FUNDAMENTALS OF ELECTRONICS

Electronic Devices and Circuit Applications

(Book 1)

In the same series:

Amplifiers: Analysis and Design
Thomas F. Schubert, Jr. and Ernest M. Kim

Active Filters and Amplifier Frequency Response
Thomas F. Schubert, Jr. and Ernest M. Kim

Oscillators and Advanced Electronics Topics
Thomas F. Schubert, Jr. and Ernest M. Kim

FUNDAMENTALS OF ELECTRONICS

Electronic Devices and Circuit Applications

(Book 1)

Thomas F. Schubert, Jr. and Ernest M. Kim
University of San Diego

I.K. International Publishing House Pvt. Ltd.
NEW DELHI

Fundamentals of Electronics: Electronic Devices and Circuit Applications (Book 1)
(Thomas F. Schubert, Jr. and Ernest M. Kim)

Published by arrangement
I.K. International Publishing House Pvt. Ltd.
S-25, Green Park Extension
Uphaar Cinema Market
New Delhi–110 016 (India)
E-mail: info@ikinternational.com
Website: www.ikbooks.com

ISBN 978-93-85909-21-4
ISBN 978-93-85909-61-0 (Set)

First Indian Reprint 2017

Published by Krishan Makhijani for I.K. International Publishing House Pvt. Ltd., S-25, Green Park Extension, Uphaar Cinema Market, New Delhi–110 016 and Printed by Rekha Printers Pvt. Ltd., Okhla Industrial Area, Phase II, New Delhi–110 020.

Synthesis Lectures on Digital Circuits and Systems

Editor
Mitchell A. Thornton, ***Southern Methodist University***

The *Synthesis Lectures on Digital Circuits and Systems* series is comprised of 50- to 100-page books targeted for audience members with a wide-ranging background. The Lectures include topics that are of interest to students, professionals, and researchers in the area of design and analysis of digital circuits and systems. Each Lecture is self-contained and focuses on the background information required to understand the subject matter and practical case studies that illustrate applications. The format of a Lecture is structured such that each will be devoted to a specific topic in digital circuits and systems rather than a larger overview of several topics such as that found in a comprehensive handbook. The Lectures cover both well-established areas as well as newly developed or emerging material in digital circuits and systems design and analysis.

Fundamentals of Electronics: Book 1 Electronic Devices and Circuit Applications
Thomas F. Schubert, Jr. and Ernest M. Kim
2015

Applications of Zero-Suppressed Decision Diagrams
Tsutomu Sasao and Jon T. Butler
2014

Modeling Digital Switching Circuits with Linear Algebra
Mitchell A. Thornton
2014

Arduino Microcontroller Processing for Everyone! Third Edition
Steven F. Barrett
2013

Boolean Differential Equations
Bernd Steinbach and Christian Posthoff
2013

Bad to the Bone: Crafting Electronic Systems with BeagleBone and BeagleBone Black
Steven F. Barrett and Jason Kridner
2013

Introduction to Noise-Resilient Computing
S.N. Yanushkevich, S. Kasai, G. Tangim, A.H. Tran, T. Mohamed, and V.P. Shmerko
2013

Atmel AVR Microcontroller Primer: Programming and Interfacing, Second Edition
Steven F. Barrett and Daniel J. Pack
2012

Representation of Multiple-Valued Logic Functions
Radomir S. Stankovic, Jaakko T. Astola, and Claudio Moraga
2012

Arduino Microcontroller: Processing for Everyone! Second Edition
Steven F. Barrett
2012

Advanced Circuit Simulation Using Multisim Workbench
David Báez-López, Félix E. Guerrero-Castro, and Ofelia Delfina Cervantes-Villagómez
2012

Circuit Analysis with Multisim
David Báez-López and Félix E. Guerrero-Castro
2011

Microcontroller Programming and Interfacing Texas Instruments MSP430, Part I
Steven F. Barrett and Daniel J. Pack
2011

Microcontroller Programming and Interfacing Texas Instruments MSP430, Part II
Steven F. Barrett and Daniel J. Pack
2011

Pragmatic Electrical Engineering: Systems and Instruments
William Eccles
2011

Pragmatic Electrical Engineering: Fundamentals
William Eccles
2011

Introduction to Embedded Systems: Using ANSI C and the Arduino Development Environment
David J. Russell
2010

Arduino Microcontroller: Processing for Everyone! Part II
Steven F. Barrett
2010

Arduino Microcontroller Processing for Everyone! Part I
Steven F. Barrett
2010

Digital System Verification: A Combined Formal Methods and Simulation Framework
Lun Li and Mitchell A. Thornton
2010

Progress in Applications of Boolean Functions
Tsutomu Sasao and Jon T. Butler
2009

Embedded Systems Design with the Atmel AVR Microcontroller: Part II
Steven F. Barrett
2009

Embedded Systems Design with the Atmel AVR Microcontroller: Part I
Steven F. Barrett
2009

Embedded Systems Interfacing for Engineers using the Freescale HCS08 Microcontroller II: Digital and Analog Hardware Interfacing
Douglas H. Summerville
2009

Designing Asynchronous Circuits using NULL Convention Logic (NCL)
Scott C. Smith and JiaDi
2009

Embedded Systems Interfacing for Engineers using the Freescale HCS08 Microcontroller I: Assembly Language Programming
Douglas H.Summerville
2009

Developing Embedded Software using DaVinci & OMAP Technology
B.I. (Raj) Pawate
2009

Mismatch and Noise in Modern IC Processes
Andrew Marshall
2009

Asynchronous Sequential Machine Design and Analysis: A Comprehensive Development of the Design and Analysis of Clock-Independent State Machines and Systems
Richard F. Tinder
2009

An Introduction to Logic Circuit Testing
Parag K. Lala
2008

Pragmatic Power
William J. Eccles
2008

Multiple Valued Logic: Concepts and Representations
D. Michael Miller and Mitchell A. Thornton
2007

Finite State Machine Datapath Design, Optimization, and Implementation
Justin Davis and Robert Reese
2007

Atmel AVR Microcontroller Primer: Programming and Interfacing
Steven F. Barrett and Daniel J. Pack
2007

Pragmatic Logic
William J. Eccles
2007

PSpice for Filters and Transmission Lines
Paul Tobin
2007

PSpice for Digital Signal Processing
Paul Tobin
2007

PSpice for Analog Communications Engineering
Paul Tobin
2007

PSpice for Digital Communications Engineering
Paul Tobin
2007

PSpice for Circuit Theory and Electronic Devices
Paul Tobin
2007

Pragmatic Circuits: DC and Time Domain
William J. Eccles
2006

Pragmatic Circuits: Frequency Domain
William J. Eccles
2006

Pragmatic Circuits: Signals and Filters
William J. Eccles
2006

High-Speed Digital System Design
Justin Davis
2006

Introduction to Logic Synthesis using Verilog HDL
Robert B.Reese and Mitchell A.Thornton
2006

Microcontrollers Fundamentals for Engineers and Scientists
Steven F. Barrett and Daniel J. Pack
2006

ABSTRACT

This book, *Electronic Devices and Circuit Application*, is the first of four books of a larger work, *Fundamentals of Electronics*. It is comprised of four chapters describing the basic operation of each of the four fundamental building blocks of modern electronics: operational amplifiers, semiconductor diodes, bipolar junction transistors, and field effect transistors. Attention is focused on the reader obtaining a clear understanding of each of the devices when it is operated in equilibrium. Ideas fundamental to the study of electronic circuits are also developed in the book at a basic level to lessen the possibility of misunderstandings at a higher level. The difference between linear and non-linear operation is explored through the use of a variety of circuit examples including amplifiers constructed with operational amplifiers as the fundamental component and elementary digital logic gates constructed with various transistor types.

Fundamentals of Electronics has been designed primarily for use in an upper division course in electronics for electrical engineering students. Typically such a course spans a full academic years consisting of two semesters or three quarters. As such, *Electronic Devices and Circuit Applications*, and the following two books, *Amplifiers: Analysis and Design* and *Active Filters and Amplifier Frequency Response*, form an appropriate body of material for such a course. Secondary applications include the use in a one-semester electronics course for engineers or as a reference for practicing engineers.

KEYWORDS

operational amplifiers, amplifiers, modeling, gain, semiconductor diodes, load lines, zener diodes, rectifiers, logic gates, transistors, bipolar junction transistors, TTL, ECL, transistor biasing, bias stability, field effect transistors, BJT, FET, MOSFET, SPICE modeling

Preface

It is expected that the reader of this text is familiar with the common passive elements of linear circuit analysis (resistors, inductors, capacitors, and transformers) as well as the idealized linear active elements (independent and dependent voltage and current sources). Unfortunately, the field of electronics makes great use of active elements that do not necessarily fall into either of the above categories. These active elements may behave in either a linear or non-linear fashion depending on their circuit application.

The study of electronic circuit behavior traditionally begins with three active semiconductor electronic elements:

- The Semiconductor Diode
- The Bipolar Junction Transistor (BJT)
- The Field Effect Transistor (FET)

To this trio of fundamental devices has been added an additional electronic circuit building block, the Operational Amplifier (OpAmp). While the OpAmp is composed of tens of transistors (usually either BJTs or FETs, but sometimes a mixture of both types) and often a few diodes, its easily understood terminal properties, high use in industry, and commercial availability make it a good companion for study with the other devices.

Quasistatic analysis explores the potentially non-linear action of each of these four elements (or any other similar element) in a variety of applications. The fundamental assumption in this exploration is that voltage and current transitions take place slowly and that the circuit is always in equilibrium: hence the term quasistatic.

The authors have chosen to begin the study of electronics with a chapter on the operational amplifier for several reasons, among which are:

- in most simple applications, the OpAmp behaves in a near-ideal fashion.
- typical analysis of OpAmp circuitry provides a good review of basic circuit analysis techniques.
- discussion of the OpAmp provides a good framework for understanding of electronic circuitry.

While many readers will find much in this chapter on OpAmps a review, the chapter presents several concepts fundamental to the study of electronic circuitry. Most significant among these concepts are:

- undistorted amplification
- gain
- device modeling
- conditions under which device models, particularly linear models, fail

Of particular importance is the concept that a device with extremely complex interior working mechanisms can be modeled simply by its terminal characteristics.

The remaining three chapters in this book present the semiconductor diode, the BJT and the FET. Each chapter follows the same basic framework and has the same goals:

- To present each device through real experimental data and through theoretical functional relationships.
- To use the above presented relationships to observe the action of the device in relatively simple circuits.
- To devise a progression of realistic piecewise-linear models for the devices. The theoretical basis for each model is presented and the appropriate use of these models is explored. Only when a model fails to properly predict device behavior will new, more complex, models be introduced. This simple-to-complex route provides for progressively more detailed analysis using the newly introduced models.
- To use realistic applications to demonstrate the usefulness of the device models.
- To provide a solid foundation for the linear and non-linear modeling and applications found in later books of this series.

Upon completing Book 1, the reader will have a good foundation in the operation of these four basic active, non-linear devices. The fundamental regions of operation for each device will have been explored: both linear and non-linear device models will be available for further investigations.

Thomas F. Schubert, Jr. and Ernest M. Kim
May 2015

Acknowledgments

In the development of any book, it seems that an infinite number of people provide and incalculable amount of guidance and help. While our thanks goes out to all those who helped, us, we can only mention a few of our many benefactors here. Special thanks go to Lynn Cox, who sparked our interest in writing an electronics text and, of course, to the staff at Morgan and Claypool Publishers, specifically Joe Claypool and Dr. C.L. Tondo.

Thomas F. Schubert, Jr. and Ernest M. Kim
October 2015

Contents

CHAPTER 1

Operational Amplifiers and Applications

The Operational Amplifier (commonly referred to as the OpAmp) is one of the primary active devices used to design low and intermediate frequency analog electronic circuitry: its importance is surpassed only by the transistor. OpAmps have gained wide acceptance as electronic building blocks that are useful, predictable, and economical. Understanding OpAmp operation is fundamental to the study of electronics.

The name, operational amplifier, is derived from the ease with which this fundamental building block can be configured, with the addition of minimal external circuitry, to perform a wide variety of linear and non-linear circuit functions. Originally implemented with vacuum tubes and now as small, transistorized integrated circuits, OpAmps can be found in applications such as: signal processors (filters, limiters, synthesizers, etc.), communication circuits (oscillators, modulators, demodulators, phase-locked loops, etc.), Analog/Digital converters (both A to D and D to A), and circuitry performing a variety of mathematical operations (multipliers, dividers, adders, etc.).

The study of OpAmps as circuit building blocks is an excellent starting point in the study of electronics. The art of electronics circuit and system design and analysis is founded on circuit realizations created by interfacing building block elements that have specific terminal characteristics. OpAmps, with near-ideal behavior and electrically good interconnection properties, are relatively simple to describe as circuit building blocks.

Circuit building blocks, such as the OpAmp, are primarily described by their terminal characteristics. Often this level of modeling complexity is sufficient and appropriately uncomplicated for electronic circuit design and analysis. However, it is often necessary to increase the complexity of the model to simplify the analysis and design procedures. These models are constructed from basic circuit elements so that they match the terminal characteristics of the device. Resistors, capacitors, and voltage and current sources are the most common elements used to create such a model: an OpAmp can be described at a basic level with two resistors and a voltage-controlled voltage source.

OpAmp circuit analysis also offers a good review of fundamental circuit analysis techniques. From this solid foundation, the building block concept is explored and expanded throughout this text. With the building block concept, all active devices are treated as functional blocks with specified input and output characteristics derived from the device terminal behavior. Circuit design is the process of interconnecting active building blocks with passive components to produce a wide variety of desired electronic functions.

1.1 BASIC AMPLIFIER CHARACTERISTICS

One of the fundamental characteristics of an amplifier is its gain.[1] Gain is defined as the factor that relates the output to the input signal intensities. As shown in Figure 1.1, a time dependent input signal, $x(t)$, is introduced to the "black box" which represents an amplifier and another time dependent signal, $y(t)$, appears at the output.

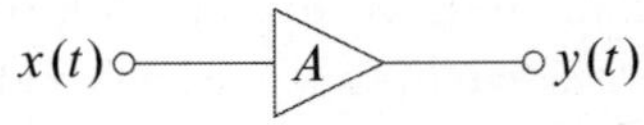

Figure 1.1: "Black box" representation of an amplifier with input $\mathbf{x}(t)$ and output $\mathbf{y}(t)$.

In actuality, $\mathbf{x}(t)$ can represent either a time dependent or time independent signal. The output of a good amplifier, $\mathbf{y}(t)$, is of the same functional form as the input with two significant differences: the magnitude of the output is scaled by a constant factor, A, and the output is delayed by a time, t_d. This input-output relationship can be expressed as:

$$\mathbf{y}(t) = A\mathbf{x}(t - t_d) + \alpha \tag{1.1}$$

where

A is the gain of the amplifier,
α is the output DC offset, and
t_d is the time delay between the input and output signals.

The signal is "amplified" by a factor of A. Amplification is a ratio of output signal level to the input signal level. The output signal is amplified when $|A|$ is greater than 1. For $|A|$ less than 1, the output signal is said to be attenuated. If A is a negative value, the amplifier is said to invert the input. Should $x(t)$ be sinusoidal, inversion of a signal is equivalent to a phase shift of 180°: negative A implies the output signal is $\pm 180°$ out of phase with the input signal.

For time-varying signals, it may be convenient to find the amplification (ratio) by comparing either the root-mean-squared (RMS) values or the peak values of the input and output signals. Good measurement technique dictates that amplification is found by measuring the input and output RMS values since peak values may, in many instances, be ambiguous and difficult to quantify.[2] Unfortunately, in many practical instances, RMS or power meters are not available dictating the measurement of peak amplitudes. The delay time is an important quantity that is

[1]Other amplifier specifications of interest include input and output impedances, power consumption, frequency response, noise factor, Mean Time to Failure (MTTF), and operational temperature range. An understanding of the basis for these specifications and their impact on design will be developed in the chapters that follow. The discussion in this chapter will, for the most part, be restricted to gain and time-domain effects.

[2]Peak values are also strongly affected by the presence of noise.

often overlooked in electronic circuit analysis and design.[3] The signal encounters delay between the input and output of an amplifier simply because it must propagate through a number of the internal components of the amplifying block.

In Figure 1.1, $\mathbf{x}(t)$ and $\mathbf{y}(t)$ are time dependent signals. Depending on the amplifier, $\mathbf{x}(t)$ and $\mathbf{y}(t)$ can be either current or voltage signals. Every amplifier draws power from a power supply, typically in the form of current from a DC voltage source. As will be shown in later sections of this text, the maximum possible output signal level is determined largely by the power supply voltage and current limitations. For instance, assume that the amplifier in Figure 1.1 is powered by a DC voltage source with output equal to V_{CC}. If the output signal, $\mathbf{y}(t)$, is a voltage signal, the maximum output voltage attainable under ideal conditions for the gain block is V_{CC}.[4] The phenomenon of limiting output voltage levels to lie within the limits set by the power supplies is called *saturation*. Should the power supply be unable to provide sufficient current to the gain block, the output will also be limited, although in a manner that is not as simple as in saturation.

In order to discuss terminal characteristics of commercially available OpAmps, a specific amplifier must be selected. The μA741 (or LM741, MC1741) is a good choice since it is the most commonly used and studied OpAmp available. The prefixes {μA, OP, LM, MC} designate the manufacturer of the integrated circuit (IC): μA represents Fairchild Semiconductor, OP is used by Linear Technologies, LM by National Semiconductor, and MC by Motorola Semiconductors. The specification sheets for the three OpAmps listed above can be found on the internet. In many instances, one or two letters follow the numerical designation of the IC. These letters indicate the package type or size and package material. For example, a μA741CP is a 741 IC manufactured by National Semiconductor that is in a commercial grade plastic standard eight lead dual-in-line package (DIP) or MINI DIP. Other manufacturers, such as Texas Instruments, manufacture the μA741CP using the Fairchild part designation.

Other common OpAmps include the OP-27, LF411, and LM324. The OP-27 and LF411 OpAmps have specifications that are similar to the μA741 and come in selected packages. The OP-27 and LF411 OpAmps are, like the μA741, dual power rail amplifiers; that is, the amplifier usually operates with both a positive and negative power supply voltages. The LM324, on the other hand, is a single supply amplifier; it requires a positive voltage and a common reference (ground).

Figure 1.2 shows a top view of a μA741CP package with the terminal designations. The terminals of interest are:

- the inverting input (pin 2),
- the non-inverting input (pin 3),

[3]The implication of delay time is addressed in the transistor amplifier time domain analysis portion of the third book of this series.

[4]Note that upper case letters are used for DC signals and lower case letters for time-varying signals. Lower case letters with lower case subscripts is used for AC signals. Lower case letters with upper case subscripts are used for AC signals with DC components.

- the output (pin 6),
- the positive power supply (designated $+V_{cc}$, pin 7), and
- the negative power supply (designated $-V_{cc}$, pin 4).

The offset null pins (1 and 5) are used to compensate for minor fabrication imperfections as well as degradation due to aging. Although commonly left disconnected by the circuit designer, these pins are sometimes utilized in applications that require the amplification of very small level signals. The μA741 OpAmp is a compensated amplifier. The performance implications of compensated and uncompensated amplifiers are related to frequency response and stability: they will be discussed in detail in the third book of this series.

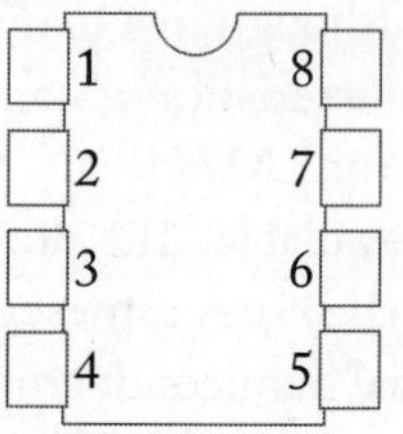

Figure 1.2: The top view of the μA741CP package with pin numbers.

A conventional simplified OpAmp schematic representation is shown in Figure 1.3. This representation shows two input terminals designated (−) and (+) corresponding to the inverting and non-inverting inputs, respectively, the output terminal, and the positive and negative power supply terminals labeled V^+ and V^-, respectively. Not shown are the offset null pins. Unless used, these pins are usually not included in schematic representations.

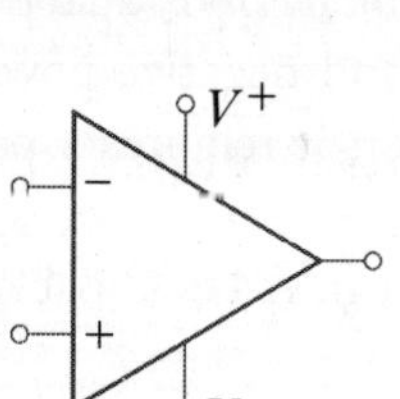

Figure 1.3: OpAmp schematic representation.

Notice that the schematic symbol of the OpAmp does not have a ground pin. In many ways, the lack of a ground pin on the OpAmp is the key to its operation. Ideally, only the differential voltage between the two input pins affects the output voltage of an OpAmp. A ground reference is provided external to the chip package.

1.2 MODELING THE OPAMP

Terminal voltages and currents are used to characterize OpAmp behavior. In order to unify all discussions of OpAmp circuitry, it is necessary to define appropriate descriptive conventions. All voltages are measured relative to a common reference node (or ground) which is external to the chip as is shown in Figure 1.4. The voltage between the inverting pin and ground is denoted as v_1: the voltage between the non-inverting pin and ground is v_2. The output voltage referenced to ground is denoted as v_o. Power is typically applied to an OpAmp in the form of two equal-magnitude supplies, denoted V_{CC} and $-V_{CC}$, which are connected to the V^+ and V^- terminals of the OpAmp, respectively.

The reference current directions are shown in Figure 1.4. The direction of current flow is always into the nodes of the Op Amp. The current into the inverting input terminal is i_1; current into the non-inverting input terminal is i_2; current into the output terminal is i_o; and the currents into the positive and negative power supply terminals are I_{C-} and I_{C+}, respectively.

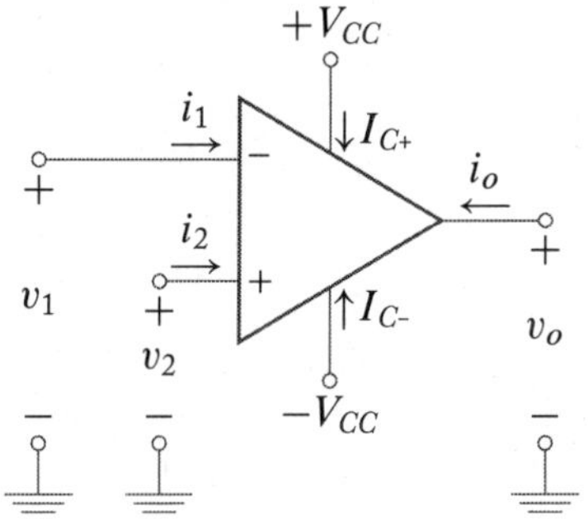

Figure 1.4: Terminal voltages and currents.

The voltage and current constraints inherent to the input and output terminals of an OpAmp must be understood prior to connecting external circuit elements. The OpAmp is considered as a building block element with specific rules of operation. A short discussion of these rules of operation follows.

The terminal voltages are constrained by the following relationships[5]

$$v_o = A(v_2 - v_1) \tag{1.2}$$

and

$$-V_{CC} \leq v_o \leq V_{CC}. \tag{1.3}$$

The first of the two voltage constraints states that the output voltage is proportional to the difference between the non-inverting and inverting terminal inputs, v_2 and v_1, respectively. The

[5]For introductory purposes, the time delay factor t_d has been assumed to be zero. Delay time considerations will be discussed at length in the third book of this series.

proportionality constant A is the called the *open loop gain*, whose significance will be detailed later. The second voltage constraint states that the output voltage is limited to the power supply rails.[6] That is, v_o must lie between $\pm V_{CC}$. If the output reaches the limiting values, the amplifier is said to be saturated. In reality, the amplifier saturates at voltages slightly shy of $\pm V_{CC}$ due to device characteristics within the OpAmp. So long as $|v_o| < V_{CC}$, the amplifier is operating in the linear region. Between the limiting values lies the "linear region" where the output voltage is related to the input voltage by the proportionality constant A. Figure 1.5a shows an idealized voltage transfer characteristic[7] of an OpAmp. A more realistic voltage transfer characteristic is shown in Figure 1.5b where the amplifier response gradually tapers toward saturation at higher input voltages due to the characteristics of the circuit design internal to the chip.

From the data sheet for the μA741C, the typical open loop gain (designated as Large Signal Voltage Gain), A, is 200k. It is reasonable to assume that all OpAmps have very large voltage gain and that a first-order approximation of the voltage gain is:

$$A \approx \infty. \tag{1.4}$$

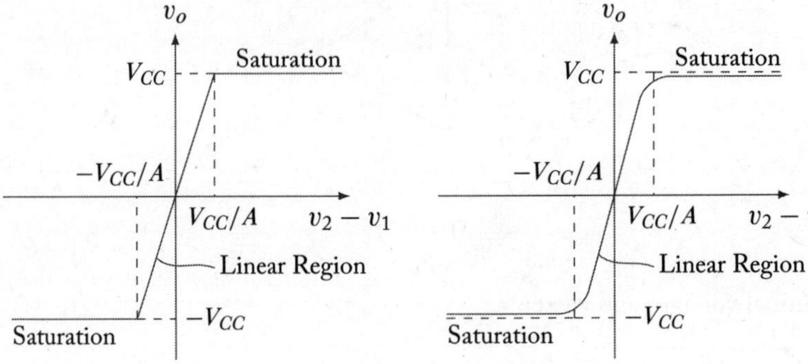

Figure 1.5: Transfer characteristics of (a) an ideal OpAmp and (b) more realistic "soft-limiting" OpAmp transfer characteristic.

In the 741 OpAmp, the absolute maximum supply voltages (V_{CC} and $-V_{CC}$) are ± 18V. Therefore, the output cannot exceed ± 18V. Knowing the maximum value of the output voltage v_o and the typical A, the maximum difference between v_2 and v_1 is found to be:

$$(v_2 - v_1)_{\text{max}} = \frac{18\text{V}}{200,000} = 0.09\,\text{mV}.$$

[6] The power supply voltages are commonly called *rails:* they limit the output voltages of a functional electronic block.
[7] A *transfer characteristic* is a graphical representation of the output as a function of the input. In this instance, the voltage input-output relationship is shown. A *transfer function* is usually a mathematical description of the output as a function of the input.

In most OpAmp applications this voltage level can be considered negligible. Therefore, in the linear region of operation, the input voltages are assumed to be equal:

$$v_1 \approx v_2. \tag{1.5a}$$

The input terminals of an OpAmp exhibit high input resistance: the μA741 typically has an input resistance of 2 MΩ. The input resistance of an ideal OpAmp is approximated as infinite.

$$R_i \approx \infty. \tag{1.6}$$

Within the linear region of operation the maximum current flowing between the two input terminals is given by:

$$i_{\text{input(max)}} = \frac{(v_2 - v_1)_{\max}}{2\,\text{M}\Omega} = \frac{0.09\,\text{mV}}{2\,\text{M}\Omega} = 45\,\text{pA}.$$

The input currents, i_1 and i_2, are extremely small and are considered to be approximately zero:

$$i_1 = i_2 \approx 0. \tag{1.5b}$$

The two relationships given in Equations (1.5a) and (1.5b) together form what is referred to as a "virtual short" between the inverting and non-inverting input terminals of the OpAmp. A virtual short implies that two terminals act in a *voltage sense* as if they were *shorted*, but ***no current flows between the terminals***.

Kirchhoff's current law (KCL) can be used to sum the currents of an OpAmp. Since the input currents are very small, the resulting relationship is,

$$i_o = -(I_{C+} + I_{C-}). \tag{1.7}$$

Equation (1.7) indicates that although the input currents are negligible, the output current is substantial. That is, $i_o \neq 0$.

For completeness, the typical output resistance of a 741 OpAmp is 75 Ω. All OpAmps have low output resistance: Ideal OpAmps are considered to have zero-value output resistance.

$$R_o \approx 0. \tag{1.8}$$

Equations (1.2) to (1.8) define the ***ideal OpAmp model***. These defining properties are summarized in Table 1.1.

The SPICE macromodel[8] for the μA741C OpAmp yields a value of A of approximately 195 k. The input and output impedances are complex and vary with input signal frequency. The real part of the input impedance dominates with a value of approximately 996 kΩ. The output impedance is essentially 50 Ω at frequencies above 100 Hz. The component parameters specified in the SPICE macromodel of the μA741C OpAmp and the typical specifications found in

[8]A SPICE macromodel is a complex subcircuit model of a device intended to correctly model all terminal performance characteristics of a device. The SPICE macromodel for the μA741C OpAmp was provided by MicroSim Corp.

Table 1.1: OpAmp characteristic property values

Property	OpAmp Value Typical	Ideal
Gain, A	$> 200{,}000$	∞
Input Resistance, R_i	$> 2\,\text{M}\Omega$	∞
Output Resistance, R_o	$< 75\,\Omega$	0
Input Voltage Difference, $v_2 - v_1$	$< 0.1\,\text{mV}$	0 (virtual short)
Input Current, i_1 or i_2	$< 50\,\text{pA}$	0 (virtual short)
Output Voltage Limits	$\lvert v_o \rvert < V_{CC}$	$\lvert v_o \rvert < V_{CC}$

the data sheet both lie within the acceptable range of parameter values found in manufactured components. Therefore, using either specification will yield acceptable results when designing a circuit using the μA741C.

The voltage and current constraints given are the parameters used to describe an *ideal* OpAmp model. By attaching external components to an OpAmp, a functional circuit can be designed. The significance of this exercise is to demonstrate the power of modelling active elements (in this case, an OpAmp) in terms of its current and voltage characteristics at the input and output ports. Although an understanding of the internal operation of the active device is desirable, circuits can be designed using the device's input and output port parameters. The terminal characteristics of the active device in conjunction with the Kirchhoff's current and voltage laws are used to analyze the circuit.

A simple application of an OpAmp is the unity gain buffer, which is primarily used to isolate electronic signals. A unity gain buffer or voltage follower is shown in Figure 1.6. The circuit can be analyzed using the voltage and current constraints given by Equations (1.5a) and (1.5b), respectively,

$$v_1 \approx v_2$$

and

$$i_1 = i_2 \approx 0.$$

The input signal voltage v_i is equal to v_2 since negligible current flows into the positive input terminal of the OpAmp ($i_2 = 0$): there is no voltage drop across R_s. Owing to the virtual short between the inverting and non-inverting terminals, the non-inverting terminal voltage is also at the same voltage level, v_i, that is,

$$v_i = v_2 = v_1.$$

Since the inverting terminal is directly connected to the output of the OpAmp,

$$v_o = v_1 = v_i,$$

or

$$\frac{v_o}{v_1} = 1. \tag{1.9}$$

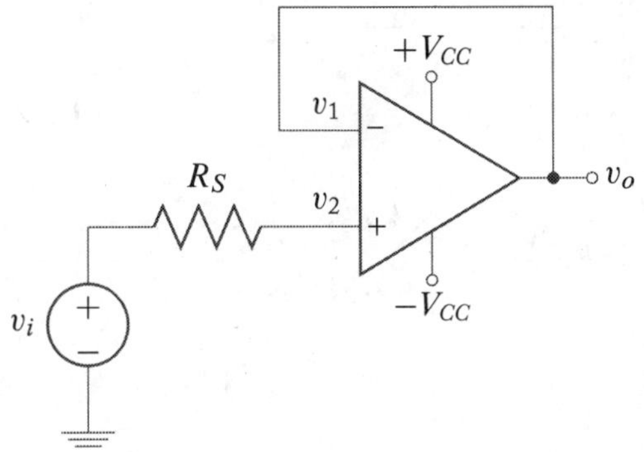

Figure 1.6: Unity gain buffer.

The voltage follower is called a unity gain *buffer* since it is an ideal impedance transformer. The input impedance of the voltage follower is very high and its output impedance is, for all practical purposes, zero. Verification of these impedance characteristics provides a useful exercise in the study of OpAmp properties.

The input and output resistances of the voltage follower can be found by using the simplified equivalent circuit for the OpAmp. The simplified equivalent circuit, shown in Figure 1.7, differs from the ideal OpAmp model in that the equivalent circuit is a lumped parameter (resistors and sources) model of the OpAmp in the linear region of operation. It is a functional equivalent of the OpAmp which is not the actual circuitry in the OpAmp chip, but behaves functionally as an OpAmp to external circuitry (the principle is much the same as Thévenin or Norton equivalent circuits). The simplified OpAmp equivalent model shown here assumes frequency independent behavior: that is, the response of the amplifier is independent of signal frequency.[9]

The current and voltage constraints in Equations (1.5a) and (1.5b) assumed in the ideal OpAmp model must be discarded when using the simplified equivalent model. Equation (1.2) is no longer valid due to the non-zero output resistance, R_o.

The voltage follower circuit is analyzed using the simplified equivalent circuit as shown in Figure 1.8. In the model, $R_i = 2\,\text{M}\Omega$, $R_o = 75\,\Omega$, and $A = 200$ k. Let $R_S = 1\,\text{k}\Omega$.

The current i is found by Kirchhoff's Voltage Law (KVL),

$$i = \frac{v_i - A\,(v_2 - v_1)}{R_i + R_O + R_S}, \tag{1.10}$$

but i is also given by:

$$i = \frac{v_2 - v_1}{R_i}. \tag{1.11}$$

[9]In reality this frequency independence is not true. The frequency dependent nature of OpAmps will be discussed in 9 (Book 3).

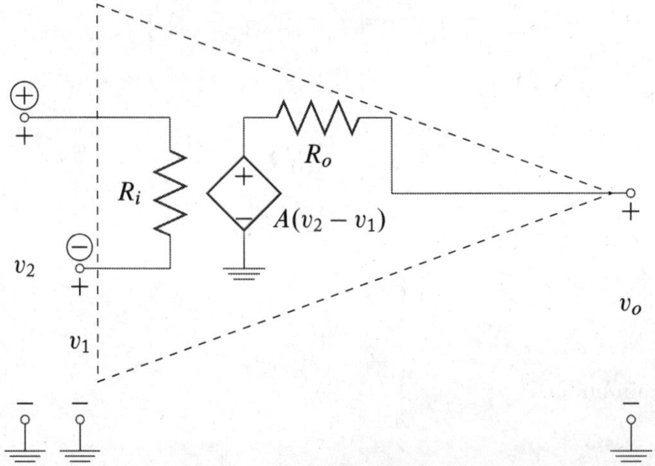

Figure 1.7: Simplified equivalent circuit model of the OpAmp.

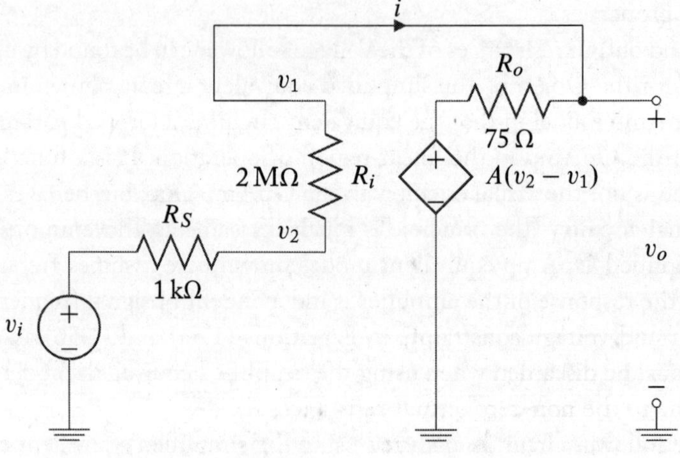

Figure 1.8: Voltage follower analysis with simplified equivalent OpAmp model.

Since the output terminal of a voltage follower is connected to the negative input terminal of the OpAmp,

$$v_o = v_1 \tag{1.12}$$

Equating Equations (1.10) and (1.11), substituting v_i for v_2, and solving for v_o/v_i yields the expression for voltage gain for the voltage follower:

$$\frac{v_o}{v_i} = 1 - \frac{R_i}{R_S + R_o + R_i\,(1 + A)}. \tag{1.13}$$

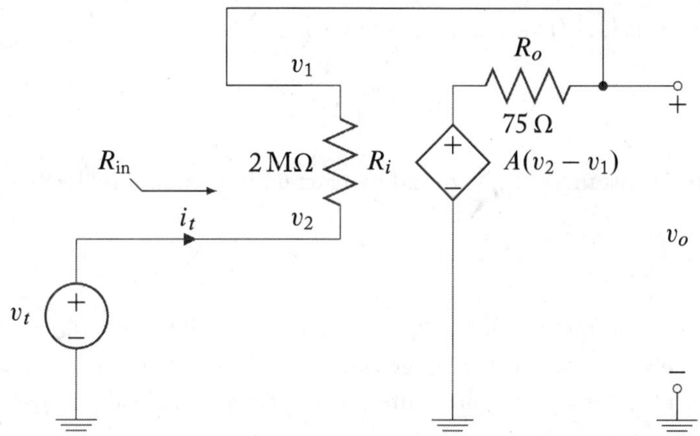

Figure 1.9: Using a test source to find the Thévenin equivalent input resistance, R_{in}.

It can easily be seen that large values of the OpAmp gain, A, will lead to a voltage gain closely approximating unity. The parameter values for this example, $R_i = 2\,\text{M}\Omega$, $R_o = 75\,\Omega$, $R_S = 1\,\text{k}\Omega$ and $A = 200\,\text{k}$, result in:

$$\frac{v_o}{v_i} = 1 - 2.6875 \times 10^{-9} \approx 1. \tag{1.14}$$

The input and output resistances of the voltage follower can be calculated using the simplified equivalent model. Thévenin equivalent resistances are found using the standard "test source" method. To calculate the input resistance, a test source, v_t is used to excite the circuit shown in Figure 1.9. Note that v_t is applied directly to v_2. The input resistance directly from the voltage source v_i in Figure 1.8 is found by adding R_S to the Thévenin resistance found at v_2. The Thévenin equivalent input resistance, R_{in}, is the ratio of the test voltage over the current delivered by the test source,

$$R_{in} = \frac{v_t}{i_t}. \tag{1.15}$$

Using KVL, the current delivered by the test voltage, v_t, is

$$i_t = \frac{v_t - A\,(v_2 - v_1)}{R_i + R_o}. \tag{1.16}$$

But

$$v_2 = v_t, \tag{1.17}$$

and, knowing the voltage drop across R_i,

$$v_1 = v_t - i_t R_i. \tag{1.18}$$

Substituting Equations (1.17) and (1.18) into (1.16) yields

$$i_t = \frac{v_t - A i_t R_i}{R_i + R_o}. \tag{1.19}$$

The Thévenin input resistance, R_{in}, is found by rearranging Equation (1.19),

$$R_{in} = \frac{v_t}{i_t} = R_i\,(1 + A) + R_o. \tag{1.20}$$

For the given typical parameter values ($R_i = 2\,\text{M}\Omega$, $A = 200\,\text{K}$, and $R_o = 75\,\Omega$), the input resistance can be calculated to be the very large value: $R_{in} = 400 \times 10^9\,\Omega$. It is reasonable to assume that the input resistance of a unity gain buffer, R_{in} is, for all practical purposes, infinite.

Example 1.1
Determine the output resistance of an OpAmp voltage follower.

Solution:
To find the output resistance, R_{out}, a test voltage source is connected to the output of the voltage follower to find the Thévenin equivalent resistance at the output. Note also that all independent sources must be zeroed. That is, all independent voltage sources are short circuited and all independent currents are open circuited. The circuit used to find R_{out} is shown in Figure 1.10.[10] To find the Thévenin equivalent output resistance, a test voltage source, v_t is connected at the

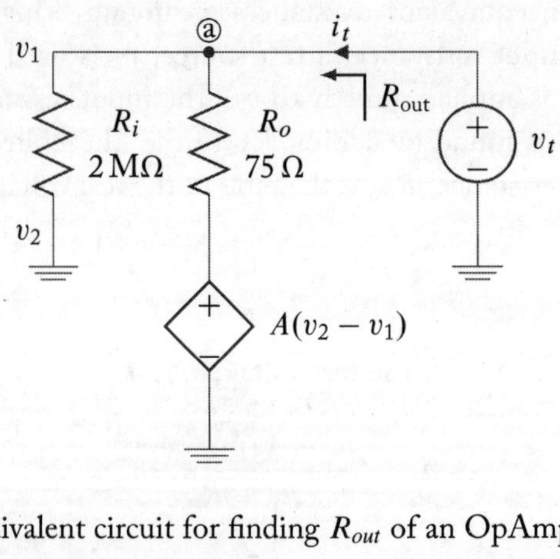

Figure 1.10: Simplified equivalent circuit for finding R_{out} of an OpAmp voltage follower.

output. The circuit draws i_t source current. The input at v_2 has been short circuited to ground to set independent sources to zero.

[10]In this example the resistance of the source connected to the input was considered to be zero: that is, $R_S = 0$. When the resistance of the source is not zero, it appears in series with the input resistance of the OpAmp, R_i. The effect of considering non-zero source resistance in the results of this example is insignificant.

Using the node voltage method of analysis by summing all currents flowing into node a yields:

$$i_t + \frac{v_2 - v_1}{R_i} + \frac{A\,(v_2 - v_1) - v_t}{R_o} = 0. \tag{1.21}$$

But

$$v_2 = 0 \quad \text{and} \quad v_1 = v_t. \tag{1.22}$$

Then equation (1.21) is simplified to

$$i_t = \frac{R_o + (A+1)\,R_i}{R_o R_i} v_t, \tag{1.23}$$

and R_{out} is found to be:

$$R_{out} = \frac{v_t}{i_t} = \frac{R_o R_i}{R_o + (A+1)\,R_i}. \tag{1.24}$$

A typical value for the output resistance of the unity gain buffer, R_{out}, can be calculated using the given typical OpAmp values of $R_i = 2\,\mathrm{M}\Omega$, $R_o = 75\,\Omega$, and $A = 200\,\mathrm{k}$. R_{out} is found to be $375\,\mu\Omega$ which in most circuit applications can be considered essentially zero.

A SPICE simulation to determine the output resistance of a unity gain buffer using the circuit of Figure 1.10 is shown in Example 1.2. Solution #1 of Example 1.2 uses a SPICE-engine simulator (National Instruments MultiSim) solution using the Transfer Function Analysis command in conjunction with a source voltage to yield the gain and input and output resistances. Although the Transfer Function Analysis command is adequate for this particular example, a more general approach is to use a test source at the output as shown in Solution #2. This is particularly important when the input and output resistances are complex and are dependent on frequency. For instance, the input and output impedances of real OpAmps are complex and frequency dependent.

Example 1.2
Find the output resistance for a unity gain buffer using SPICE and the simplified OpAmp equivalent circuit.

Solution #1:

The National Instruments MultiSim solution using the Transfer Function Analysis command with a source voltage.

The Transfer Function Analysis command yields[11] $R_{in} = 400\ \mathrm{G}\Omega$ and $R_{out} = 0\,\Omega$. The input resistance found here exactly matches the calculations using Equation (1.20). The output resistance is functionally the same value as found in Example 1.1.

[11]Many SPICE-based circuit simulators, as a default, insert a shunt resistance from analog nodes to ground [RSHUNT] to help eliminate problems such as "singular matrix" errors. Multisim™ uses a default value of $10^{12}\,\Omega$ for RSHUNT: that value is not large enough to produce correct results for the circuit under consideration in this example. RSHUNT was removed from the list of default parameters in order to produce the results shown here.

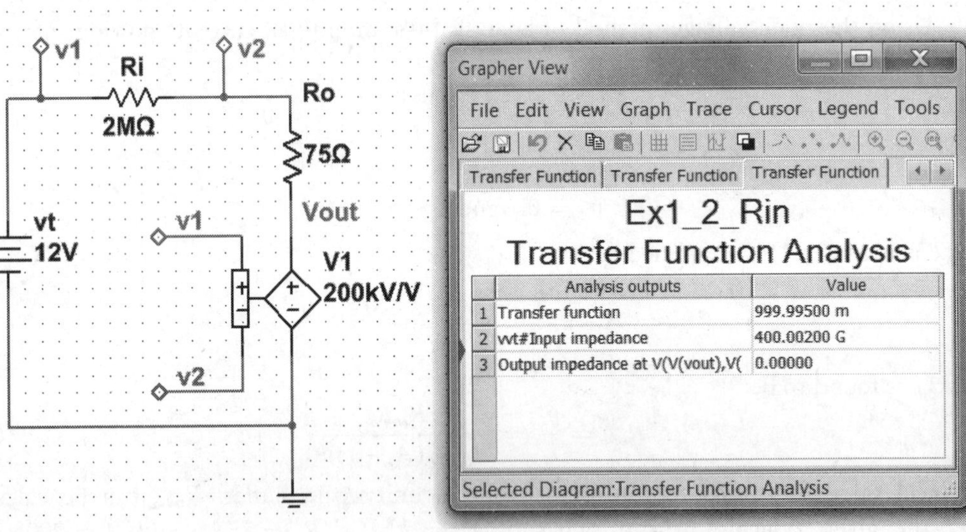

Solution #2:

SPICE solution using a test source at the output terminal of the equivalent model. A test source at the output of the equivalent model yields $R_{out} = 375\ \mu\Omega$ which exactly matches the hand-calculated results of Example 1.1 and is, for all practical purposes, zero.

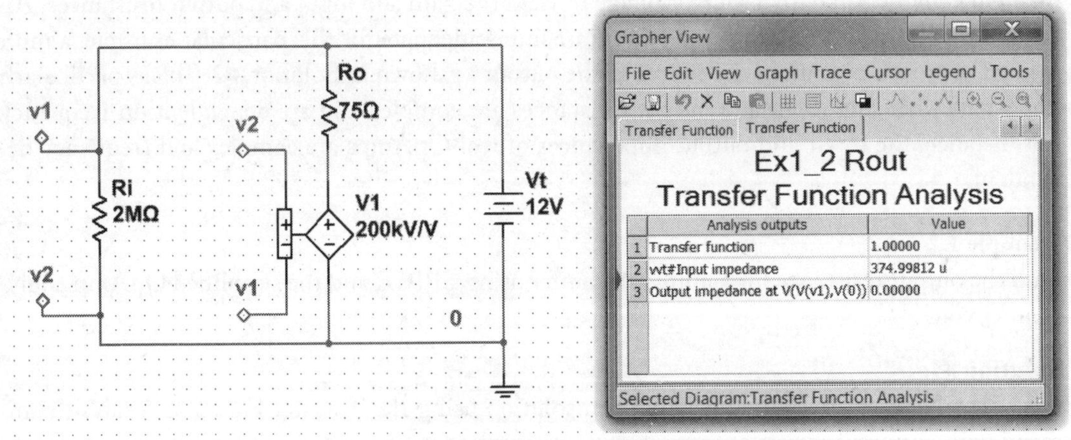

1.3 BASIC APPLICATIONS OF THE OPAMP

Although the OpAmp can be used in infinite circuit configurations, several configurations have become basic electronic building blocks. These commonly-found OpAmp circuit configurations are the inverting amplifier, summing amplifier, non-inverting amplifier, difference amplifier, integrator, and differentiator. All five configurations can be analyzed using the voltage and current constraints, and the ideal model of the OpAmp discussed in Sections 1.1 and 1.2.

1.3.1 INVERTING AMPLIFIER

The inverting amplifier configuration shown in Figure 1.11 amplifies and inverts the input signal in the linear region of operation. The circuit consists of a resistor R_S in series with the voltage source v_i connected to the inverting input of the OpAmp. The non-inverting input of the OpAmp is short circuited to ground (common). A resistor R_f is connected to the output and provides a negative feedback path to the inverting input terminal.[12] Because the output resistance of the OpAmp is nearly zero, the output voltage v_o will not depend on the current that might be supplied to a load resistor connected between the output and ground.

For most OpAmps, it is appropriate to assume that their characteristics are approximated closely by the ideal OpAmp model of Section 1.2. Therefore, analysis of the inverting amplifier can proceed using the voltage and current constraints of Equations (1.5a) and (1.5b),

$$v_1 = v_2,$$

and

$$i_1 = i_2 \approx 0.$$

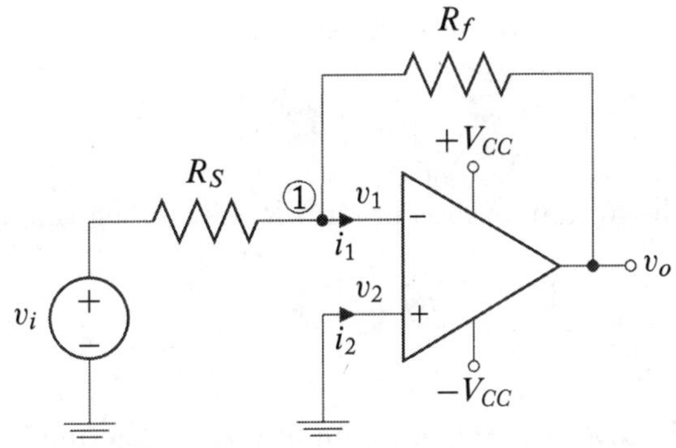

Figure 1.11: The inverting amplifier configuration.

Since v_2 is connected to the common or ground terminal,

$$v_2 = 0.$$

[12]Further detailed discussion of feedback theory and the implication of negative feedback is found in 8 (Book 2). Analysis of the OpAmp circuits in this chapter will rely on standard circuit analysis techniques.

Node 1 is said to be a virtual ground due to the virtual short circuit between the inverting and non-inverting terminals (which is grounded) as defined by the voltage constraint,

$$v_1 = v_2 = 0. \tag{1.25}$$

The node voltage method of analysis is applied at node 1,

$$0 = \frac{v_i - v_1}{R_S} + \frac{v_O - v_1}{R_f} + i_1. \tag{1.26}$$

By applying Equation (1.25), obtained from the virtual short circuit, and the constraint on the current i_1 as defined in Equation (1.5b), Equation (1.26) is simplified to

$$0 = \frac{v_i}{R_S} + \frac{v_o}{R_f}. \tag{1.27}$$

Solving for the voltage gain, v_o/v_i,

$$\frac{v_o}{v_i} = -\frac{R_f}{R_S} \tag{1.28}$$

Notice that the voltage gain is dependent only on the ratio of the resistors external to the OpAmp, R_f and R_S. The amplifier increases the amplitude of the input signal by this ratio. The negative sign in the voltage gain indicates an inversion in the signal.

The output voltage is also constrained by the supply voltages V_{CC} and $-V_{CC}$,

$$|v_o| < V_{CC}.$$

Using Equation (1.28), the maximum resistor ratio R_f/R_s for a given input voltage v_i is

$$\frac{R_f}{R_S} < \left|\frac{V_{CC}}{v_i}\right|. \tag{1.29}$$

The input resistance of the inverting amplifier can be readily determined by applying the voltage constraint: $v_1 = v_2 = 0$. Therefore, the resistance that the signal source v_i encounters is simply R_S due to the virtual short to ground at the inverting terminal. Then R_S must be large for a high input resistance. If R_S is large, R_f must be very large to achieve large gain, R_f/R_s. In some instances, R_f may be prohibitively high.[13] Therefore, in most applications, the input resistance of the inverting amplifier is low to moderate.

[13]The non-ideal characteristics of OpAmps place constraints on externally connected elements. A discussion of these constraints is found in Section 1.5 of this chapter.

Example 1.3

For the circuit shown in Figure 1.12, find the gain and i_o. If $v_i = 2 \sin \omega_o t$ V, what is the output? What input voltage amplitude will cause the amplifier to saturate?

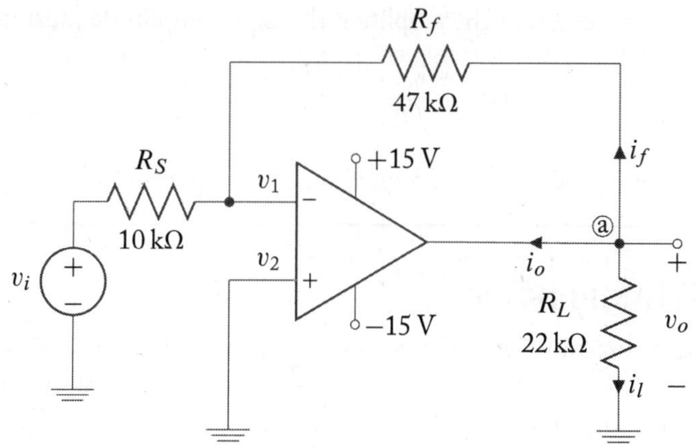

Figure 1.12: Inverting amplifier with load resistor, R_L.

Solution:

The output voltage v_o is independent of the load resistor, R_L, because of the low output resistance of the OpAmp. Therefore, the gain of the amplifier is

$$\frac{v_o}{v_i} = -\frac{R_f}{R_S}$$

$$= -\frac{47\,\text{k}\Omega}{10\,\text{k}\Omega} = -4.7.$$

Using KCL at node a,

$$0 = i_o + i_f + i_l.$$

The currents i_f and i_l are

$$i_1 = -\frac{v_o}{22000} = -\frac{4.7 \times 2 \sin(\omega_o t)}{22000}$$

and

$$i_f = -\frac{v_o}{47000} = -\frac{4.7 \times 2 \sin(\omega_o t)}{47000}.$$

Solving for i_o yields,

$$io = \left(\frac{1}{47000} + \frac{1}{22000}\right) \times 4 \times 4.7 \sin(\omega_o t)$$

$$= 1.255 \sin \omega_o t \text{ mA}.$$

For an input voltage signal of $v_i = 2 \sin \omega_o t$ V,

$$v_o = \frac{v_o}{v_i} \times 2 \sin(\omega_o t) = -4.7 \times 2 \sin(\omega_o t) = -9.4 \sin(\omega_o t) \text{ V}.$$

For operation in the linear region of the amplifier, the input amplitude must not exceed,

$$|v_i| < \frac{+V_{CC}}{R_f / R_S} < \frac{15}{4.7} = 3.19 \text{ V}.$$

Input signal amplitudes greater than or equal to 3.19 will cause the amplifier to saturate.

1.3.2 SUMMING AMPLIFIER

The output voltage of a summing amplifier is an inverted, amplified sum of the input voltages. A summing amplifier can theoretically have a large number of input voltages. Figure 1.13 shows a summing amplifier with three inputs, v_{i1}, v_{i2}, and v_{i3}.

Using the node voltage method by summing the current entering node 1 gives

$$0 = \frac{v_{i1} - v_1}{R_1} + \frac{v_{i2} - v_1}{R_2} + \frac{v_{i3} - v_1}{R_3} + \frac{v_o - v_1}{R_f} \tag{1.30}$$

The virtual short between input terminals of the OpAmp leads to:

$$v_1 = v_2 = 0.$$

Therefore, Equation (1.30) simplifies to,

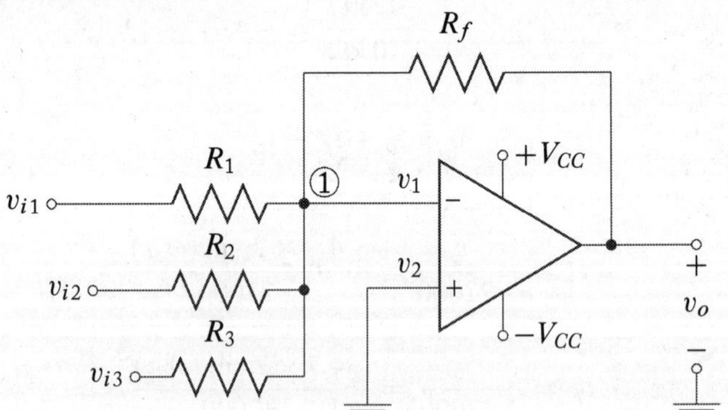

Figure 1.13: Summing amplifier with three input signals.

$$0 = \frac{v_{i1}}{R_1} + \frac{v_{i2}}{R_2} + \frac{v_{i3}}{R_3} + \frac{v_o}{R_f}. \tag{1.31}$$

Solving for the output voltage v_o yields,

$$v_o = -\left(\frac{R_f}{R_1}v_{i1} + \frac{R_f}{R_2}v_{i2} + \frac{R_f}{R_3}v_{i3}\right). \tag{1.32}$$

The output voltage is an inverted sum of scaled input voltages.

A particularly useful case occurs when $R_1 = R_2 = R_3 = R_S$. In this case, Equation (1.32) is simplified to,

$$v_o = -\frac{R_f}{R_s}\left(v_{i1} + v_{i2} + v_{i3}\right). \tag{1.33}$$

The number of input signal voltages may be increased to meet the requirements of the application. For n input signals,

$$v_o = -\frac{R_f}{R_s}\sum_{j=1}^{n} v_{ij}. \tag{1.34}$$

Example 1.4

Two voltage signals,

$$v_{i1} = 2\cos(\omega_o t + 25°)\,\text{V} \quad \text{and} \quad v_{i2} = 1.5\cos(\omega_o t - 35°)\,\text{V}$$

are added by the summing amplifier in Figure 1.14.

Find the output voltage, v_o.

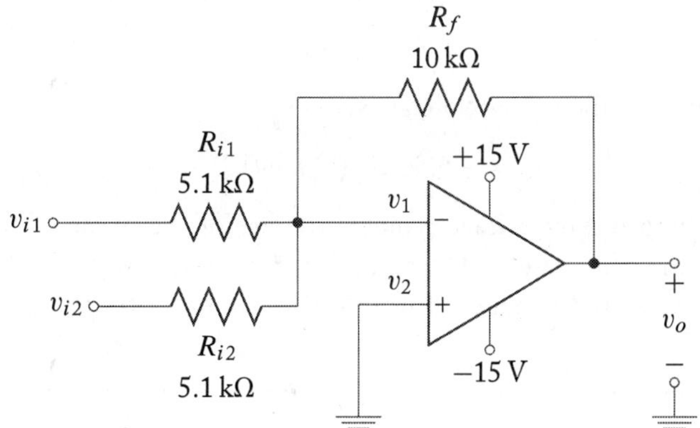

Figure 1.14: Summing amplifier with two input voltages.

Solution:

Since $R_1 = R_2$, the expression for the output voltage is,

$$v_o = -\frac{R_f}{R_s}\left(v_{i1} + v_{i2}\right).$$

The input voltages in this example are sinusoids of the same frequency. Therefore, the two input voltages can be combined using phasor representation,

$$\mathbf{V_i} = \mathbf{V_{i1}} + \mathbf{V_{i2}}.$$

The sum of the voltages is,

$$\begin{aligned} v_i &= v_{i1} + v_{i2} \\ &= 2\cos(\omega_o t + 25°) + 1.5\cos(\omega_o t - 35°). \end{aligned}$$

The sum of the voltages in phasor representation is,

$$\begin{aligned} \mathbf{V_i} &= \mathbf{V_{i1}} + \mathbf{V_{i2}} \\ &= 2\angle 25° + 1.5\angle - 35° \\ &= (1.81 + j0.845) + (1.23 - j0.860) \\ &= 3.04 - j0.015 \\ &= 3.04\angle - 0.285°\text{ V}. \end{aligned}$$

The output voltage in phasor notation is,

$$\begin{aligned} \mathbf{V_o} &= -\frac{R_f}{R_1}\mathbf{V_i} \\ &= -\left(\frac{10k}{5.1k}\right)(3.04\angle - 0.285°) \\ &= -5.96\angle - 0.285°\text{ V}. \end{aligned}$$

In time domain notation, the output voltage is,

$$v_o = -5.96\cos(\omega_o t - 0.285°)\text{ V}.$$

Note that the resulting output voltage requires the use of the phase of the two input signals. The output voltage in this case is the amplifier gain, $-R_f/R_1 = -1.96$ multiplied by the phasor sum of the two input voltages. This example demonstrates that proper attention to the phase and frequency of the input signals is required when designing and analyzing circuits.

1.3.3 NON-INVERTING AMPLIFIER

A non-inverting amplifier is shown in Figure 1.15 where the source is represented by v_S and a series resistance R_S.

The analysis of the non-inverting amplifier in Figure 1.15 assumes an ideal OpAmp operating within its linear region. The voltage and current constraints at the input to the OpAmp yield the voltage at node 1,

$$v_1 = v_2 = v_S,$$

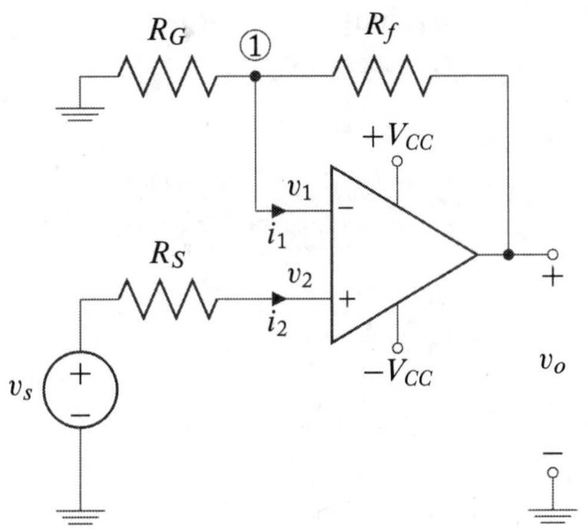

Figure 1.15: Non-inverting amplifier configuration.

since $i_1 = i_2 = 0$. Using the node voltage method of analysis, the sum of the currents flowing into node 1 is,

$$0 = \frac{0 - v_1}{R_G} + \frac{v_o - v_1}{R_f}. \tag{1.35}$$

Solving for the output voltage v_o using the voltage constraints, $v_1 = v_S$

$$v_o = v_i \left(1 + \frac{R_f}{R_G}\right). \tag{1.36}$$

The gain of the non-inverting amplifier is,

$$\frac{v_o}{v_i} = 1 + \frac{R_f}{R_G}. \tag{1.37}$$

Unlike the inverting amplifier, the non-inverting amplifier gain is positive. Therefore, the output and input signals are ideally in phase. The amplifier will operate in its linear region when,

$$1 + \frac{R_f}{R_G} < \left|\frac{V_{CC}}{v_s}\right|. \tag{1.38}$$

Note that, like the inverting amplifier, the gain is a function of the external resistors R_f and R_G.

1.3.4 DIFFERENCE AMPLIFIER

The output voltage signal of a difference amplifier is proportional to the difference of the two input voltage signals. A schematic of a difference amplifier is shown in Figure 1.16.

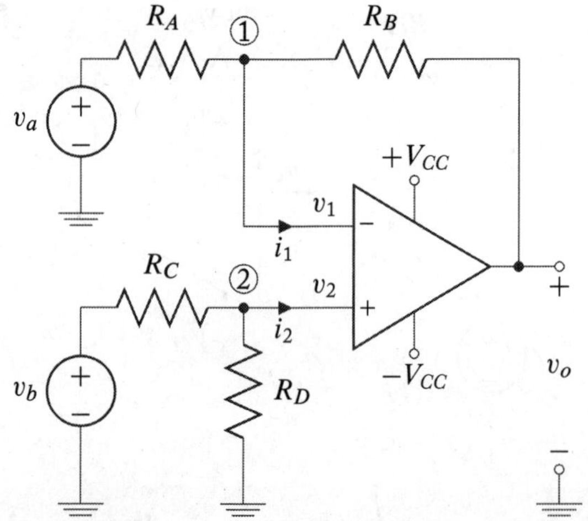

Figure 1.16: Difference amplifier with input voltages v_a and v_b.

By assuming an ideal OpAmp operating in the linear region, the current constraints can be used to yield the voltages at nodes 1 and 2 as a simple voltage division at the non-inverting input:

$$v_1 = v_2 = v_b\left(\frac{R_D}{R_C + R_D}\right). \tag{1.39}$$

The node voltage method of analysis is used to determine the output voltage v_o with respect to the input voltages v_a and v_b,

$$0 = \frac{v_a - v_1}{R_A} + \frac{v_o - v_1}{R_B}. \tag{1.40}$$

Solving for the output voltage v_o yields,

$$v_o = \frac{R_B}{R_A}(v_1 - v_a) + v_1. \tag{1.41}$$

Substituting Equation (1.39) into (1.41) provides the output voltage as a function of the input voltages,

$$v_o = \frac{R_D}{R_C + R_D}\left(\frac{R_B}{R_A} + 1\right)v_b - \frac{R_B}{R_A}v_a. \tag{1.42}$$

The expression for the output voltage in Equation (1.42) can be simplified for the particular case where the resistor ratios are given by:

$$\frac{R_A}{R_B} = \frac{R_C}{R_D}. \tag{1.43}$$

By applying the ratio of Equation (1.43), the output voltage in Equation (1.42) is reduced to a scaled difference of the input voltages,

$$v_o = \frac{R_B}{R_A}(v_b - v_a)\,. \tag{1.44}$$

The difference amplifier is commonly used in circuits that require comparison of two signals to control a third (or output) signal. For instance, v_a could be a voltage reading representing temperature from a thermistor (a resistor that changes values with temperature) circuit and v_b a reference voltage representing a temperature setting. The output of the difference amplifier would then be the deviation of the measured temperature from the reference temperature setting.

Example 1.5

The difference amplifier in Figure 1.17 has an input voltage $v_a = 3$ V. What values of v_b will result in operation in the linear region?

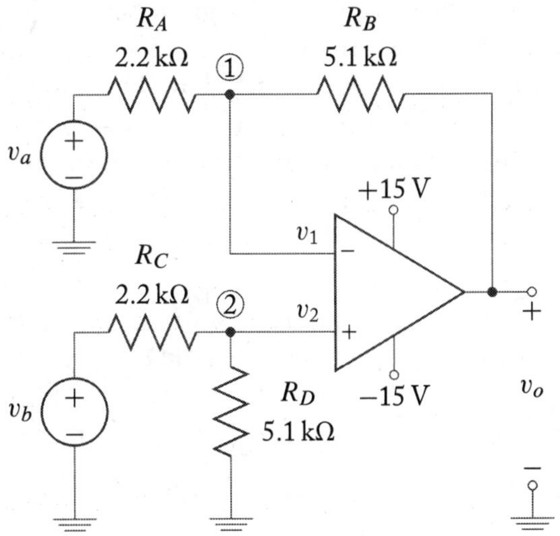

Figure 1.17: Difference amplifier of Example 1.5.

Solution:

The limits on the output voltage are determined by the power supply rail voltages. In this example, the supply voltages are +15 V and −15 V. Therefore, the output voltage must be,

$$-15\text{ V} < v_o < +15\text{ V}.$$

Since $\frac{R_A}{R_B} = \frac{R_C}{R_D}$ the input voltage v_b from Equation (1.44) can be calculated,

$$v_b = \frac{R_A}{R_B} v_o + v_a.$$

Substituting $R_A = 2.2\,\text{k}\Omega$ and $R_B = 5.1\,\text{k}\Omega$ into the equation for v_b yields for the upper and lower limits of the output voltage,

$$v_o = +15\,\text{V} : \qquad v_b = 9.47\,\text{V},$$

and

$$v_o = -15\,\text{V} : \qquad v_b = -3.47\,\text{V}.$$

Then the input voltage range for v_b to insure linear operation of the amplifier is,

$$-3.47\,\text{V} < v_o < 9.47\,\text{V}.$$

The Multisim schematic and simulation results are given below. The simplified model of the OpAmp is used. The voltage v_b is swept from -15 V to 15 V. The output voltage is at node V_o. Note that the node voltage at the output extends well above 15 V and below -15 V. The excursion occurs because the simplified OpAmp model assumes operation in the linear region. Therefore, care must be taken when using this model to take into account the limits on the output voltage. As shown in the output plot, the range of values for v_b is from -3.47 V to 9.47 V as indicated by the cursors for output voltages of -15 V and 15 V, respectively.

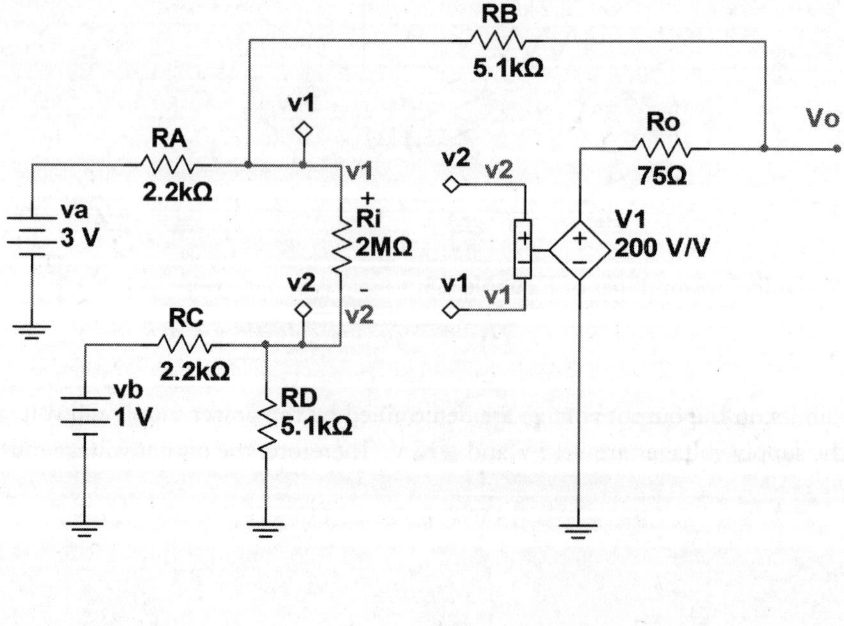

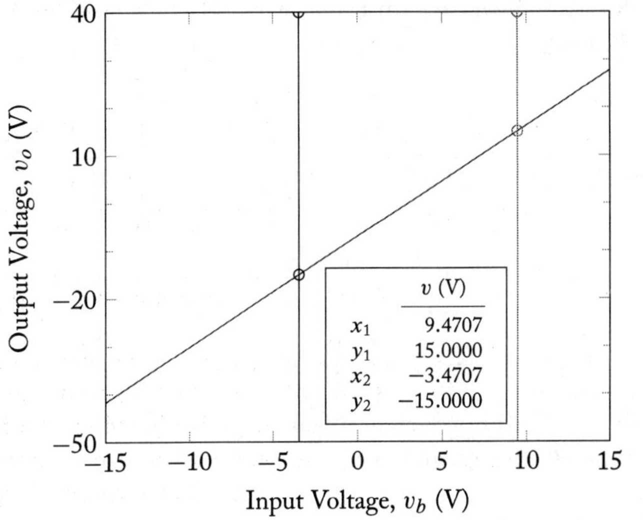

1.3.5 INTEGRATOR

The integrator is commonly used in signal generation or processing applications. The name of the circuit is accurately descriptive: the integrator performs an integration operation on the input signal. An integrator is shown in Figure 1.18. The circuit is similar to the inverting amplifier with the feedback resistor R_f replaced by a capacitor C.

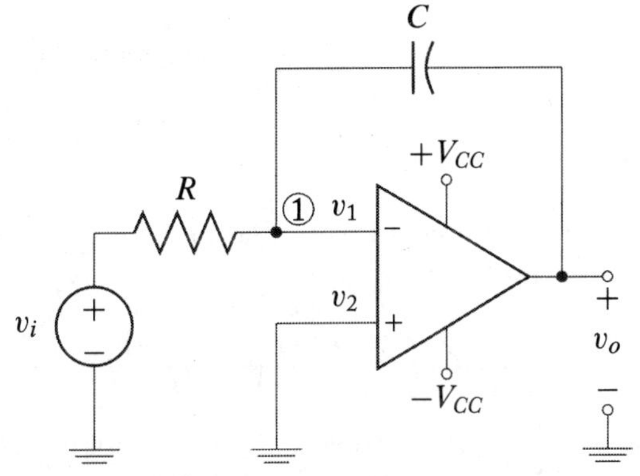

Figure 1.18: Integrator circuit.

With an ideal OpAmp operating in its linear region, the node voltage method of analysis can be applied at node 1 using the OpAmp voltage and current constraints,

$$0 = \frac{v_i - v_1}{R} + C\frac{d}{dt}(v_o - v_1). \tag{1.45}$$

But $v_1 = 0$ due to the virtual ground so,

$$0 = \frac{v_i}{R} + C\frac{dv_o}{dt}. \tag{1.46}$$

Solving for v_o yields,

$$v_o = -\frac{1}{RC}\int v_i\, dt. \tag{1.47}$$

Equation (1.47) shows that the output voltage of an integrator circuit is a product of the reciprocal of the RC time constant and the integral of the inverted input signal.

1.3.6 DIFFERENTIATOR

If the capacitor and resistor positions in the integrator schematic are switched, the circuit performs a differentiation operation on the input signal. The resulting circuit is shown in Figure 1.19.

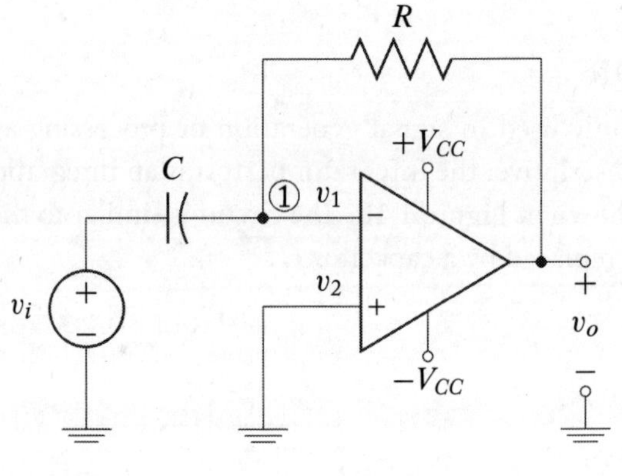

Figure 1.19: Differentiator circuit.

The analysis of the differentiator circuit is similar to the integrator. Apply the node voltage method of analysis at node 1 assuming ideal OpAmp characteristics to yield,

$$0 = C\frac{d(v_i - v_1)}{dt} + \frac{v_o - v_1}{R} \tag{1.48}$$

Using the voltage constraint $v_1 = v_2 = 0$, Equation (1.48) simplifies to,

$$0 = C\frac{dv_i}{dt} + \frac{v_o}{R} \tag{1.49}$$

The output voltage v_o is therefore,

$$v_o = -RC\frac{dv_i}{dt}. \tag{1.50}$$

Equation (1.50) shows that the output voltage of a differentiator circuit is a product of the RC time constant and the derivative of the inverted input signal.

1.4 DIFFERENTIAL AMPLIFIERS

A differential amplifier is any two-input amplifier that has an output proportional to the difference of the inputs. The defining equation for a differential amplifier is then:

$$y_o = A\left(x_{i1} - x_{i2}\right), \tag{1.51}$$

where the output, y_o, and the inputs $\{x_i\}$ could be either voltages or currents. Previous discussions in this chapter have explored two differential amplifiers: the difference amplifier (shown in Figure 1.20) and the basic OpAmp itself. Each of these two examples has an output voltage that is proportional to the difference of two input voltages. In the case of the difference amplifier, the output expression was derived to be:

$$v_o = \frac{R_B}{R_A}\left(v_{i2} - v_{i1}\right), \tag{1.52}$$

if the resistor values were chosen so that

$$\frac{R_A}{R_B} = \frac{R_C}{R_D}. \tag{1.53}$$

Ideally this amplifier (or any differential amplifier) is sensitive only to the difference in the two

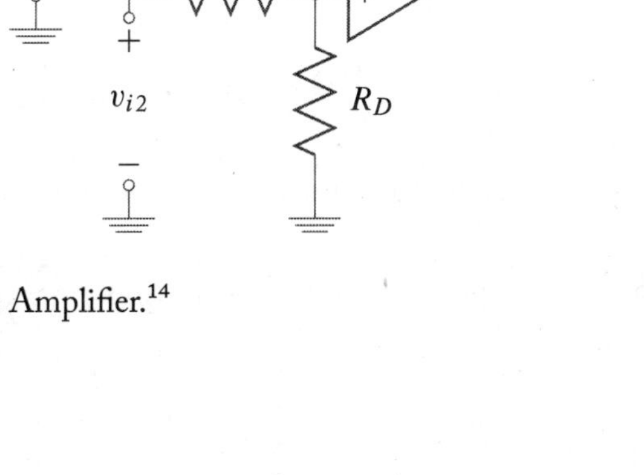

Figure 1.20: A Difference Amplifier.[14]

input signals, and is completely insensitive to any common component of the two signals. That is, if the difference in inputs remains constant, the output should not vary if the average value of the two inputs changes. Unfortunately, a differential amplifier rarely meets this goal, and the output has a slight dependence on the average of the input signals. The output for this type of imperfect differential amplifier is given by:

$$v_o = A_{DM} v_{iDM} + A_{CM} v_{iCM} = A_{DM}\left(v_{i2} - v_{i1}\right) + A_{CM}\left(\frac{v_{i2} + v_{i1}}{2}\right), \tag{1.54}$$

where

$$A_{DM} = \text{the amplification of the input signal difference, } v_2 - v_1,$$

and

$$A_{CM} = \text{the amplification of the input signal average, } \frac{(v_2+v_1)}{2}.$$

The *quality* of a differential amplifier is displayed in its ability to amplify the differential signal while suppressing the common signal. A measure of this quality is the ratio of the differential gain to the amplification of the average (or common) part of the input signals. The measure of quality is named *Common-mode rejection ratio* (CMRR) and is usually expressed in decibels (dB). The defining equation for CMRR is:

$$\text{CMRR} = 20\log_{10}\left|\frac{A_{DM}}{A_{CM}}\right|. \tag{1.55}$$

Unfortunately, usual analysis procedures do not produce an expression in the form of Equation (1.54): the differential-mode gain, A_{DM}, and the common-mode gain, A_{CM}, are not the usual results of analysis. A more typical result of analysis procedures is:

$$v_o = A_1 v_{i1} + A_2 v_{i2}. \tag{1.56}$$

A conversion between the two output expressions can be obtained by realizing that:

$$v_{i1} = \frac{(v_{i2} + v_{i1})}{2} - \frac{(v_{i2} - v_{i1})}{2} = v_{iCM} - \frac{1}{2} v_{iDM}, \tag{1.57a}$$

and

$$v_{i2} = \frac{(v_{i2} + v_{i1})}{2} + \frac{(v_{i2} - v_{i1})}{2} = v_{iCM} + \frac{1}{2} v_{iDM}. \tag{1.57b}$$

If Equations (1.57a) and (1.57b) are combined with Equation (1.56) the result is:

$$v_o = \left(\frac{A_2 - A_1}{2}\right)\left(v_{i2} - v_{i1}\right) + \left(A_1 + A_2\right)\left(\frac{(v_{i2} + v_{i1})}{2}\right), \tag{1.58}$$

The conclusions easily drawn from Equations (1.58) and (1.54) are:

$$A_{DM} = \left(\frac{A_2 - A_1}{2}\right) \qquad \text{and} \qquad A_{CM} = (A_1 + A_2)\,. \tag{1.59}$$

A good differential amplifier has $A_1 \approx -A_2$: the differential-mode gain will be large and the common-mode gain small. The CMRR will be large for a good differential amplifier: as an example, the μA741 OpAmp has a typical CMRR of 90 dB with a guaranteed minimum CMRR of 70 dB.[15]

Example 1.6

The difference amplifier of Figure 1.20 is constructed with an ideal OpAmp and 1% tolerance resistors of nominal values 2.2 kΩ and 5.1 kΩ. The resistors were measured and found to have the following resistance values:

$$R_A = 2.195\,\text{k}\Omega \qquad R_C = 2.215\,\text{k}\Omega$$
$$R_B = 5.145\,\text{k}\Omega \qquad R_D = 5.085\,\text{k}\Omega\,.$$

Determine gain of the differential amplifier and its common-mode rejection ratio.

Solution:

The design gain of this amplifier is:

$$A = \frac{R_B}{R_A} = \frac{5.1\,\text{k}\Omega}{2.2\,\text{k}\Omega} = 2.318.$$

However, this value is based on the assumption of equal resistor ratios:

$$\frac{R_A}{R_B} = \frac{R_C}{R_D}.$$

The quality of this difference amplifier depends strongly on whether the resistor ratio described in Equation (1.53) is *exactly* valid. In this case, 0.42663 ≠ 0.43559. Typical resistor variation leads to the conclusion that the Equation (1.53) is slightly in error and the more exact expression for the output voltage as a function of the inputs is more correct:

$$v_o = \frac{R_D}{R_C + R_D}\left(\frac{R_B}{R_A} + 1\right) v_{i2} - \frac{R_B}{R_A} v_{i1}.$$

This input-output transfer function is of the general form of Equation (1.56):

$$v_o = A_1 v_{i1} + A_2 v_{i2}.$$

[15]CMRR is dependent on external circuitry as well as the fundamental properties of an OpAmp. The condition under which the μA741 measurements is made is: the output resistance of the source (and any series resistance between the source and the OpAmp) must be less than 10 kΩ.

where

$$A_2 = \frac{R_D}{R_C + R_D}\left(\frac{R_B}{R_A} + 1\right) \qquad \text{and} \qquad A_1 = -\frac{R_B}{R_A}.$$

The common-mode and differential-mode gains can now be evaluated using the correct gain expression and Equations (1.59).

$$A_2 = \frac{5.085}{2.215 + 5.085}\left(\frac{5.145}{2.195} + 1\right) = 2.329 \qquad \text{and} \qquad A_1 = -\frac{5.145}{2.195} = -2.344$$

therefore

$$A_{DM} = 2.337 \qquad \text{and} \qquad A_{CM} = -0.01464$$

The CMRR is then obtained using Equation (1.55):

$$\text{CMRR} = 20\log_{10}\left|\frac{A_{DM}}{A_{CM}}\right| = 20\log|-159.6| = 44.06\,\text{dB}.$$

The ratio of differential-mode gain to common-mode gain is about 160: this is only a moderately good differential amplifier. A good circuit designer would notice that the resistors in this example were paired in the worst possible manner. If the physical resistors used for R_A and R_C were exchanged, the resistor ratios in each gain path would be more nearly exact: $0.4305 \gg 0.4317$. Continuing with the gain calculations for this new configuration leads to:

$$A_2 = \frac{5.085}{2.195 + 5.085}\left(\frac{5.145}{2.215} + 1\right) = 2.321 \qquad \text{and} \qquad A_1 = -\frac{5.145}{2.215} = -2.323$$

with

$$A_{DM} = 2.322 \qquad \text{and} \qquad A_{CM} = -\ 0.00186$$

which results in a CMRR of:

$$\text{CMRR} = 20\log_{10}\left|\frac{A_{DM}}{A_{CM}}\right| = 20\log|-1248.0| = 61.92\,\text{dB}$$

Rearranging the resistors has brought an improvement in the CMRR of almost 18 dB: the ratio of gains has been improved by a factor of about 7.8. Obviously care must be taken in the choice and placement of element values to provide the optimum amplifier.

Differential amplifiers are not restricted to circuits with single OpAmps. It is possible to construct a differential amplifier without any OpAmps, while many differential amplifiers have three or more OpAmps as essential elements. Figure 1.21 shows the basic schematic representation of an instrumentation amplifier using two OpAmps. Instrumentation amplifiers are high performance voltage amplifiers that are primarily used for the initial amplification of signals from a variety of types of transducers. They are available packaged as a single item in a DIP package or can be realized with discreet components. Packaged instrumentation amplifiers usually have greater control on the factors contributing to CMRR and are often the advantageous choice for the circuit designer.

The basic topology of this particular instrumentation amplifier is that of an inverting amplifier connected in series with a summing amplifier. The extra resistors at the positive terminals of each OpAmp, R_2 and R_2', serve no obvious function if the OpAmps are considered to be ideal: their function is to reduce the effects of input parameter variations which are non-ideal properties of OpAmps.[16] The inversion of v_{i1} prior to summation allows for the output of the amplifier to be a multiple of the difference of the two inputs. This particular circuit topology also allows, with appropriate external resistor choices, for large (on the order of 100 V) input voltages.

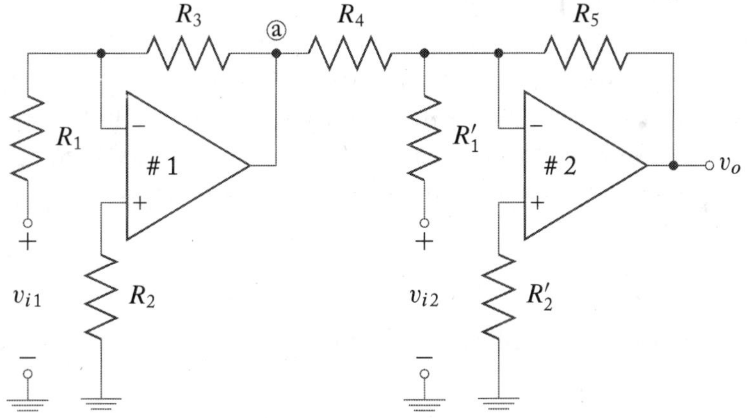

Figure 1.21: An instrumentation amplifier with high input voltage capability.

Analysis of this amplifier begins, as usual, with the assumption that each OpAmp is near-ideal: the input are virtually-shorted, the gain is infinite, the input resistance is infinite, and the output resistance is zero. Since no current flows into the inputs of either OpAmp, there is no voltage drop across the resistors, R_2 and R_2'. The OpAmp positive input terminals are therefore at ground potential: each OpAmp circuit acts in the same manner as if its positive terminal were directly connected to ground. The zero output resistance of the first OpAmp implies that the remainder of the circuit does not affect its output, and the voltage at node a is obtained from the gain equation for an inverting amplifier:

$$v_a = -\frac{R_3}{R_1} v_{i1}. \tag{1.60}$$

The zero output resistance of the first OpAmp circuit also implies that it acts as a perfect voltage source input to the summing amplifier of the second OpAmp. The output of the summing

[16]The effects of input parameter variations on OpAmp performance is discussed in Section 1.5.

amplifier is then given by:

$$v_o = -\frac{R_5}{R_4}v_a - \frac{R_5}{R_1'}v_{i2}; \tag{1.61}$$

or, with the results of Equation (1.60),

$$v_o = \frac{R_5 R_3}{R_4 R_1}v_{i1} - \frac{R_5}{R_1'}v_{i2}. \tag{1.62}$$

If proper resistor value choices are made, this instrumentation amplifier becomes a true differential amplifier. The necessary restriction on the resistor values to create a differential amplifier is:

$$\frac{R_1 R_4}{R_3} = R_1'. \tag{1.63}$$

Additional decisions are made for good design.[17] It is often important to load the input sources equally, the input resistance at each input to the instrumentation amplifier is therefore set to the same value: $R_1' = R_1$. The restriction of Equation (1.63) then requires that $R_4 = R_3$. With these restrictions, the final idealized output relationship for this instrumentation amplifier is:

$$v_o = \frac{R_5}{R_1}\left(v_{i1} - v_{i2}\right), \tag{1.64}$$

which is of the general form for a differential amplifier.

Example 1.7

Determine the common-mode and differential-mode gains and the common-mode rejection ratio for the instrumentation amplifier of Figure 1.21 with resistor values:

$$\begin{aligned} R_1 &= 50.15\,\text{k}\Omega \qquad & R_1' &= 49.80\,\text{k}\Omega \qquad & R_3 &= 10.05\,\text{k}\Omega \\ R_2 &= 8.215\,\text{k}\Omega \qquad & R_2' &= 8.250\,\text{k}\Omega \qquad & R_4 &= 9.965\,\text{k}\Omega \end{aligned}$$

Solution:

In order to determine CMRR it is necessary to use the results of Equation (1.62) to determine the gains. The input-output relationship is given by:

$$v_o = \frac{(49.85)\,(10.05)}{(9.965)\,(50.15)}v_{i1} - \frac{49.85}{49.80}v_{i2} = 1.0025v_{i1} - 1.0010v_{i2}.$$

The differential-mode gain and the common-mode gains are calculated using Equation (1.59) and found to be:

$$A_{DM} = 1.00175 \quad \text{and} \quad A_{CM} = 0.001493.$$

[17] Additional design guidelines are discussed in Section 1.5.

CMRR is calculated using Equation (1.55) and is given by:

$$\text{CMRR} = 20 \log_{10} \left| \frac{A_{DM}}{A_{CM}} \right| = 20 \log |671.07| = 56.54 \, \text{dB}.$$

1.5 NON-IDEAL CHARACTERISTICS OF OPAMPS

In this section the most significant limitations of the non-ideal Operational Amplifier are discussed. A fundamental understanding of these non-ideal properties allows the electronics designer to choose circuit topologies and parameter values so that the performance of real, practical circuitry closely approximates the ideal case. The concept of an *ideal* OpAmp has allowed the use of simplified circuit analysis techniques to determine the performance of OpAmp circuits and concentration on the design philosophy behind the various OpAmp circuit topologies. The ideal OpAmp was defined with the following properties:

- Infinite Voltage Gain
- Infinite Input Resistance
- Zero Output Resistance
- Output Independent of Power Source Characteristics
- Properties Independent of Input Frequency

A number of non-ideal characteristics have been considered briefly in prior sections of this chapter:

- Output Saturation
- Finite Input Resistance
- Finite Voltage Gain
- Non-zero Output Resistance

These characteristics will be further discussed along with the following additional non-ideal characteristics:

- Input Parameter Variations
- Output Parameter Limitations
- Supply and Package Related Parameters

In addition, the performance of an OpAmp is dependent on the frequency of the input signals. In many low-frequency applications this frequency dependence is not significant: OpAmps are commonly used in the audio frequency range and beyond without significant distortion. A discussion of frequency dependent behavior and its close relative, slew rate, is beyond the scope of this section: a discussion of the frequency dependence of OpAmp circuit performance can be found in Section 9.9 (Book 3).

1.5.1 FINITE GAIN, FINITE INPUT RESISTANCE AND NON-ZERO OUTPUT RESISTANCE

The properties of finite gain, finite input resistance and non-zero output resistance were observed in previous discussions concerning using an OpAmp to create a unity gain buffer in Section 1.2. The effect of these non-ideal properties was approached through a simple equivalent model of the OpAmp as shown in Figure 1.7: that approach will be continued here. While the effects due to these properties may vary slightly with OpAmp application, a demonstration of their typical effects on circuit performance will be illustrated using the basic inverting amplifier configuration.

An inverting amplifier can be constructed using an OpAmp as shown in Figure 1.22. While the power supply, $\pm V_{CC}$, is not shown in this figure, its presence is assumed.

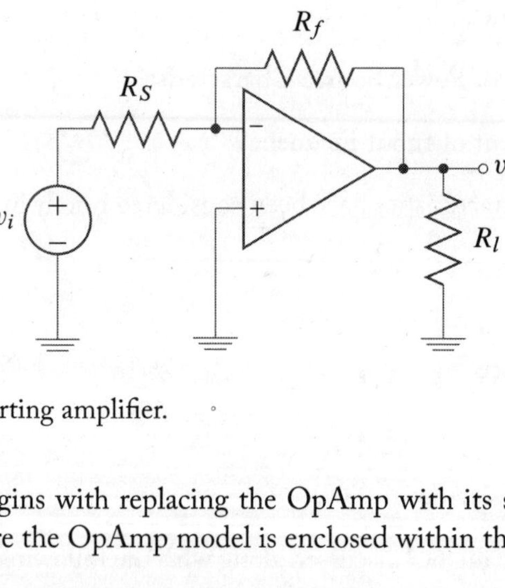

Figure 1.22: A simple inverting amplifier.

Analysis of this circuit begins with replacing the OpAmp with its simple equivalent circuit as shown in Figure 1.23. Here the OpAmp model is enclosed within the dashed box with an input voltage, $v_+ = v_2 - v_1$.

Equations expressing Kirchhoff's Current Law at the input and output terminals of the OpAmp are the first step to obtaining an expression for the voltage gain:

$$\frac{v_i - (-v_+)}{R_S} + \frac{v_+}{R_i} + \frac{v_o - (-v_+)}{R_f} = 0 \tag{1.65}$$

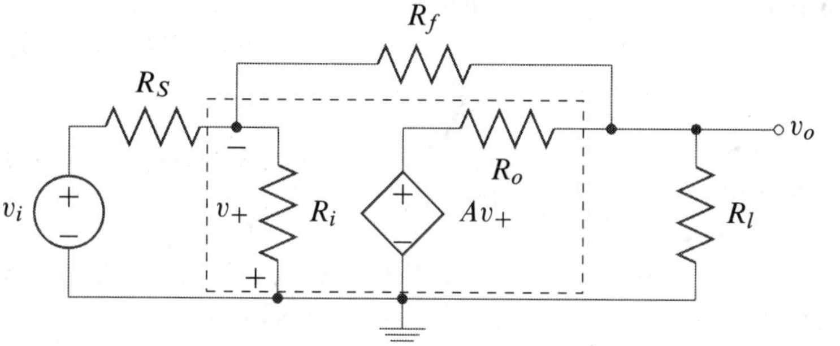

Figure 1.23: Inverting amplifier equivalent circuit.

and

$$\frac{Av_+ - v_o}{R_o} - \frac{v_o}{R_l} + \frac{(-v_+) - v_o}{R_f} = 0. \tag{1.66}$$

Solving for v_+ in Equation (1.66) yields:

$$v_+ = v_o \left(\frac{G_o + G_l + G_f}{AG_o - G_f} \right) \tag{1.67}$$

where the subscripted quantities $\{G_x\}$ are conductances corresponding to the resistances with the same subscript $\{R_x\}$. For example:

$$G_f = \frac{1}{R_f} \qquad \text{and} \qquad G_o = \frac{1}{R_o}.$$

Equations (1.67) and (1.65) can now be combined to obtain an expression for the voltage gain:

$$\frac{v_o}{v_i} = -\frac{G_S}{G_f + \frac{(G_o+G_l+G_f)(G_S+G_i+G_f)}{(AG_o-G_f)}}. \tag{1.68}$$

The expression for voltage gain using the ideal OpAmp model in Section 1.3 was:

$$\frac{v_o}{v_i} = -\frac{R_f}{R_S} = -\frac{G_S}{G_f}. \tag{1.69}$$

Consideration of the non-ideal characteristics of an OpAmp has added complexity to the gain function and increases the magnitude of the denominator of the expression: the overall gain is reduced. Good circuit design practices imply that near-ideal performance is the desired goal. With *appropriate* external element choices, the gain function can approach the ideal. A first obvious design choice to decrease the size of the additional term in the denominator of Equation (1.68) is to make the parallel combination of R_l and R_f large with respect to the output impedance of the OpAmp, R_o. This choice of resistors is equivalent to making $G_o \gg (G_l + G_f)$, which allows for the simplification of the gain expression to:

$$\frac{v_o}{v_i} \approx -\frac{G_S}{G_f + \frac{1}{A}\left(G_S + G_i + G_f\right)} = -\frac{R_f // \left\{A \times \left(R_S // R_i // R_f\right)\right\}}{R_S}. \tag{1.70}$$

If A is large, the parallel combination of R_f and $A \times (R_S // R_i // R_f)$ will be very close to R_f in value and the gain of the circuit will be near that of the idealized case.

Example 1.8

Given the following circuit parameters for an inverting amplifier,

$$R_f = 47\,\mathrm{k\Omega} \qquad R_S = 10\,\mathrm{k\Omega} \qquad R_l = 22\,\mathrm{k\Omega},$$

and non-ideal OpAmp parameters:

$$R_i = 2\,\mathrm{M\Omega} \qquad R_o = 75\,\Omega \qquad A = 200{,}000$$

Determine the voltage gain of the amplifier and compare to the ideal gain.

Solution:

The admittances are first calculated:

$$\begin{aligned} G_f &= 21.28\,\mathrm{mS} \qquad & G_i &= 0.500\,\mathrm{mS} \qquad & G_l &= 45.46\,\mathrm{mS} \\ G_S &= 100.0\,\mathrm{mS} \qquad & G_o &= 13.33\,\mathrm{mS}. \end{aligned}$$

Equation (1.68) becomes:

$$\begin{aligned} \frac{v_o}{v_i} &= -\frac{100\mu}{21.28\mu + \dfrac{(13.33\mathrm{m} + 45.46\mu + 21.28\mu)\,(100\mu + 0.5\mu + 21.28\mu)}{(200{,}000 \times 13.33\mathrm{m} - 21.28\mu)}} \\ &= -\frac{100\mu}{21.28\mu + 611.96\mathrm{p}} = -4.699865. \end{aligned}$$

This result corresponds to a –0.0029% change in value from the ideal case (–4.7). Obviously, proper choices lead to near-ideal performance. Equation (1.70) could have also been used with the given set of circuit parameter values (the resistor values fit the necessary restriction): it yields similar results (–4.699865).

The input resistance of an inverting amplifier using a non-ideal OpAmp can be obtained using many of the results from the gain calculations previously derived. In order to simplify the process, R_S is removed from the circuit, as shown in Figure 1.24, and the Thévenin input resistance of the remaining circuit is calculated: the input resistance of the total amplifier will be:

$$R_{in} = R_S + R_{th}. \tag{1.71}$$

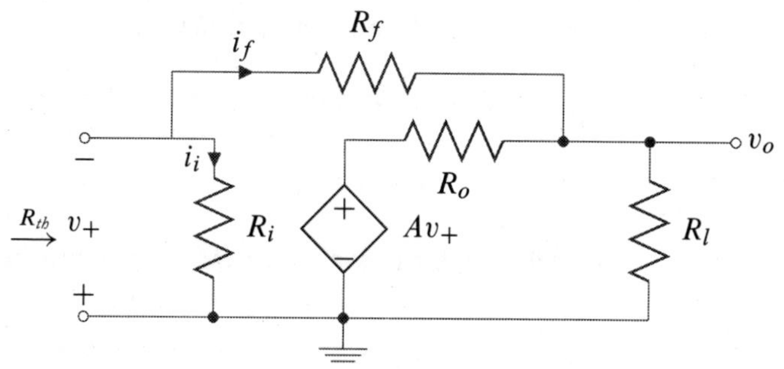

Figure 1.24: An inverting amplifier equivalent circuit with R_S removed.

Calculation of the Thévenin input resistance begins with determination of the two currents, i_i and i_f:

$$i_i = \frac{-v_+}{R_i} \tag{1.72}$$

and

$$i_f = \frac{-v_+ - v_o}{R_f}. \tag{1.73}$$

Equation (1.67) can be combined with Equation (1.73) to eliminate v_o:

$$i_f = -v_+\left[G_f\left\{1 + \frac{AG_o - G_f}{G_0 + G_l + G_f}\right\}\right]. \tag{1.74}$$

Which leads to the Thévenin resistance:

$$R_{th} = \frac{-v_+}{i_i + i_f} = \frac{1}{G_i + G_f\left\{1 + \frac{AG_o - G_f}{G_o + G_l + G_f}\right\}} \tag{1.75}$$

and the total input resistance, R_{in}:

$$R_{in} = R_S + R_{th} = R_S + \frac{1}{G_i + G_f\left\{1 + \frac{AG_o - G_f}{G_o + G_l + G_f}\right\}}. \tag{1.76}$$

The input resistance has been increased by the quantity R_{th}. In order to make the non-ideal performance mirror the ideal approximations, *appropriate* external element choices can be made. Once again, if the resistors are chosen so that $R_I//R_f \gg R_o$, the expression for the input resistance reduces to:

$$R_{in} \approx R_S + \frac{1}{G_i + G_f\,(1 + A)} = R_S + \left\{ R_i // \frac{R_f}{1 + A} \right\}. \tag{1.77}$$

If Equation (1.77) is to be a reasonable approximation to the idealized expressions ($R_{in} = R_S$), it is necessary that R_f be limited in magnitude. If R_{in} is not to vary by more than a few ohms from the ideal value, a reasonable choice for the maximum value of R_f is:

$$R_f < 1 + A\,\Omega. \tag{1.78}$$

Example 1.9

Given the circuit of Example 1.8, determine the input resistance of the inverting amplifier and compare to the ideal case.

Solution:

Equation (1.76) yields:

$$R_{in} = 10\,\text{k}\Omega + 0.2362\,\Omega.$$

The appropriate choices for external resistor values have been made for Equation (1.77) to be valid. It yields a value for the input resistance of:

$$R_{in} = 10\,\text{k}\Omega + 0.2350\,\Omega.$$

The results are deviations of less than 0.0024% from the ideal value of 10 kΩ.

Calculation of the output resistance of an inverting amplifier using a non-ideal OpAmp is accomplished using Thévenin techniques. The input source is set to zero, the load is removed (output resistance calculations rarely include the load), and the output is driven by a voltage source as shown in Figure 1.25. The ratio of the driving voltage, v_t, to the driving current, i_t, gives the output resistance, R_{out}.

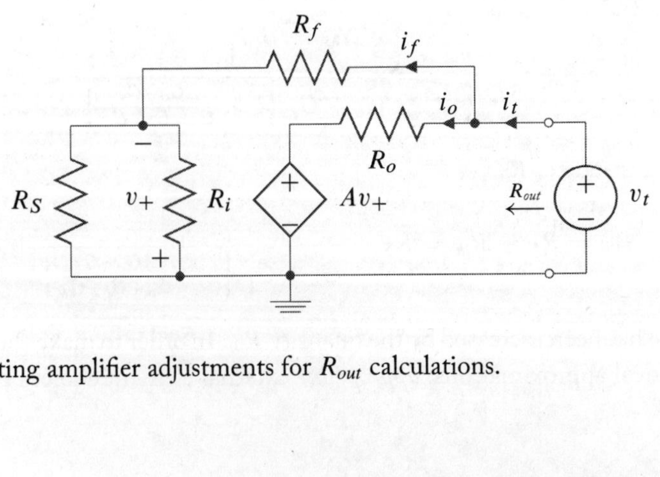

Figure 1.25: Inverting amplifier adjustments for R_{out} calculations.

The driving current is the sum of the currents through R_f and R_o:

$$i_t = i_o + i_f. \tag{1.79}$$

The current through R_f is given by:

$$i_f = \frac{v_t}{R_f + R'}, \tag{1.80}$$

where

$$R' = R_S//R_i.$$

The current through R_o is given by:

$$i_o = \frac{v_t - Av_+}{R_o}, \tag{1.81}$$

where v_+ can be obtained through a simple voltage division:

$$v_+ = -\frac{R'}{R' + R_f} v_t. \tag{1.82}$$

Combining (1.81) and (1.82) with (1.80) and (1.78) gives the total driven current:

$$i_t = \frac{R_o + R_f + (1 + A)\, R'}{R_o \left(R_f + R'\right)} v_t \tag{1.83}$$

and finally the output resistance:

$$R_{out} = \frac{v_t}{i_t} = \frac{R_o \left(R_f + R'\right)}{R_o + R_f + (1 + A)\, R'} = \frac{R_o \left(R_f + R_S//R_i\right)}{R_o + R_f + (1 + A)\left(R_S//R_i\right)}. \tag{1.84}$$

Reasonable assumptions (such as have been previously described) on the choice of external resistor values lead to an approximation of the output resistance expression:

$$R_{out} \approx \frac{R_o \left(R_f + R_S\right)}{(1 + A)\, R_S}. \tag{1.85}$$

Example 1.10

Given the circuit of Example 1.8, determine the output resistance of the inverting amplifier and compare to the ideal case.

Solution:

Equation (1.84) yields:

$$R_{out} = 0.00215\,\Omega.$$

The external resistor simplification parameters the use of Equation (1.85) have been met, therefore,

$$R_{out} \gg 0.00214\,\Omega\,.$$

The idealized value of $R_{out} = 0$ seems justified for practical circuitry given appropriate choices of external components.

It has been shown that finite input impedance, finite gain, and non-zero output resistance do have an effect on the performance of an inverting amplifier. Still, it is possible to have near-ideal performance if appropriate choices on the external circuitry are imposed. A reasonable set of restrictions on the external components has been shown to be:

- $R//R_f \gg R_o$
- $R_f < (1 + A)\,\Omega$

If resistances connected to the inputs and output of an OpAmp circuit obey these general restrictions, the OpAmp circuit performance will be near-ideal. Any non-ideal variations can be detected with a computer simulation: good circuit design practice always includes simulation.

1.5.2 INPUT PARAMETER VARIATIONS

When an ideal OpAmp has zero output, one expects that the input will have the following properties:

- the voltage difference at the input will be zero, and
- each input current will be zero.

Real OpAmps have input voltage differences and currents that vary from the ideal. These differences are described by the quantities Input Offset Voltage, Input Bias Current, and Input Offset Current.

Input Offset Voltage

The input offset voltage, V_{OS}, is defined as the difference in voltage between the OpAmp input terminals when the output voltage is zero.[18] This voltage difference is due to slightly different properties of the input circuitry at each of the input terminals.

For typical OpAmps the offset voltage is a few millivolts or less, and can often be nulled with an external three-terminal potentiometer connected between the offset null terminals of the OpAmp with the middle terminal of the potentiometer connected either to ground or one of the supply voltage terminals (the connections vary with OpAmp type and manufacturer). If the

[18] An alternate (but entirely equivalent) definition is: V_{OS} is the differential input voltage which must be applied to drive the output voltage to zero.

offset voltage is not nulled, it appears as an additional input voltage in series with the true inputs to the OpAmp (See Figure 1.26). The input offset voltage is also a function of temperature: the nulling circuitry may need to be adjusted as OpAmp temperature varies. The offset voltage and its variation with temperature place a lower limit on the magnitude of DC voltages that can act as inputs to an OpAmp without erroneous circuit operation.

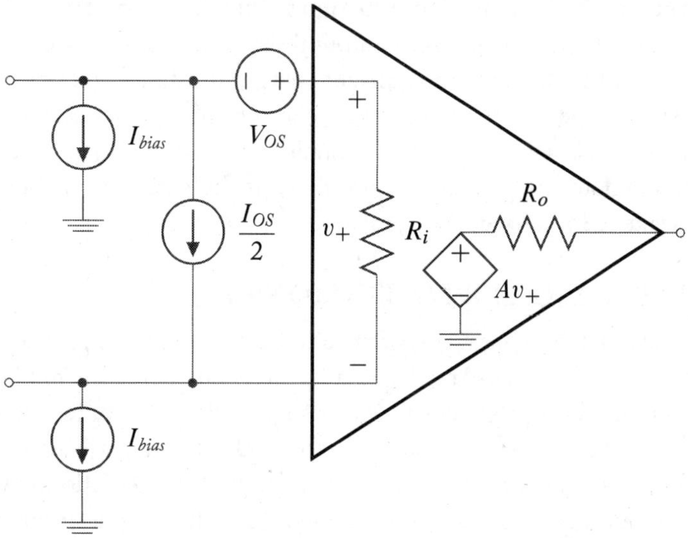

Figure 1.26: Equivalent circuit for an OpAmp including input offset voltage and current, input bias current, finite input and non-zero output resistance, and finite voltage gain.

Input Bias and Offset Current

The input circuitry of an OpAmp also draws a small amount of current: this non-zero input current is in variation to the ideal OpAmp assumptions. Input bias current is defined as the average of the two input currents when the output of the OpAmp is zero volts:

$$I_{bias} = \frac{I_{in1} + I_{in2}}{2}. \tag{1.86}$$

The magnitudes of the input bias current of an OpAmp lies in the range of a few picoamperes to tens of nanoamperes depending on the type of input circuitry. Bipolar Junction Transistor input stages tend to have larger bias currents while Field Effect Transistor input stages have smaller bias currents. In many applications the effect of a balanced bias current (both input currents equal) can be eliminated through the use of external circuit elements. Slightly different properties of the input circuitry at each of the input terminals, particularly due to random manufacturing variations, create a more serious problem. Input offset current is defined as the difference between the input currents:

$$I_{OS} = I_{in1} - I_{in2}. \tag{1.87}$$

Typical variation of input circuitry bias current is approximately 5% of the mean value. This means that the mismatch in the two currents is random (among OpAmps) and cannot be compensated by a fixed external resistor.

An equivalent circuit diagram of an OpAmp including input offset voltage and current, input bias current, finite input and non-zero output resistance, and finite voltage gain is shown in Figure 1.26. The effects of finite input and non-zero output resistance and finite voltage gain have been discussed at length: the effects of Input bias current, input offset current, and input offset voltage can be analyzed in a similar fashion. It should be noted that manufacturer specifications on these parameters denote the maximum *magnitude* of the parameter—the parameter can be either positive or negative for a particular OpAmp.

1.5.3 OUTPUT PARAMETER LIMITATIONS

The maximum output voltage swing was described in Section 1.2 as "slightly shy of $\pm V_{CC}$ due to device characteristics within the OpAmp." Manufacturers guarantee the minimum value of this parameter using a graph of output voltage swing as a function of supply voltage for a specified load resistance (often 2 kΩ or 10 kΩ). Variation in the load resistance will also alter the maximum output voltage swing: manufacturers often provide a graph of this variation as well.

The maximum output current is often specified through a graph. For protection purposes, many OpAmps have a current-limiting circuit in the output stage that will limit the maximum output current to a specified value. A μA741 OpAmp (which has a current-limiting output stage) can source, or sink, approximately 25 mA.

1.5.4 PACKAGE AND SUPPLY RELATED PARAMETERS

There are several other limitations on the operation of OpAmps that relate to the manufacturer's package and the power supply to which the OpAmp is connected. The limitations can be found in OpAmp data sheets. The primary limitations are:

- Power Dissipation
- Operating Temperature Range
- Supply Voltage Range
- Supply Current
- Power Supply Rejection Ratio

All OpAmps have a limitation on the maximum amount of power that can be dissipated safely on a continuous basis. Power dissipation is package dependent with ceramic packages having the highest rating. Metal and plastic packages have lower ratings with plastic the lowest. Typical values are in the 100–500 mW range.

OpAmps are guaranteed to operate within specifications provided the temperature of the package is within the operating temperature range specification. Commercial grade devices have a temperature range of 0°C to +70° C, the range for industrial grade devices is –25°C to +85°C, and military grade devices operate from −55°C to +125°C.

The supply voltages V_{CC} have *maximum* and *minimum* values for proper operation of the OpAmp. Typical maxima are in the range of ±18 V to ±22 V, but specialized units may operate at much higher levels. Minima are typically about ±5 V but may range as low as ±2 V. As has been mentioned before, the output voltage swing must lie within the rails set by the supply voltage. Specialized OpAmps can operate with a one-sided supply: typically the negative power supply terminal is grounded and the other power supply terminal is connected to $+V_{CC}$.

The supply current is defined as the current that an OpAmp draws from the power supply when the OpAmp output is zero. This is a particularly important parameter in battery-operated applications.

Variations in the supply voltage, $\pm V_{CC}$, can feed through to the output—typically through offset voltage variation. The ratio of the change in offset voltage to the change in power supply voltage is defined as the power-supply rejection ratio (*PSRR*):

$$PSRR = \frac{\Delta V_{OS}}{\Delta V_{CC}}$$

PSRR can be expressed in V/V or in decibels, where

$$PSRR|_{dB} = 20 \log \frac{\Delta V_{OS}}{\Delta V_{CC}}.$$

If OpAmps are used with a high-performance voltage regulator, the error due to *PSRR* can essentially be eliminated in OpAmp applications.

While the list of non-ideal OpAmp properties may seem large, each property contributes but a small error that, with careful choices of circuit topology and circuit element value, can be nearly eliminated. There is insufficient space in a text of this nature to investigate all effects in all possible circuits. While the demonstrations have been kept to a minimum, it is hoped that the reader has developed a “feel” for the most important effects and a sense of how to compensate for them. The good circuit Designer should keep all of these second-order effects in mind and act appropriately.

1.6 CONCLUDING REMARKS

The Operational Amplifier has been described in this chapter as a highly useful device with near-ideal terminal characteristics, summarized in Table 1.2. Many of common OpAmp applications can be described using only these characteristics and simple circuit analysis techniques.

Table 1.2: Idealized OpAmp characteristics

Property	Ideal OpAmp Value
Gain, A	∞
Input Resistance, R_i	∞
Output Resistance, R_o	0
Input Voltage Difference, $v_2 - v_1$	0 (virtual short)
Input Current, i_1 or i_2	0 (virtual short)
Output Voltage Limits	$\lvert v_o \rvert \leq V_{CC}$

In well-designed OpAmp applications, the properties of the application depend most strongly on the circuit elements external to the OpAmp rather than on the OpAmp itself. In order to preserve this primary dependence on the external circuit elements, certain design restrictions have been presented. In general, these restrictions relate to the resistance values connected to the terminals of the OpAmp:

- Resistors connected to the output should be large with respect to the output impedance, and

- resistors connected to the input should be less than $(1 + A)\,\Omega$.

Additional design restrictions concerning frequency response will be discussed in 9 (Book 3).

While a variety of linear applications have been examined in this chapter, the possibilities for circuitry using OpAmps extend far beyond what has been shown here. Additional OpAmp linear applications and many non-linear applications will be examined in later chapters. Later chapters will also investigate components used in the internal design of several OpAmp types and will shed light on non-ideal characteristics and the limitations these characteristics impose on OpAmp usage.

SUMMARY DESIGN EXAMPLE

In order to investigate the low-frequency volt-ampere (V-I) relationship of a two-terminal electronic device, it is often desirable to display the V-I relationship on the screen of an oscilloscope. A typical experimental circuit diagram for such a display is shown below: it consists of the series connection of a low-frequency function generator, a resistor, and the device under test (DUT).

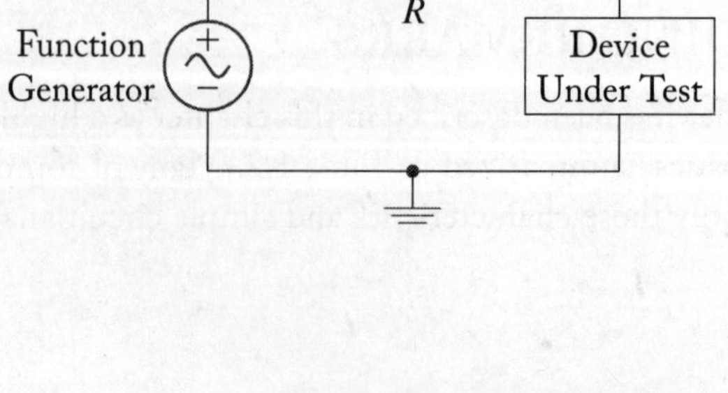

The voltage across the DUT is given by V_B and will serve as one of the inputs to the oscilloscope. The other input to the oscilloscope is the loop current. The most economical method for measuring this current is given by (current probes are quite expensive):

$$I = \frac{V_{AB}}{R}.$$

The location of the ground node in this circuit presents a problem. Safety regulations require that one terminal of the output of most function generators be at ground potential: similarly, one terminal of the input to most oscilloscopes is at ground potential. These ground connections do not pose a problem in measuring the voltage, V_B, but measuring the voltage, V_{AB}, is difficult. The differential input mode to most oscilloscopes can solve this difficulty in measurement, but this mode cannot usually be invoked simultaneously with the required x-y display mode.

The obvious solution to the measurement problem is an external differential amplifier with inputs at nodes A and B and an output to one of the oscilloscope channels. Design such a differential amplifier.

Solution:

A list of specifications is necessary for good design. The connection of the differential amplifier across the resistor, R, must not significantly disturb the measurements: It should have very high input resistance: $R_{in} > 1\,\text{M}\Omega$ matches the input resistance of most oscilloscopes. Similarly, the output of the differential amplifier should have low output resistance so that an accurate measurement can be made, $R_{out} < 100\,\Omega$ is adequate. The amplifier differential gain should be either unity or ten (10) so that a mix of oscilloscope probes can be utilized. CMRR should be high.

If a low-frequency function generator is used, OpAmps can be used for the realization of the differential amplifier. The differential amplifier of Figure 1.20 can easily be designed to meet all the specifications except input resistance. If the resistors are chosen to be sufficiently large to meet input resistance requirements, the ideal OpAmp approximations will fail. Therefore, it is necessary connect unity-gain buffers in series with each input. The circuit topology (next page) is therefore chosen.

The input and output resistance of this circuit automatically meets specifications using all common, commercial OpAmps. The requirement for two distinct values of the differential gain is accomplished with a double-pole, single-throw switch. When both indicated switches (each is a pole of the actual switch) are open, the differential gain is unity: when both are closed the differential gain is ten. After these topological design decisions, all that remains in the design is the choice of resistor values.

For unity magnitude gain in the v_B path, $R_A = R_B$ (Equation 1.42). For a gain magnitude of ten in that path,

$$R_A//R_a = 0.1R_B = 0.1R_A \quad \Rightarrow \quad R_A = 9R_a.$$

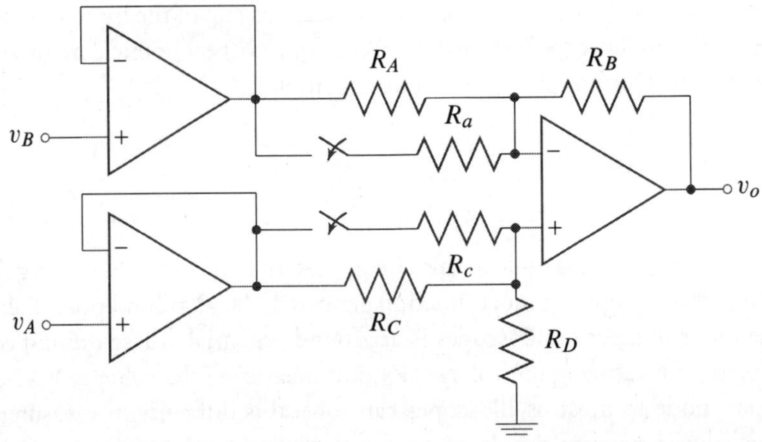

Assuming the above results, unity magnitude gain in the v_A path implies $R_C = R_D$ (Equation 1.43). Similarly, a gain magnitude of ten in that path implies,

$$R_C//R_c = 0.1R_D = 0.1R_C \quad \Rightarrow \quad R_C = 9R_c$$

While many choices will fulfill these ratios with adequate accuracy, one reasonable choice is:

$$R_A = R_B = R_C = R_D = 90.9\,\text{k}\Omega$$
$$R_a = R_c = 10.1\,\text{k}\Omega.$$

Both these resistor values are available as 0.5% resistors. Resistors with such small tolerances will ensure high CMRR.

1.7 PROBLEMS

1.1. A sinusoidal input is applied to the input of a linear amplifier. The input and the output voltage signals are displayed on the screen of an oscilloscope as in the figure. The oscilloscope vertical scale is set at 2 V/div. and the horizontal scale is set at 1 ms./div. It is known that the gain of the amplifier is greater than unity. Find the following:

a) The frequency of the signals.

b) The gain of the amplifier.

c) The delay time.

d) The phase shift.

e) A mathematical expression for the input and output voltages.

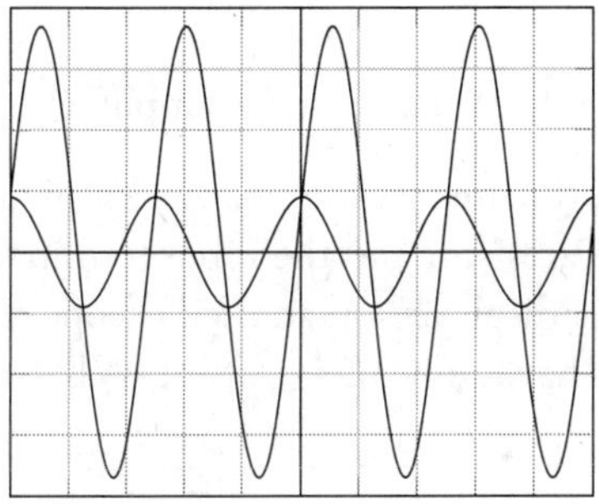

1.2. A sinusoidal input is applied to the input of a linear amplifier. The input and the output voltage signals are displayed on the screen of an oscilloscope in its x-y display mode as shown. The input signal is displayed on the horizontal axis and the output on the vertical scale. Both input amplifiers for the oscilloscope are adjusted so that the scales are 1 V/div. It is known that the output signal lags the input signal by a phase angle between 0° and 90°. Determine:

a) The gain of the amplifier.
b) The phase shift.

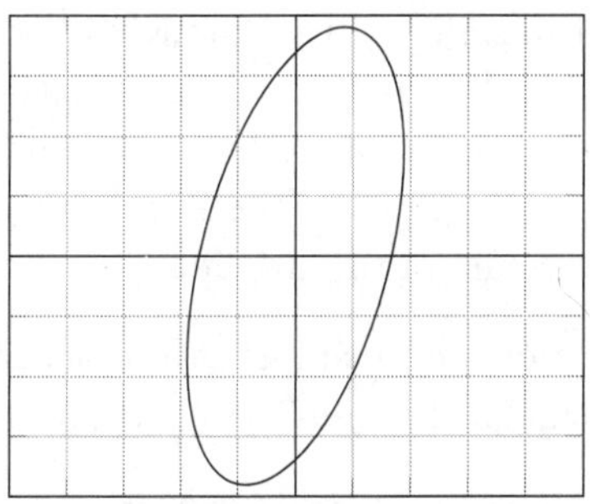

1.3. In order to measure the output resistance of an amplifier, an engineer connects a variable resistor to the amplifier output. The engineer then carefully measures the peak-to-peak output voltage for several settings of the variable resistor. The following experimental data results:

Measurement Number	Resistor Value, R (Ω)	Output Voltage, $v_{p\text{-}p}$ (V)
1	20	1.68
2	50	3.20
3	100	4.57
4	200	5.82
5	500	6.96

Determine the output resistance of the amplifier.

1.4. In order to measure the input resistance of a linear amplifier, a voltage source is connected in series with a 1 kΩ resistor and the input terminal of the amplifier. The voltage at each end of the resistor is displayed on an oscilloscope as shown with the vertical scales set at 1 V/div. The signals have zero voltage offset. What is the input resistance of the amplifier?

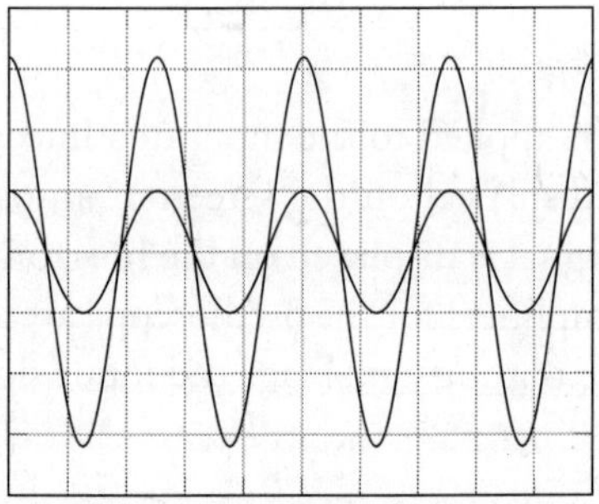

1.5. In order to measure the input resistance and voltage gain of a linear amplifier, a voltage source is connected in series with a true RMS ammeter and the input terminal of the amplifier. The input and output signals are displayed on an oscilloscope. Peak-to-peak readings of the input and output voltages are found to be 1.8 V_{p-p} and 6.7 V_{p-p}, respectively. Each has zero voltage offset. The ammeter reads 1 mA RMS. Determine the input resistance and voltage gain of the amplifier.

1.6. The design specifications for a simple inverting amplifier require an input resistance of 10 kΩ and a voltage gain, $A = -6.2$.

a) Prepare a design using an ideal OpAmp.

b) If a real OpAmp with the properties:

- $R_i = 2\,\text{M}\Omega$
- $A = 200\,\text{k}$
- $R_o = 75\,\Omega$

is used, what error in the gain and input resistance will result due to the non-ideal properties of the OpAmp?

1.7. The design specifications for a simple inverting amplifier require an input resistance of 10 kΩ and a voltage gain, $A = -6.2$.

a) Prepare a design using an ideal OpAmp.

b) What is the maximum error in the gain and input resistance that will result due to resistor variation if:

- 5% resistors are used?
- 1% resistors are used?

1.8. An ideal OpAmp can be modeled using the SPICE statement:

Eopamp 101 0 103 102 10MEG

where the nodes are related to the OpAmp shown.

Explain why this one statement can be used as a very simple model an OpAmp. Draw appropriate circuit diagrams.

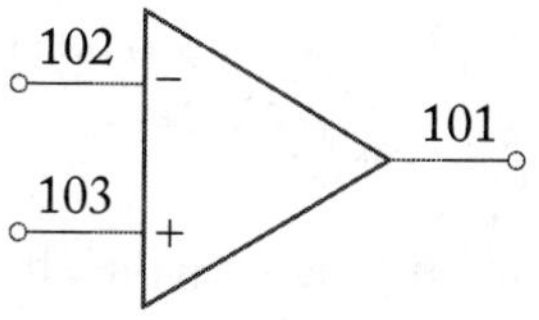

1.9. Design a non-inverting amplifier with a voltage gain of approximately 12 with an input resistance of 12 kΩ. Use SPICE to confirm the result using a simple model of the OpAmp.

1.10. For the circuit shown determine the voltage gain v_o/v_i and the input resistance R_{in}.

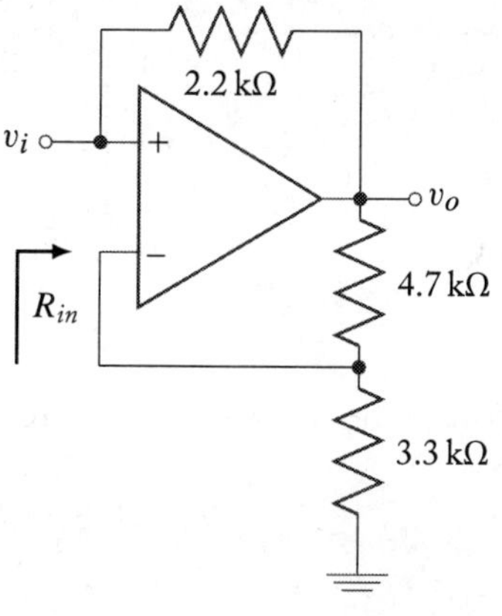

1.11. Determine the current through the load resistor, R_L, as a function of the input voltage, v_i, for the given circuit.

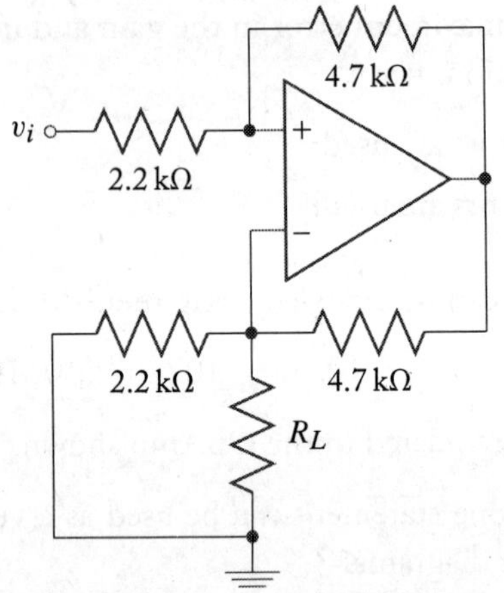

1.12. Design an operational amplifier circuit to meet the following specifications:

- $v_o = 3.0\,v_2 - 5.0\,v_1$
- $R_{i1} = 15\,\text{k}\Omega$ (the input resistance seen by source v_1)
- $R_{i2} = 25\,\text{k}\Omega$ (the input resistance seen by source v_2)

1.13. In an attempt to create a non-inverting, summing amplifier the circuit topology shown is chosen. Complete the design so that the output voltage is given by:

$$v_o = 5v_{i1} + 3v_{i2}$$

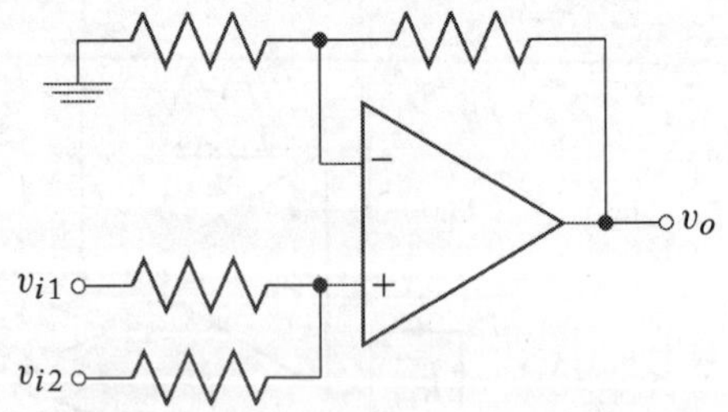

1.14. What is the expression of the output voltage for the circuit shown for

$$v_{i1} = \cos\left(\omega_o t\right)$$

and

$$v_{i2} = 1.5\cos\left(\omega_o t + 30°\right)\,?$$

Use SPICE to verify the result.

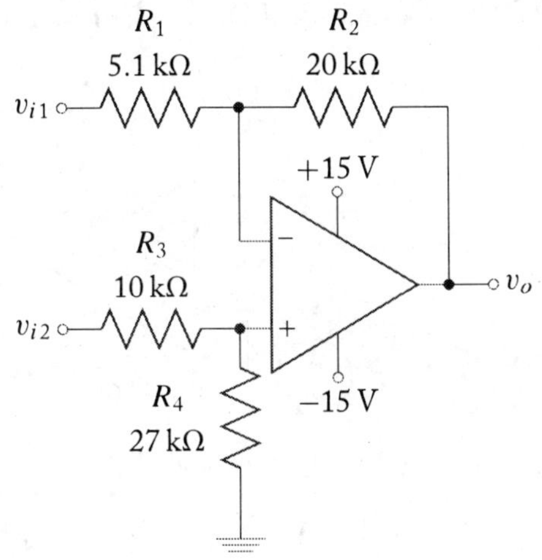

1.15. For the circuit shown, what are the range of values of a and b for the output to remain in the linear region of operation when

$$v_{i1} = a\cos(\omega_o t)$$

and

$$v_{i2} = b\cos\left(\omega_o t + 45^\circ\right),$$

for $a = 0.3b$?

Use SPICE to confirm the result.

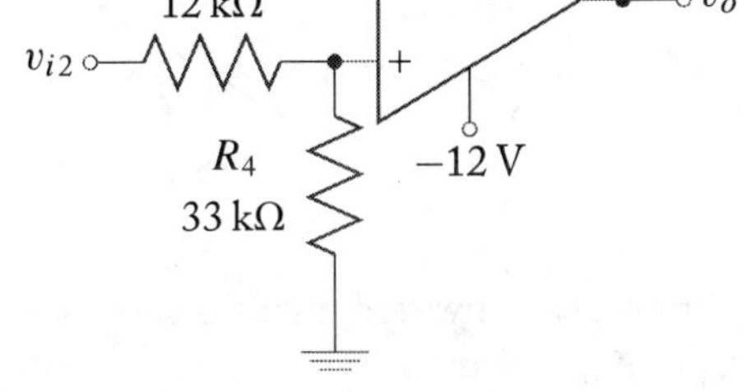

1.16. In applications where the output of an amplifier is no longer ground referenced (as in a D'Arsonval meter), an amplifier like that shown below may be used. Determine the current gain, $A_i = \frac{i_i}{i_o}$. Assume ideal OpAmp characteristics. Use SPICE to confirm the result.

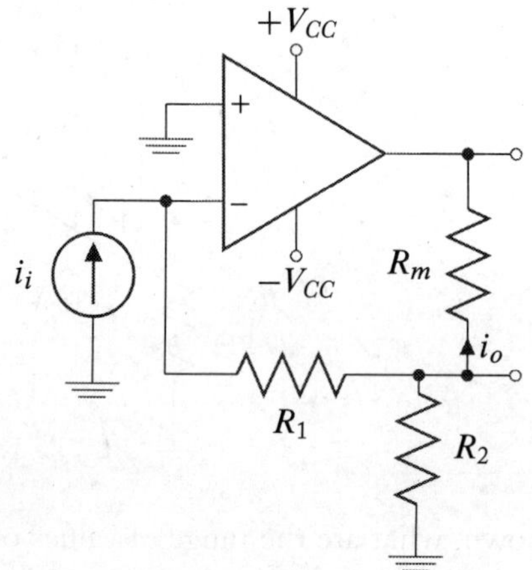

1.17. Determine the voltage gain and output resistance of the amplifier shown. Use a simplified equivalent model of an OpAmp for the analysis. Use SPICE to confirm the result.

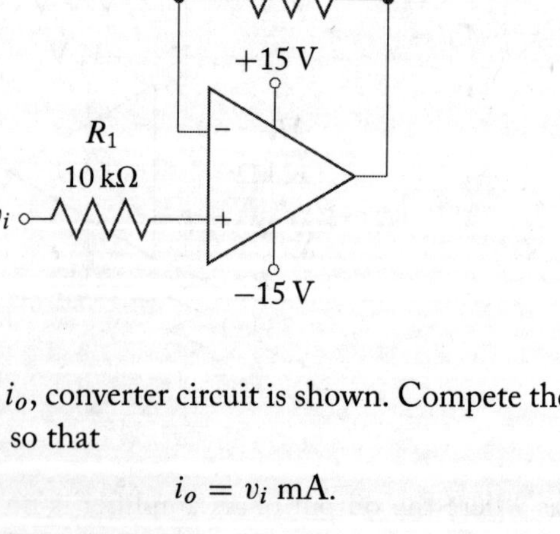

1.18. A voltage, v_i, to current, i_o, converter circuit is shown. Compete the design of the circuit by determining $R_4 - R_1$ so that

$$i_o = v_i \text{ mA}.$$

Let $2R_1 = 2R_2 = R_3 + R_4 = 1020\,\Omega$.

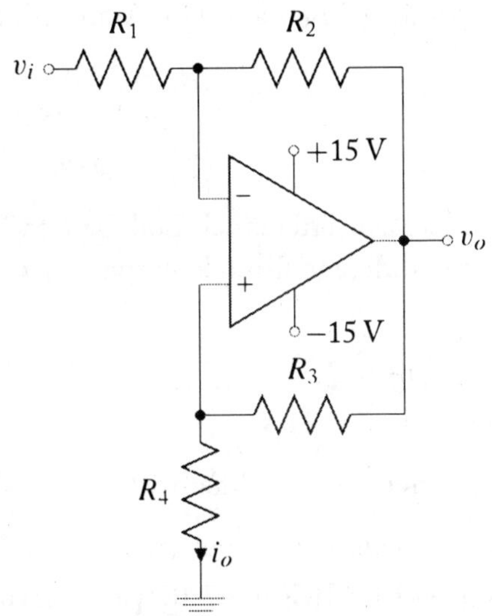

1.19. For the amplifier shown, confirm that v_o is half of v_i. Find the output resistance, R_o. Use SPICE to confirm the result.

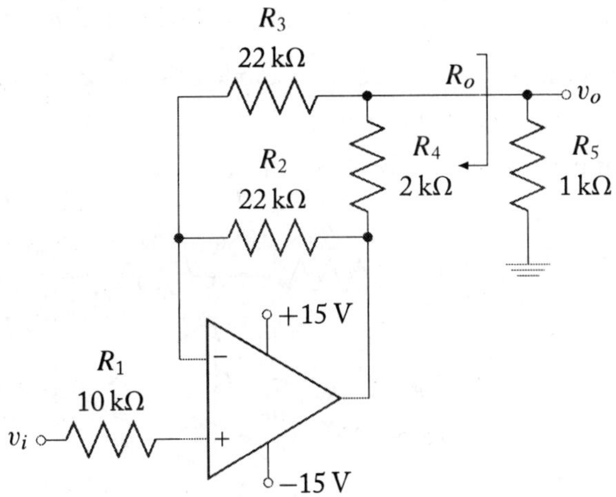

1.20. Design an OpAmp circuit to implement the following equation:

$$v_o = v_1 - 5.1\,v_2.$$

1.21. A difference amplifier with a gain of 10 is to be designed.

a) Prepare a design that will not have a saturated output with input voltages of −0.5 V and +0.2 V.

b) What is the error in gain for a real OpAmp with the following specifications?

$$R_i = 2\,\text{M}\Omega$$
$$A = 200,000$$
$$R_o = 75\,\Omega$$

1.22. A differential amplifier has a differential-mode gain of 92 dB and a CMRR of 80 dB. Find the magnitude of the differential-mode output $v_{o(DM)}$ and the common-mode output $v_{o(CM)}$ if:

a) $v_1 = 1.6\,\mu\text{V}$ and $v_2 = 2\,\mu\text{V}$
b) $v_1 = -1.6\,\mu\text{V}$ and $v_2 = 2\,\mu\text{V}$

1.23. Design an OpAmp differential amplifier with:

a) A gain of 67 and a minimum input resistance of 22 kΩ for each input.
b) For an OpAmp with CMRR = 67 dB with a maximum common-mode input signal of 0.08 V, find the differential input signal for which the differential-mode output is greater than 90 times the common-mode output.

1.24. A differential amplifier is constructed with an ideal OpAmp and resistors of nominal values 1.0 kΩ and 4.7 kΩ (i.e., $R_A \approx R_C \approx 1.0\,\text{k}\Omega$ and $R_B \approx R_D \approx 4.7\,\text{k}\Omega$).

a) What is the worst case common-mode gain using resistors with 5% tolerance?
b) What is the CMRR for that case?
c) Repeat parts a) and b) for 0.5% tolerance resistors.

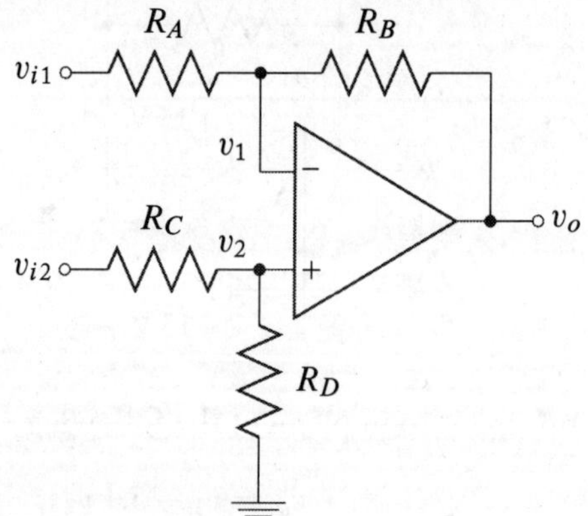

1.25. A differential amplifier shown has a differential-mode gain, $A_{DM} = 5000$, and a CMRR of 56 dB. Let $v_o = v_{o(CM)} + v_{o(DM)} = 1.2\,\text{V}$. Construct a graph of v_2 vs. v_1 showing the locus of all possible inputs that provide this output. Compare significant graph points to resistor ratios. Assume that $v_2 \geq v_1$ and the outputs add, and maintain $|v_2| \leq 5\,\text{V}$.

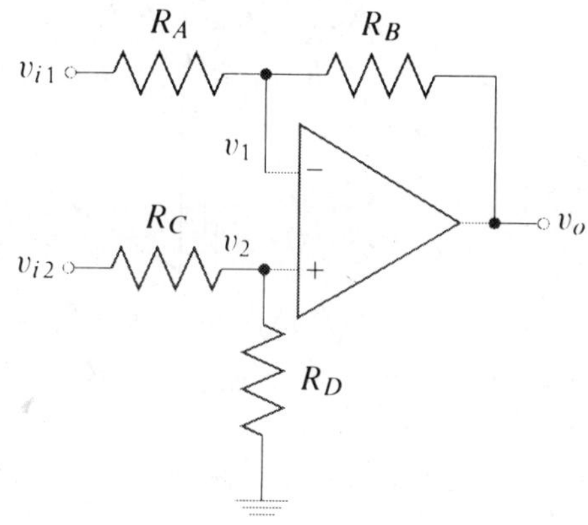

1.26. For the differential amplifier shown, the common-mode and differential-mode inputs are $v_{CM} = 150\,\text{mV}$ and $v_{DM} = 25\,\text{mV}$, respectively. Let $A_{DM} = 10\,\text{k}$ and $A_{CM} = 3$. Assume that $v_{o(DM)}$ and $v_{o(CM)}$ add. Find

a) CMRR in dB,

b) v_{i1},

c) v_{i2},

d) $v_{o(DM)}$,

e) $v_{o(CM)}$, and

f) v_o.

1.27. Determine the output voltage, v_o, as a function of the input voltage, v_i, for the given circuit. (Hint at answer: this is an integrator of some sort.)

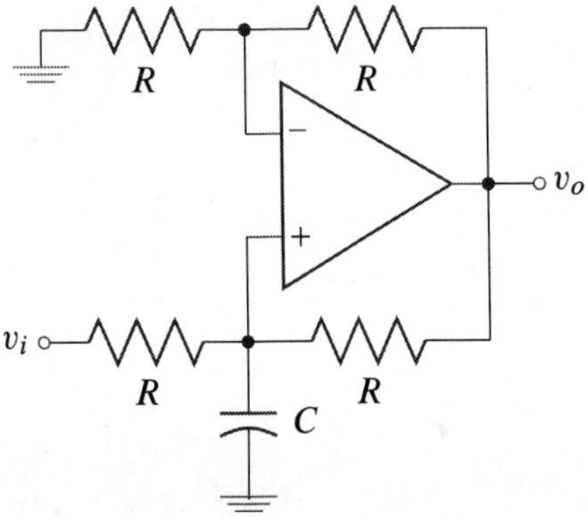

1.28. Assuming that the OpAmp is ideal and the capacitor is uncharged, find:

a) $v_o(t)$ for $t > 0$,

b) The time for the output voltage, $v_o(t)$ to reach 3 V.

c) Use SPICE to confirm the result.

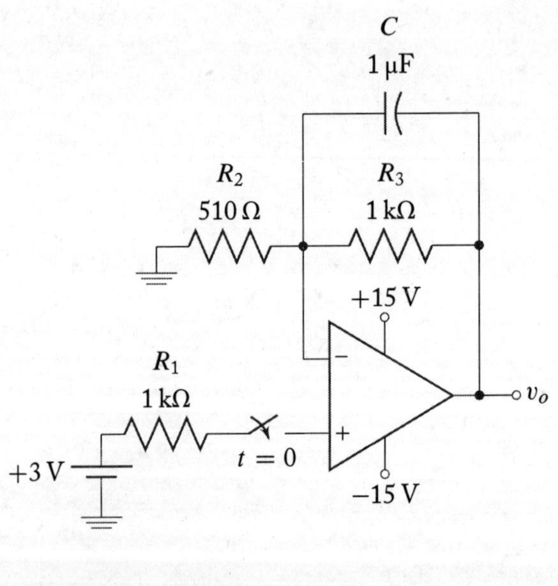

1.29. For the circuit below, graph the output signal for the square wave input signal as shown. Show amplitude and time scales. Simulate the circuit using SPICE.

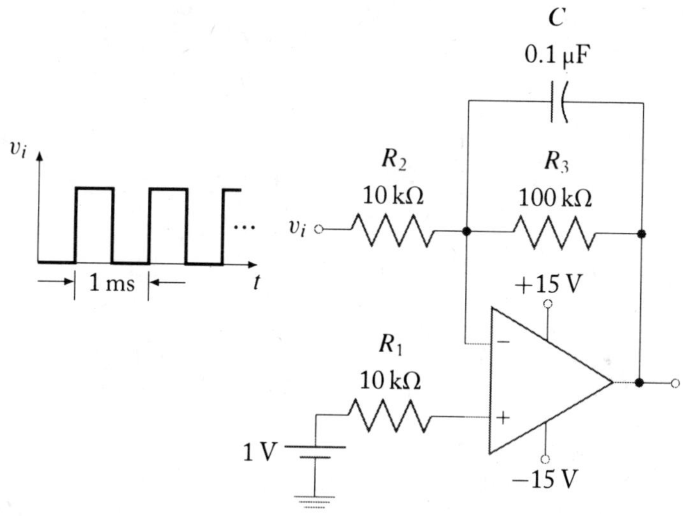

1.30. Given the attached circuit constructed with ideal OpAmps.

a) Determine the output voltage as a function of the input voltage in the given circuit.

b) At what input voltages will the output saturate?

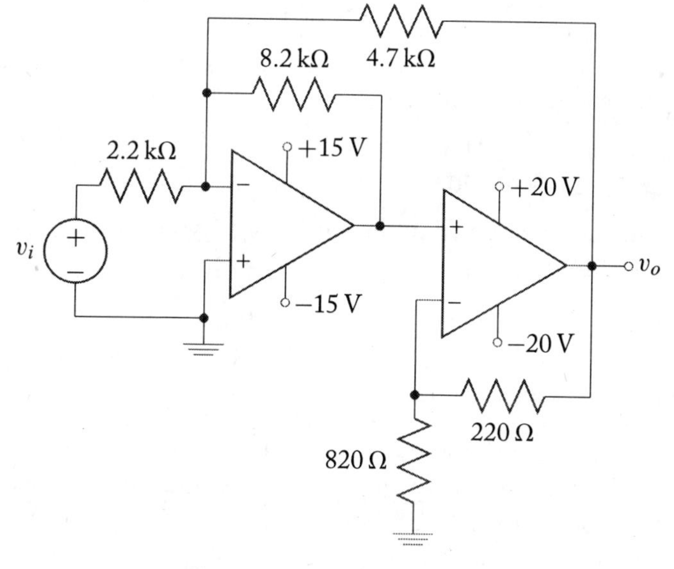

1.31. Determine the output voltage, v_o, as a function of the two input voltages, v_{i1} and v_{i2}, for the given circuit.

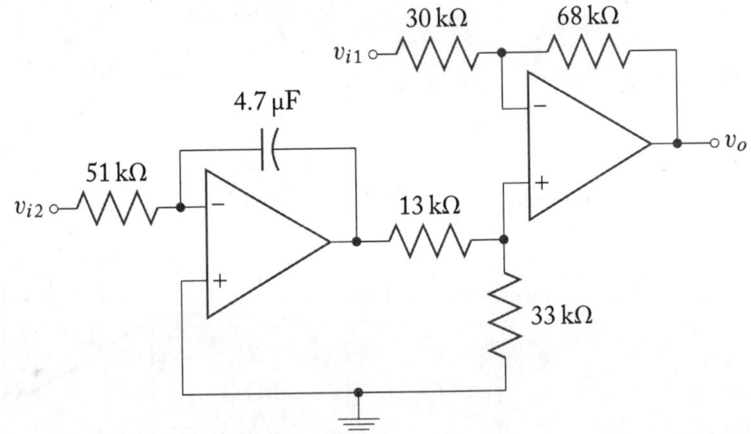

1.32. Design an inverting amplifier with voltage gain ≈ -5.1 and an input resistance $\approx$ 10 kΩ using a μA741 OpAmp. The μA741 has the following typical performance parameters:

Input resistance – 2 MΩ
Voltage gain – 200 kV/V
Output resistance – 75 Ω

Determine the variation from the ideal design goals due to the non-ideal properties of the μA741 OpAmp.

1.33. Design an inverting amplifier with voltage gain ≈ -5.6 and an input resistance 10 kΩ using an OP27 OpAmp. The OP27 has the following typical performance parameters:

Input resistance – 2 GΩ
Voltage gain – 1.5 MV/V
Output resistance – 70 Ω

Determine the variation from the ideal design goals due to the non-ideal properties of the OP27 OpAmp.

1.34. Find an expression for the input resistance of the non-inverting amplifier shown in Figure 1.15 if a non-ideal OpAmp is used. What conditions must be met for the input resistance to be approximated by the ideal?

1.35. Find an expression for the output resistance of the non-inverting amplifier shown in Figure 1.15 if a non-ideal OpAmp is used. What conditions must be met for the output resistance to be approximated by the ideal?

1.36. In high-gain applications, it is useful to replace the unity gain buffers of the circuit described in the Summary Design Example with an alternate input buffer. In the circuit shown, the new input buffer increases the differential-mode gain without altering the common-mode gain: an improvement in the CMRR of the total circuit results. Determine expressions for the common-mode and differential-mode gain of the new input buffer stage.

Note: in applications where there are two outputs, the gain quantities are defined as follows:

$$A_{DM} = \frac{\{v_a - v_b\}}{\{v_A - v_B\}}, \qquad A_{CM} = \frac{1/2\,\{v_a + v_b\}}{1/2\,\{v_A + v_B\}}$$

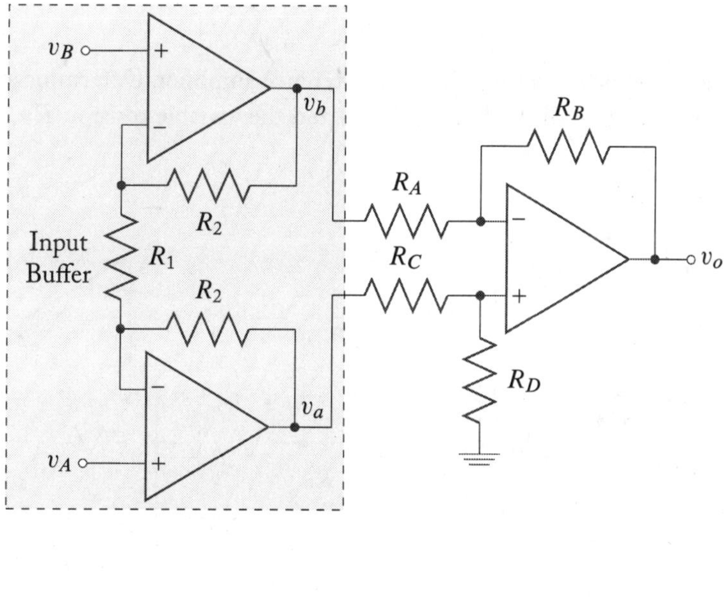

1.37. For the circuit shown in the above problem, the CMRR of the input buffer is 20 dB while the CMRR of the differential amplifier is 54 dB. What is the CMRR of the total amplifier (consisting of the buffer in series with the differential amplifier). Present theoretical validation of the results.

1.38. The circuit shown is a variable-gain difference amplifier. Determine an expression for the gain as a function of the fixed resistors and the variable resistor, R_V.

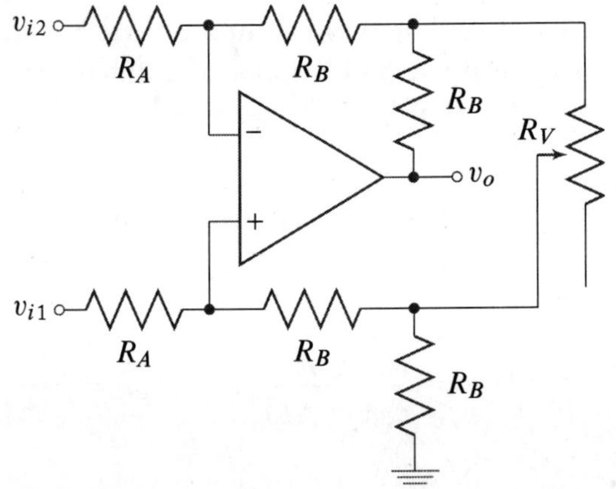

1.39. The circuit shown is a variable-gain difference amplifier. Determine an expression for the gain as a function of the fixed resistors and the variable resistor, R_V.

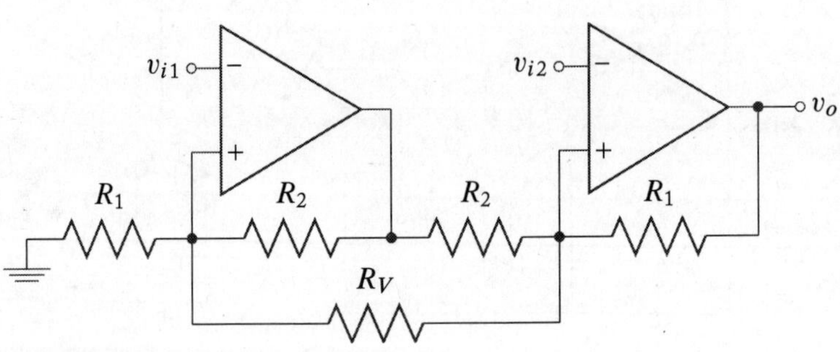

1.40. For the amplifier shown below, determine the values of the load resistor, R_L, that will lead to gain that deviates from the ideal value by –0.01%. Assume the OpAmp has the following properties:

$$A_v = 500,000$$
$$R_i = 1\,\text{M}\Omega$$
$$R_o = 75\,\Omega$$

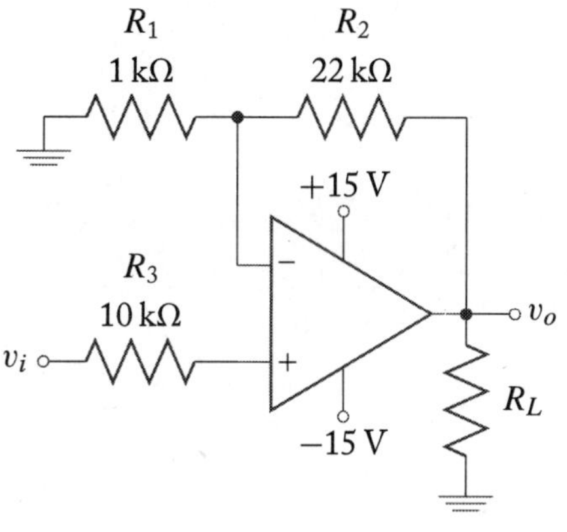

1.8 REFERENCES

[1] Ghausi, M. S., *Electronic Devices and Circuits: Discrete and Integrated,* Holt, Rinehart and Winston, New York, 1985.

[2] Gray, P. R., and Meyer, R. G., *Analysis and Design of Analog Integrated Circuits*, 2nd. Ed., John Wiley & Sons, Inc., New York, 1984.

[3] Millman, J., *Microelectronics, Digital and Analog Circuits and Systems*, McGraw-Hill Book Company, New York, 1979.

[4] Nilsson, J. W., *Electric Circuits,* 3rd. Ed., Addison-Wesley Publishing Co., Reading, 1989.

[5] Soclof, S., *Analog Integrated Circuits*, Prentice-Hall, Inc., Englewood Cliffs, 1985.

[6] Wojslaw, C. F., and Moustakas, E. A., *Operational Amplifiers*, John Wiley & Sons, Inc., New York, 1986.

CHAPTER 2

Diode Characteristics and Circuits

Simple electronic circuit elements can be divided into two fundamental groups by their terminal characteristics:

- Linear devices – devices that can be described by linear algebraic equations or linear differential equations;
- Non-linear devices – those devices that are described by non-linear equations.

Resistors, capacitors, and inductors are examples of passive circuit elements that are basically linear.[1] Operational amplifiers, when functioning within certain operational constraints (as described in Chapter 1), are linear, active devices.

The diode is the most basic of the non-linear electronic circuit elements. It is a simple two-terminal device whose name is derived from the vacuum tube technology device with similar characteristics: a tube with two electrodes (*di* - two; *ode* - path), the anode and the cathode. Vacuum tube devices have largely been superseded in electronic applications by semiconductor junction diodes. This chapter will restrict its discussion to semiconductor diodes, diode characteristics, and simple electronic diode applications.

2.1 BASIC FUNCTIONAL REQUIREMENTS OF AN IDEAL DIODE

There are many applications in electronic circuitry for a one-way device: that is, a device that provides zero resistance to current flowing in one direction, but infinite resistance to current flowing the opposite direction. Protection against misapplied currents or voltages, converting alternating current (AC) into direct current (DC), demodulating Amplitude Modulated (AM) radio signals, and limiting voltages to specified maxima or minima are but a few of the many possible applications. While a device with such ideal characteristics may be impossible to manufacture, its study is still instructive and, in many applications, ideal devices closely approximate real devices and provide insight into real-device circuit operation.

[1]All linear electronic devices can become non-linear if input currents or voltages are allowed to become too large. Thermal effects, dielectric breakdown, magnetic saturation, and other physical phenomena can cause non-linearities in device transfer characteristics. Still, devices that are categorized as linear have a region of operation, usually specified by the manufacturer, in which the transfer characteristics are extremely linear.

An ideal diode is a true one-way electronic device. Its volt-ampere (V-I) transfer relationship is shown in Figure 2.1.[2] The two terminals of such a diode retain the names first used in the vacuum-tube diode:

A - the anode (Greek, *ana* up + *hodos* way)

K - the cathode (Greek, *kata* down + *hodos* way).

Analytically the transfer relationship can be described as:

$$\begin{aligned} I &= 0 \quad \text{for } V < 0 \\ V &= 0 \quad \text{for } I \geq 0 \end{aligned} \tag{2.1}$$

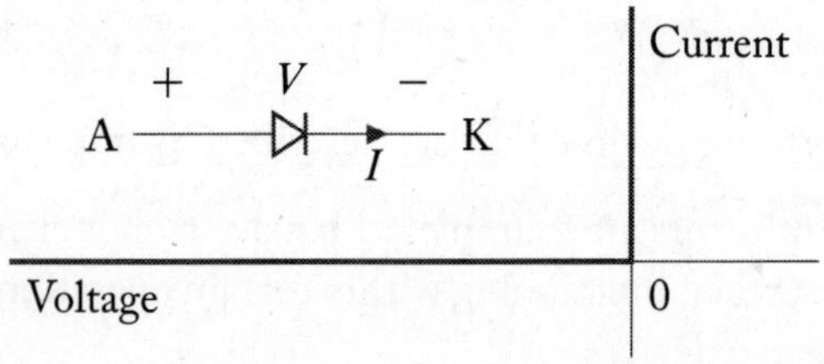

Figure 2.1: The Volt-ampere transfer relationship for an ideal diode.

It is important to notice that the definition of the sign convention of Figure 2.1 is extremely important. For many devices that are linear (for example, a resistor), reversing the polarity of both the voltage and current yields the same V-I relationship (Ohm's Law) as long as the passive sign convention[3] is followed. Reversing the polarity of the voltage and current (still keeping the passive sign convention) can yield, in general, a drastically different V-I relationship for a non-linear devices: thus the sign convention takes special significance.

The functional relationships of Equation (2.1) are a piece-wise linearization of the V-I transfer relationship for an ideal diode and lead to two piece-wise linear models that are often used to replace a diode for analysis purposes:

$$\begin{aligned} &\text{A}\text{———}\text{K} \qquad I \geq 0 \\ &\text{A}- \quad -\text{K} \qquad V < 0 \end{aligned}$$

The two linear models are:

[2]Ideal diodes will symbolically be shown using the symbol in Figure 2.1: the triangle portion of the symbol will be empty. Real diodes will be shown with a triangle that is filled as seen in Figure 2.4.

[3]The passive sign convention allows consistent equations to be written to characterize electronic devices. In two terminal devices, it simply states that, when describing the device, positive reference current enters the positive voltage reference node and exits the negative voltage node.

- a *short circuit* when the applied *current* is *positive*, and
- an *open circuit* when the applied *voltage* is *negative*.

While at first it is not always obvious which model will accurately predict the state of a diode in an electronic circuit, analysis using one model will produce results consistent with model assumptions: the other model will produce a result that contradicts the assumptions upon which that model is based.

Example 2.1
For the simple ideal diode circuit shown, determine the current in the diode if:

(a) $V_S = 1\text{V}$

(b) $V_S = -1\text{V}$

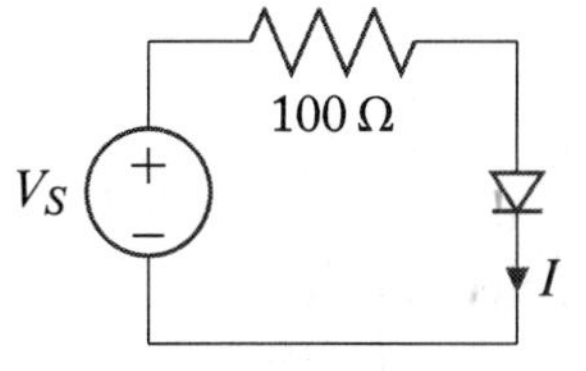

Solution:
a) $V_S = 1$
Choosing the short circuit model to replace the diode, the current, I, is found to be:

$$I = 1/100 = 10\,\text{mA}$$

If the open circuit model is used to replace the diode, the current is found to be:

$$I = 1/(100 + \infty) = 0\,\text{mA}$$

In the first case (the short circuit) the diode current (here, the same as I) is within the restrictions of the model assumptions ($I \geq 0$) and there is no contradiction to that model's assumptions. In the second case the diode voltage violates the second model assumptions ($V = 1$ violates the model assumption, $V < 0$). Thus, the diode appears to act as a short circuit and the true value of the current given by:

$$I = 10\,\text{mA}.$$

b) If the diode is replaced by its short-circuit model, the current is calculated to be:

$$I = -1/100 = -10\,\text{mA}$$

This result violates the defining constraint for the model ($I > 0$). Therefore, the open circuit model must apply:

$$I = -1/(100 + \infty) = -0\,\text{mA}.$$

Here the voltage across the diode is −1 V which fulfills the defining assumption for the model. Consequently, the diode current is zero valued.

While the studying the action of an ideal diode often provides useful insight into the operation of many electronic circuits, real diodes have a more complex V-I relationship. The fundamental operation of a real semiconductor diode in its conducting and the non-conducting regions is discussed in Section 2.2. When large reverse voltages are applied to a real diode (in what should be the far extremes of non-conducting region) the diode will enter a region of reverse conduction (the Zener region) due to one or more of several mechanisms. This sometimes useful - sometimes destructive region of reverse conduction is discussed in Section 2.7.

2.2 SEMICONDUCTOR DIODE VOLT-AMPERE RELATIONSHIP

Semiconductor Diodes are formed with the creation of a *p-n* junction. This junction is a transition region between a semiconductor region that has been injected (doped) with acceptor atoms (a *p*-region) and one that has been injected with donor atoms (an *n*-region).[4] The *p*-region becomes the anode and the *n*-region becomes the cathode of a semiconductor diode. Semiconductor diodes are real diodes and have volt-ampere relationships that are in many ways similar to the V-I relationship for an ideal diode. There are, however, distinct differences:

- In the non-conducting region (when the *p-n* junction is reverse biased) the diode current is not exactly zero: the diode exhibits a small reverse leakage current.

- The diode requires a small positive voltage to be applied before it enters the conducting region (when the *p-n* junction is forward biased). When in the conducting region the diode has a non-zero dynamic resistance.

- For large input voltages and/or currents the diode enters breakdown regions. In the forward direction, power dissipation restrictions leads to thermal destruction of the diode. In the reverse direction, the diode will first enter a Zener region of conduction then thermal destruction.

The similarities between the semiconductor diode and the ideal diode allow the semiconductor diode to be used for the applications mentioned in Section 2.1. The differences mean circuit designers and engineers must be careful to avoid an oversimplification of the analysis of diode circuitry. Other differences allow for a few applications not possible with an ideal diode.

[4]Discussions of the atomic semiconductor physics that lead to a *p-n* junction forming a diode are not within the scope of this electronics text. The authors suggest several texts in semiconductor physics and electronic engineering materials at the end of this chapter for those readers interested in these aspects of physical electronics.

In the region near the origin of the V-I relationship for a semiconductor diode, the V-I curve can be described analytically by two equivalent expressions:

$$I = I_S \left(e^{\frac{qV}{\eta k T}} - 1 \right) = I_S \left(e^{\frac{V}{\eta V_t}} - 1 \right) \tag{2.2a}$$

or

$$V = \eta\, V_t \ln \left(\frac{I}{I_s} + 1 \right). \tag{2.2b}$$

The physical constants fundamental to the diode V-I relationship are given by:

$$\begin{aligned} q &= \text{electronic charge } (160 \times 10^{-21}\text{C}) \\ k &= \text{Boltzmann's constant } (13.8 \times 10^{-24}\text{J/°K}) \\ V_t &= \text{voltage equivalent temperature of the diode} \\ &= kT/q \approx T/11600 \approx 26\,\text{mV @ room temperature}(\approx 300\,°\text{K}) \end{aligned}$$

It is difficult to describe the non-linear behavior of this mathematical expression for the diode V-I relationship throughout the entire range of possible values, however discussion of the behavior in its two extremes is useful. In the *strongly reverse-biased region*, i.e., when

$$V \ll -V_t,$$

the exponential term of Equation (2.2) is much smaller than unity and the diode current is very nearly constant with the value:

$$I \approx -I_S.$$

The non-zero value of the current implies that a reverse leakage current of value I_S is present for the diode when it is in its non-conducting region. This leakage is very small: typically in the range of a few hundredths of a nanoampere to several nanoamperes.

When the diode is in its *strongly forward biased region*, $V \gg V_t$, the current experiences an exponential growth and the diode appears to have near-zero *dynamic resistance*. Dynamic resistance is defined as the *incremental* change in voltage with respect to an *incremental* change in current. For the diode the dynamic resistance is given by:

$$r_d = \frac{\partial V}{\partial I} = \frac{\eta\, V_t}{I_s + I} = \frac{\eta\, V_t}{I_s}\, e^{-\frac{V}{\eta V_t}}. \tag{2.3}$$

In the strongly forward-biased and strongly reversed-biased regions the dynamic resistance is then:

$$r_d \approx 0, \text{ when } V \gg V_t \text{ or equivalently } I \gg I_S$$
$$\text{(strongly forward biased)}$$

and

$$r_d \approx \infty \text{, when } V \ll -V_t \text{ or equivalently } I \approx -I_S.$$
$$\text{(strongly reverse biased)}$$

These dynamic resistance values are, asymptotically, the values for forward and reverse resistance of an ideal diode. Thus, the behavior of a real diode is similar to that of an ideal diode.

A plot of the V-I relationship for a typical diode with $I_S = 1$ nA and $\eta = 1$ & 2 at a temperature of 300°K appears in Figure 2.2. Notice that the basic shape of this relationship is similar to that of the ideal diode with the following exceptions:

- In the reverse-biased region, the current is not exactly zero, it is instead a small leakage value, I_S.
- The forward-biased region exhibits real, non-zero resistance. The vertical portion of the curve is displaced to the right. There appears to be a voltage at which the diode begins to conduct. This voltage is often called a *threshold voltage* which for silicon diodes lies in the range of 0.6 V to 0.9 V. Threshold voltage will be more thoroughly discussed in Section 2.5.

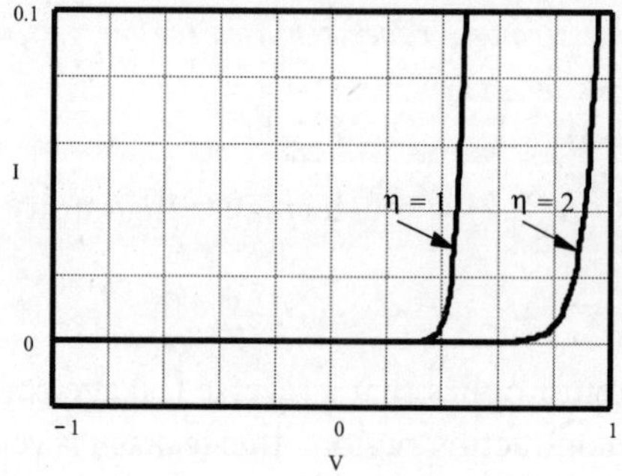

Figure 2.2: Typical diode V-I relationships.

The quantity η is an empirical scaling constant which for typical devices lies in the range:

$$1 \leq \eta \leq 2.$$

This scaling constant, η, is dependent on the semiconductor material, the doping levels of the p and n regions and the physical geometry of the diode. Typical germanium diodes have a scaling constant near unity while silicon diodes have $\eta \approx 2$.

It is important to note the temperature dependence of Equation (2.2). While the temperature dependence of V_t is evident, the temperature dependence of I_S is not explicit. It can be shown through basic principles of semiconductor physics that I_S is strongly temperature dependent. In silicon, I_S approximately doubles for every 6°K increase in temperature in the temperature range near 300°K (room temperature). That is:

$$I_s\,(T_2) = I_s\,(T_1)\;\times\;2^{\frac{T_2 - T_1}{6}}. \tag{2.4}$$

Other semiconductor materials exhibit similar variation of I_S with temperature. A graphical demonstration of the change in the diode V-I characteristic with temperature is given in Figure 2.3 .

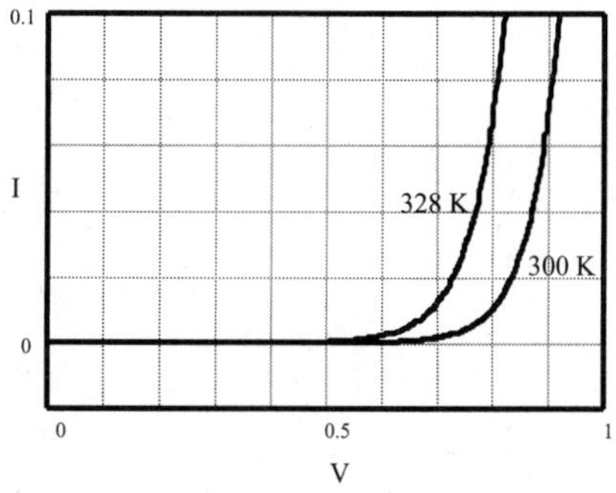

Figure 2.3: Diode V-I characteristics at two temperatures.

Example 2.2
The saturation current of a Si diode is 2.0 nA and the empirical scaling constant, $\eta = 2$. Calculate the following:

(a) At room temperature (300°K) the diode current for the following voltages.

$$-5\,\text{V}, -1\,\text{V}, 0.5\,\text{V}, 0.9\,\text{V}$$

(b) The above quantities if the temperature is raised to 55°C.

Solution:

(a) From (2.1), at room temperature.

$$I = I_S\left(e^{\frac{V}{\eta V_t}} - 1\right) = 2 \times 10^{-9}\left(e^{\frac{V}{2\times 0.026}} - 1\right) = 2 \times 10^{-9}\left(e^{\frac{V}{0.052}} - 1\right).$$

Substituting the values of V yields:

$$\begin{array}{llll} V = -5 & I = -2.00\ \text{nA} & V = 0.5 & I = 29.98\ \mu\text{A} \\ = -1 & = -2.00\ \text{nA} & = 0.9 & = 65.72\ \text{mA} \end{array}$$

(b) If the temperature is raised to 55°C, the equivalent temperature is 328°K. This is a change in temperature of 28°K. Thus,

$$V_t = 328/11600 = 28.28\,\text{mV}$$

and

$$I_S(328°\text{K}) = 2^{(28/6)} I_S(300°\text{K}) = 50.8\,\text{nA}$$

then

$$I = 50.8 \times 10^{-9}\left(e^{(V/(2\times 0.02828)} - 1\right) = 50.8 \times 10^{-9}\left(e^{(V/0.05656)} - 1\right)$$

Substituting the values of V yields:

$V = -5$	$I = -50.80\,\text{nA}$
$= -1$	$= -50.80\,\text{nA}$
$= 0.5$	$= 351.2\,\mu\text{A}$
$= 0.9$	$= 414.4\,\text{mA}$

Significant differences occur. Notice also that the warm diode with 0.9 V across it is now dissipating 0.38 W while the cool diode dissipates only 0.059 W in the same circumstances. Heating leads to increased power dissipation which, in turn, leads to heating: this cyclic process can lead to thermal run-away and eventual destruction of the diode.

2.3 THE DIODE AS A CIRCUIT ELEMENT

Due to the non-linear nature of the V-I relationship for a semiconductor diode, analysis techniques for circuits containing diodes are complex. This section deals with exact analytical solutions and lays the foundation for the graphical techniques of Section 2.4. Simplified piece-wise linear modeling techniques that allow for the use of linear analysis techniques are discussed in Section 2.5.

One of the simplest circuits involving a diode is shown in Figure 2.4. This circuit consists of an independent voltage source, a resistor, and a diode in series. Simple Thévenin extensions of this circuit show that the discussion presented here can be extended to any linear circuit connected in series with a single diode.

Kirchhoff's Voltage Law taken around the closed loop yields:

$$V - IR - V_d = 0 \tag{2.5}$$

where,

$$V_d = \eta\, V_t \ln\left(\frac{I}{I_s} + 1\right) \tag{2.6}$$

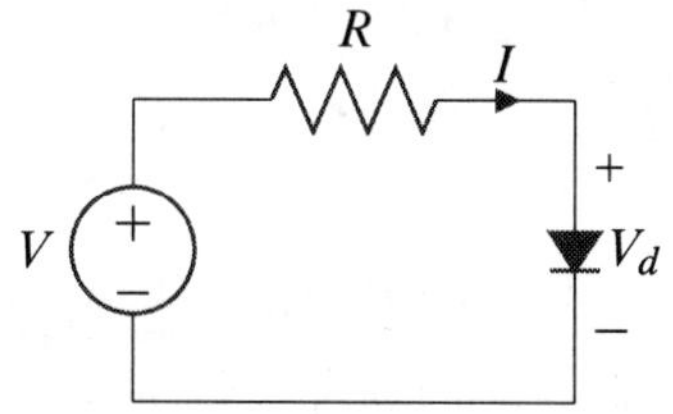

Figure 2.4: A simple diode circuit.

which leads to the non-linear equation,

$$V - IR - \eta\, V_t \ln\left(\frac{I}{I_s} + 1\right) = 0 \tag{2.7}$$

that must be solved for I. Equation (2.6) is the diode V-I relationship expressed as the voltage across the diode as a function of the diode current.

2.3.1 NUMERICAL SOLUTIONS

A simple closed-form solution for Equation (2.7) does not exist. The best technique for solution is usually a structured numerical search. Structured searches can be easily performed using mathematical software packages such as Mathcad, Matlab or similar programs. Another common technique uses programmable calculators with a built-in numerical equation solver (root-finder).

Example 2.3
For the circuit of Figure 2.4, assume the following values:

$$V = 5\,\text{V}$$
$$R = 1\,\text{k}\Omega$$

and diode parameters:

$$I_S = 2\,\text{nA}$$
$$\eta = 2$$

Find, at room temperature, the diode current, the voltage across the diode, and the power dissipated by the diode.

Mathcad can determine the solution through the use of a "solve block" as shown:

Circuit Parameters

$V := 5 \qquad I_S := 2 \bullet 10^{-9} \qquad V_t := 0.026$

$R := 1000 \qquad \eta := 2$

Guess Values $V_d := 1$ $I := 0.001$

Solve Block

Given

$$V_d = \eta \cdot V_t \cdot \ln\left(\frac{I}{I_S} + 1\right) \qquad V - I \cdot R - V_d = 0$$

$$\begin{pmatrix} V_d \\ I_d \end{pmatrix} := \text{find}\,(V_d, \text{I}) = \begin{pmatrix} 0.758 \\ 4.242 \times 10^{-3} \end{pmatrix}$$

$$P_d := V_d \cdot I_d = 3.214 \times 10^{-3}$$

Similar techniques applied to a hand-held calculator yield:

$V_d = 0.758\,\text{V}$

$I = 4.242\,\text{mA}$

2.3.2 SIMULATION SOLUTIONS

In addition to mathematical equation solving computer programs, there exists a variety of electronic circuit simulators that perform similar solutions more efficiently. Most common among these simulators is SPICE[5] and its many derivatives. These simulation programs easily solve the type of problem described in Example 2.3 and with little effort can provide solutions to more complex problems. A simple extension to Example 2.3 with a time-varying voltage source is described below.

Example 2.4

Assume the voltage source given in Figure 2.4 is a time-dependent voltage source:

$$V = 2.0 + 4\sin\{2\pi(80)t\}.$$

Determine the diode current and voltage as a function of time.

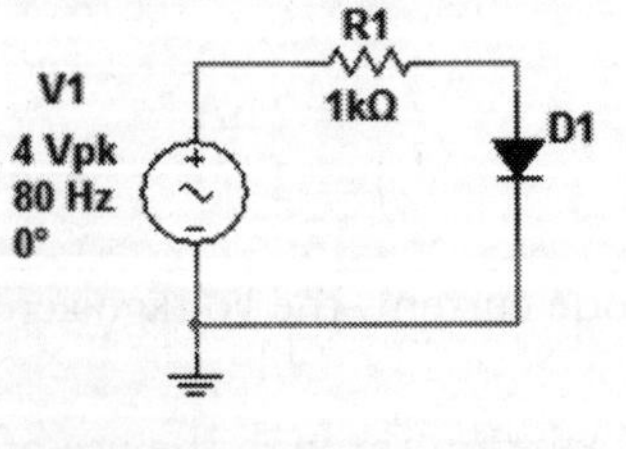

[5]SPICE, System Program with Integrated Circuit Emphasis, was developed at the University of California at Berkeley. In the development of this text, the authors used two of its derivatives: PSpice, developed by MicroSim Corporation and Multisim, developed by National Instruments.

Solution (using MultiSim)

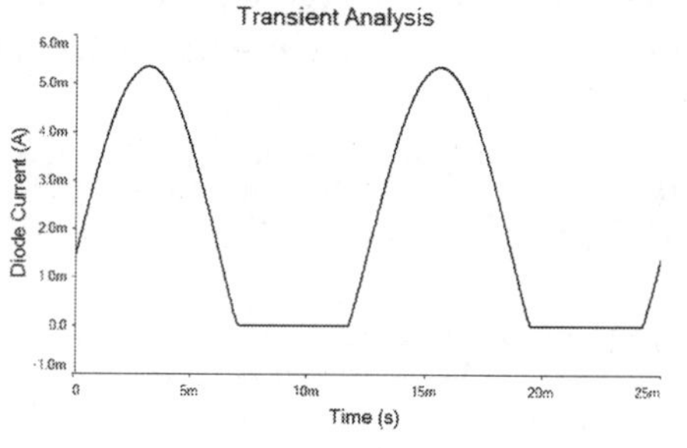

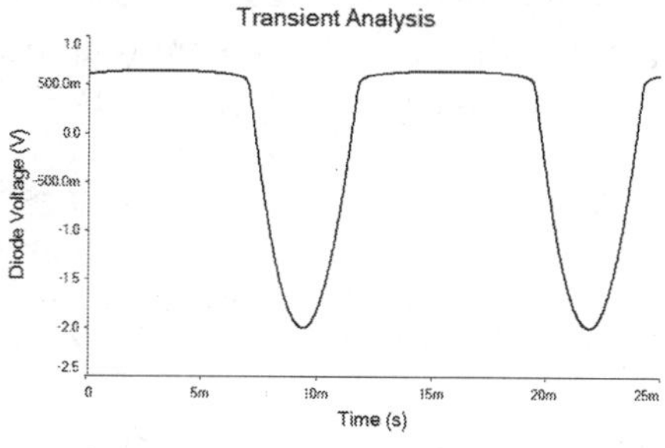

2.4 LOAD LINES

While numerical techniques are often quite useful to solve electronic circuit problems, they require the use of calculators or computers and in many circumstances provide no insight into the operation of the circuit. It is also necessary to know the parameters of the non-linear circuit elements with a reasonable degree of accuracy. Diode V-I curves are often obtained through experimental procedures and only then are the parameters derived from the experimental data. A more direct approach to working with non-linear elements that has traditionally been taken is the use of graphical techniques. The process of constructing solutions through these graphical techniques often provides useful insight into the operation of the circuit.

2.4.1 GRAPHICAL SOLUTIONS TO STATIC CIRCUITS

Graphical Solutions to the simple diode circuit of Figure 2.4 involve the use of the graphical representation of the diode V-I relationship. The graph of the V-I relationship can be obtained from Equation (2.6) or determined experimentally. The other elements in the circuit are then combined to create another relationship between the diode current and voltage. In the simple case of a DC Thévenin equivalent source driving a single diode, this additional relationship can be easily obtained by rewriting Equation (2.5) to become:

$$V_d = V - I\,R. \tag{2.8}$$

Equation (2.8) is called a load line and gives its name to this type of graphical analysis: *load line analysis*. Notice that a plot of the load line crosses the horizontal (V_d) axis at the value of the Thévenin voltage source and the vertical (I) axis at V_d/R. The slope of the load line is the negative of the inverse of the Thévenin resistance ($-1/R$).

The two relationships involving the diode current and voltage (the diode V-I relationship and the load line) can now be plotted on the same set of axes. The intersection of the two curves yields the value of the diode current and voltage for the particular values of the Thévenin source.

For circuits involving static (DC) sources the intersection set of values is called the *quiescent point* or *Q-point* of the circuit. Figure 2.5 demonstrates the load line technique applied to the circuit of Figure 2.4 using the diode V-I relationship using diode and circuit parameters of Example 2.3. The Q-point for this example appears to be:

$$V_d \approx 0.76\,\text{V}$$
$$I \approx 0.0042\,\text{A}$$

which compares nicely with the analytic solution previously obtained.

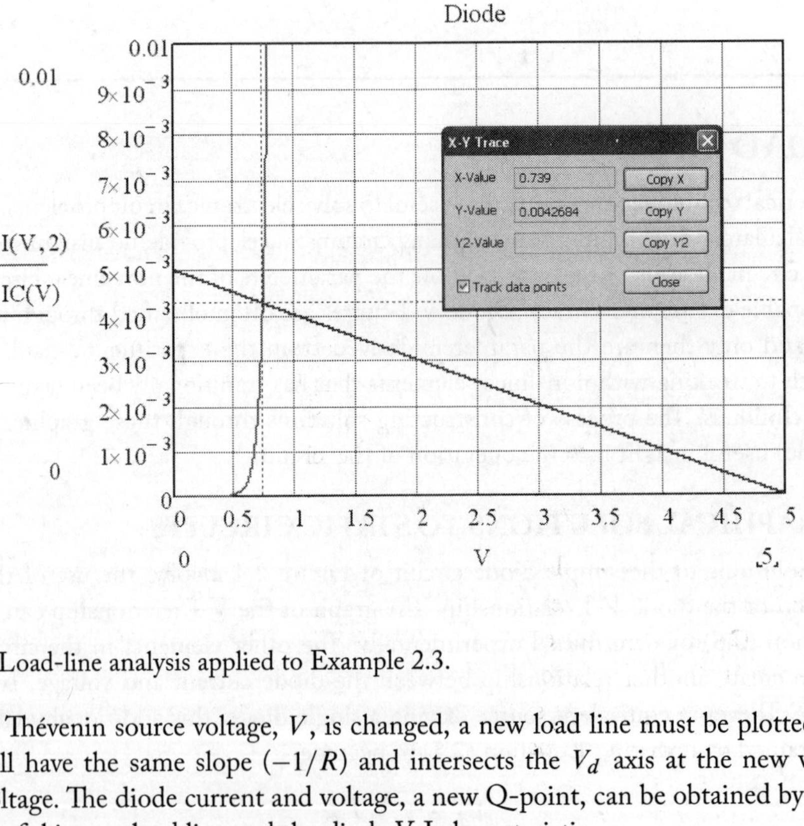

Figure 2.5: Load-line analysis applied to Example 2.3.

If the Thévenin source voltage, V, is changed, a new load line must be plotted. This new load line will have the same slope ($-1/R$) and intersects the V_d axis at the new value of the Thévenin voltage. The diode current and voltage, a new Q-point, can be obtained by finding the intersection of this new load line and the diode V-I characteristic.

2.4.2 GRAPHICAL SOLUTIONS TO CIRCUITS WITH TIME VARYING SOURCES

The load line technique outlined above can be expanded to provide solutions to circuits with time varying sources. Here the Thévenin equivalent voltage source will be time varying. Graphical load line analysis is performed as above for several instants of time $\{t_i\}$. At each instant of time the Thévenin voltage $\{v_i\}$ is known, a load line can be plotted and values for the diode voltage and/or current can be obtained. These intersection values $\{Q_i\}$ obtained from the load line analysis are then plotted against the time variables $\{t_i\}$ to obtain the time varying output of the circuit.

Example 2.5

Load lines applied to time-varying sources.

For the circuit shown, assume the following values:

$$V = 2 + 4\sin\{2\pi(80t)\}$$
$$R = 1\,\text{k}\Omega$$

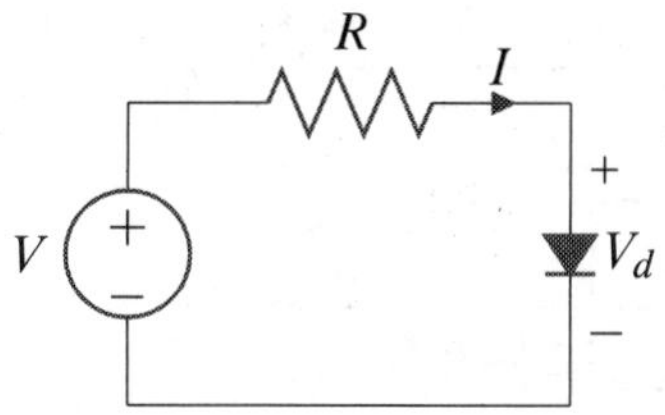

and diode parameters:

$$I_S = 2\,\text{nA}$$
$$\eta = 2.$$

Find, at room temperature, the diode current as a function of time.

Solution:

The diode current is plotted as a function of the diode current. On this plot the time dependent source voltage is also plotted as a function of time with the voltage axis corresponding to the diode voltage axis and the time axis parallel to the diode current axis. The time varying voltage is sampled and a load line is created for each sample value. The Q-point for each of these lines is determined and a plot of current as a function of time is created. The graphical representation of these steps is shown in Figure 2.6.

Notice the output waveform matches the simulation results of Example 2.4 nicely. The same load line principles can be applied to other non-linear devices and circuits.

2.5 SIMPLIFIED PIECEWISE LINEAR MODELS OF THE DIODE

The previous sections of this chapter have treated a diode as either an extremely simple device (the ideal diode) or as a complex non-linear device represented by non-linear equations or curves. While each of these treatments has its place in the analysis of electronic circuits, it is often useful to

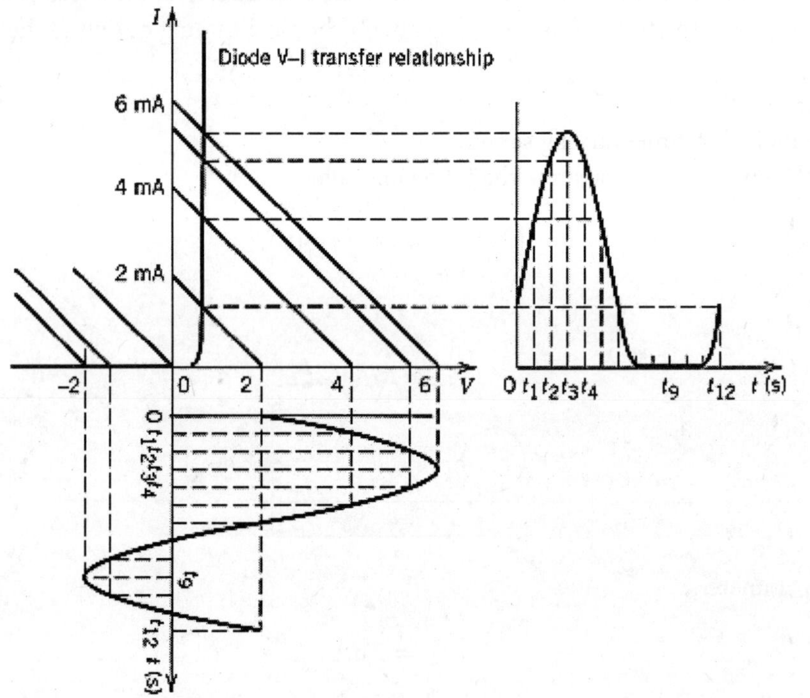

Figure 2.6: Load-line analysis applied to time-varying sources.

find a technique in the middle ground between these two extremes. One such technique involves the regional linearization of circuit element V-I characteristics.

When this technique is applied to a diode there are two basic regions.

- the region when the diode is basically conducting (Forward Bias)
- the region when the diode is basically non-conducting (Reverse Bias)

In Example 2.1 the ground work for the two-region linearization was laid with the boundary between the two regions being at the origin of the ideal diode V-I transfer relationship. With real diodes the transition between the two regions lies at a positive diode voltage, hereafter called the *threshold voltage*, V_γ.

2.5.1 FORWARD BIAS MODELING

In the region where the diode is conducting, a good linear model of the diode is a straight line tangent to the diode V-I relationship at a Q-point. The slope of this line is the dynamic resistance of the diode at the Q-point, which was derived in Section 2.2:

$$r_d = \frac{\partial V}{\partial I} = \frac{\eta\, V_t}{I_s + I} = \frac{\eta\, V_t}{I_s}\, e^{-\frac{V}{\eta V_t}}. \tag{2.9}$$

The value of the threshold voltage can be derived by finding the intersection of this tangent line with the diode voltage axis:

$$V_\gamma = V - I\, r_d = V - \eta\, V_t \left(1 - e^{-\frac{V}{\eta V_t}}\right). \tag{2.10}$$

Figure 2.7 demonstrates this principle showing a line tangent to a diode curve at the Q-point of: $I = 3\,\text{mA}$ and $V = 0.776\,\text{V}$.

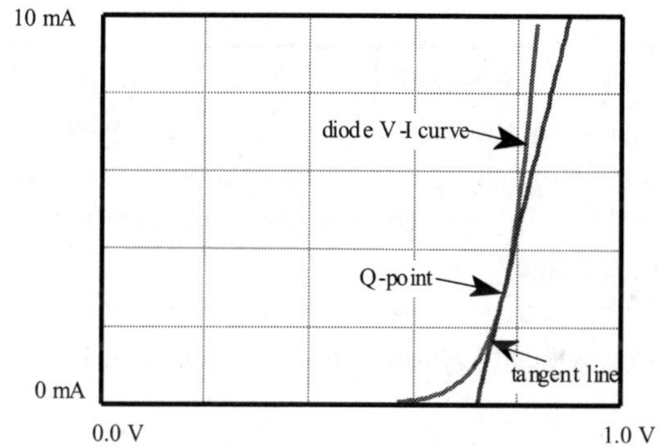

Figure 2.7: Modeling a diode with a tangent line at the Q-point.

Example 2.6

Determine a linear forward-bias model for a diode with the following parameters:

$$I_s = 1\,\text{nA}, \qquad \eta = 2$$

near the region where the diode current is 3 mA.

Solution:

The diode voltage at this Q-point is given by:

$$V_d = \eta\, V_t \ln\left(\frac{I}{I_s} + 1\right)$$

$$V = 2(0.026) \ln\left(\frac{3\,\text{mA}}{1\,\text{nA}} + 1\right) = 0.776\,\text{V}.$$

The diode dynamic resistance can then be calculated as:

$$r_d = \frac{\eta \, V_t}{I + I_S} = \frac{2\,(0.026)}{1\,\text{nA} + 3\,\text{mA}} = 17.33\,\Omega .$$

And finally the threshold voltage is calculated as:

$$V_\gamma = V - I r_d = 0.776 - (3\,\text{mA})(17.33\,\Omega) = 0.724\,\text{V}.$$

The linear forward-bias model of the diode therefore has a V-I relationship:

$$V = V_\gamma + I r_d = 0.724 + 17.33 I$$

as is shown in Figure 2.7.

The linear model of a forward-biased diode can simply be modeled with linear circuit elements as a voltage source (typically shown as a battery) in series with a resistor. The voltage source takes the value V and the resistor becomes r_d. The approximate model is shown in Figure 2.8. Care should be taken as to the polarity of the voltage source: forward biased diodes experience a voltage drop as is shown in the figure.

While this technique gives an accurate approximation of the diode V-I characteristic about a Q-point, it is not always clear what Q-point should be chosen. In practice, one usually chooses a Q-point by:

- using an ideal diode model to get an approximate Q-point for static cases, *or*
- choosing a Q-point that approximately bisects the expected range of diode currents within the application of interest.

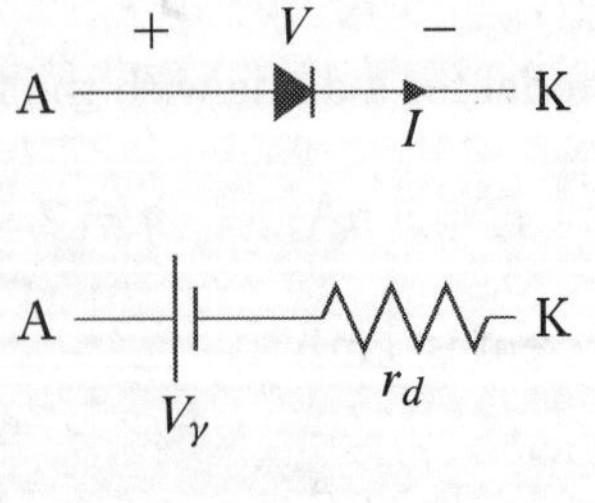

Figure 2.8: Linear modeling of a forward-biased diode.

If ideal diode modeling is excessively difficult or if the range of diode currents is not easily determined, then less accurate approximations must be made. Figure 2.9 is a plot of the threshold voltage as a function of diode quiescent current for a typical Silicon diode with reverse saturation current of 1 nA. The threshold voltage increases sharply for small diode quiescent currents and then becomes relatively constant at a value between 0.7 V and 0.8 V: approximate values should lie in that range for this Silicon diode. The dynamic resistance of this diode as a function of threshold voltage is given in Figure 2.10. This resistance, while not constant, has value of only a few Ohms (here seen to be $27\,\Omega > r_d > 4\,\Omega$). A reasonable guess at the dynamic resistance, without prior knowledge of the diode state might be $r_d \approx 15\,\Omega$ for this diode (obviously, diodes with different defining parameters will have other dynamic resistance values). Approximate models create an error in the calculation of solutions, but allow for the use of simple linear algebraic solution techniques.

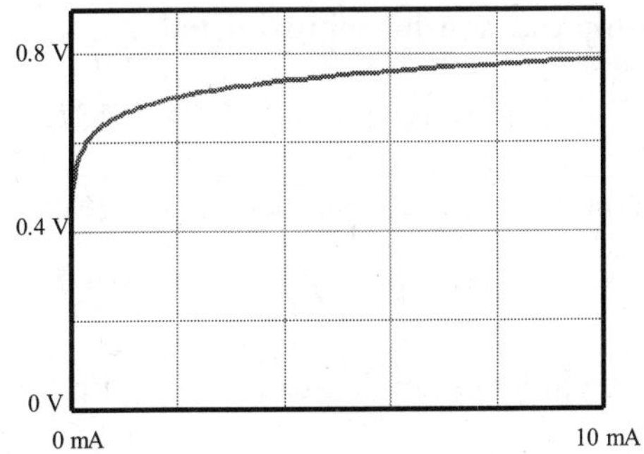

Figure 2.9: Threshold voltage, V_γ, as a function of diode current.

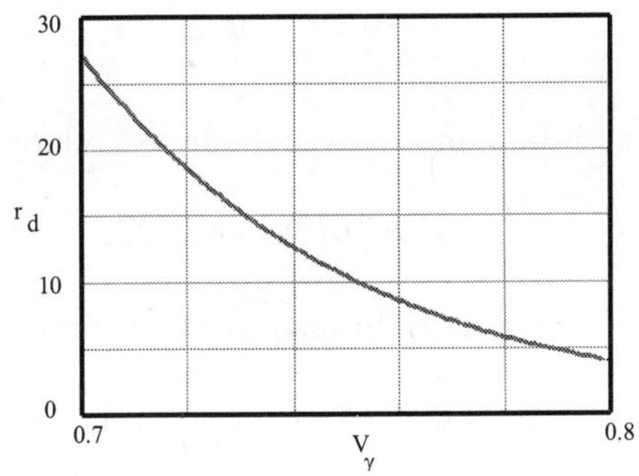

Figure 2.10: Diode resistance, r_d, as a function of threshold voltage.

Example 2.7
Assume the diode of Example 2.6 is connected in series with a 4 V source and a resistance of $820\,\Omega$ so that the diode is forward biased.

(a) Calculate the diode current with an approximate diode model.

(b) Calculate the diode current and voltage using the model derived in Example 2.6.

Solution:

(a) Using an ideal model for the diode, one would expect the diode current to be somewhat less than 5 mA. Since curves for the diode threshold voltage as a function of diode current exist, choose $V_\gamma = 0.74$ V. Again, since these curves exist, Figure 2.10 implies a diode resistance, $r_d = 13\,\Omega$ for that threshold voltage.

The diode V-I relationship can now be approximated by:

$$V = V_\gamma + r_d I = 0.74 + 13I.$$

The load line derived from the other circuit elements is given by:

$$V = V_s - RI = 4 - 820I.$$

Simple linear algebraic techniques applied to two equations with two unknowns lead to a solution of:

$$V = 0.791\text{ V} \qquad I = 3.91\text{ mA}.$$

If the diode curves for threshold voltage and resistance didn't exist other approximate values could be chosen: for example $V_\gamma = 0.7, r_d = 15\,\Omega$ lead to:

$$V = 0.759\text{ V} \qquad I = 3.95\text{ mA}.$$

(b) With the model derived in Example 2.6, the diode V-I relationship is given by:

$$V = 0.724 + 17.33I.$$

The load line is the same as in part (a). Similar linear algebraic techniques give the solution:

$$V = 0.792\text{ V} \qquad I = 3.91\text{ mA}.$$

Notice that all the approximate models give solutions that are within $\approx 1\%$ in the diode current and $\approx 5\%$ in the diode voltage. Numerical solution of this problem as outlined in Example 2.3 (using the theoretical non-linear diode V-I relationship) give the exact solution to be:

$$V = 0.789\text{ V} \qquad I = 3.92\text{ mA}.$$

The models have all given results within $\approx 0.8\%$ in current and $\approx 4\%$ in voltage of the exact theoretical values.

When accuracy of the order seen in Example 2.7 is not necessary, a diode can be represented as simply a voltage drop of approximately V_γ. This model assumes that the small series resistance is assumed to be negligible with respect to the circuit Thévenin resistance as seen by the diode, and forms an intermediate linear model between the ideal diode model and the two-element linear model. Use of this model leads to inaccuracies much larger than seen in previous linear models. It does, however, simplify circuit analysis greatly.

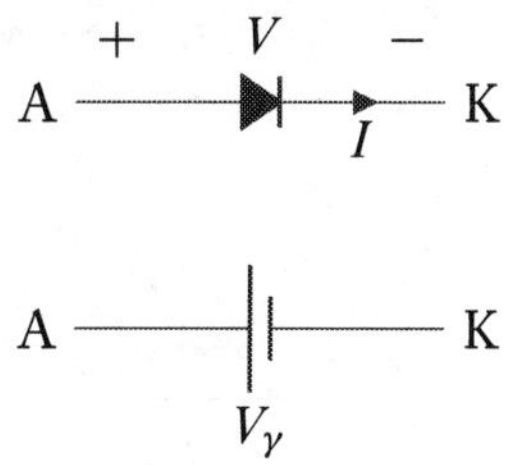

Figure 2.11: Simplified forward bias diode model.

Example 2.8

Assume the diode of Example 2.6 is connected in series with a 4 V source and a resistance of 820 Ω so that the diode is forward biased.

Calculate the diode current and voltage using a simplified forward bias diode model.

Solution:

The diode voltage for this simple model is just a voltage source, V_γ. For this simple model choose the approximate value:

$$V_\gamma = 0.7\,\text{V}.$$

Kirchhoff's voltage law applied to the loop yields:

$$V_\gamma = V_s - RI.$$

Thus,

$$0.7 = 4 - 820I$$

and

$$I = 4.02\,\text{mA}.$$

This approximate solution has an error of approximately 2.7%: the diode voltage (which was just guessed at) is in error by approximately 11.3%.

2.5.2 REVERSE BIAS MODELING

In the region where the diode is basically not conducting there are several possible linear models from which to choose. Each of these is based on the principle that the diode has a small leakage current that is fairly constant in the reverse bias region: that is when the diode voltage is between a few negative multiples of V_t and the Zener breakdown voltage[6] the diode current is constant at $-I_S$. The two common models are: a current source of value $-I_S$; or a large resistor. These models are shown in Figure 2.12.

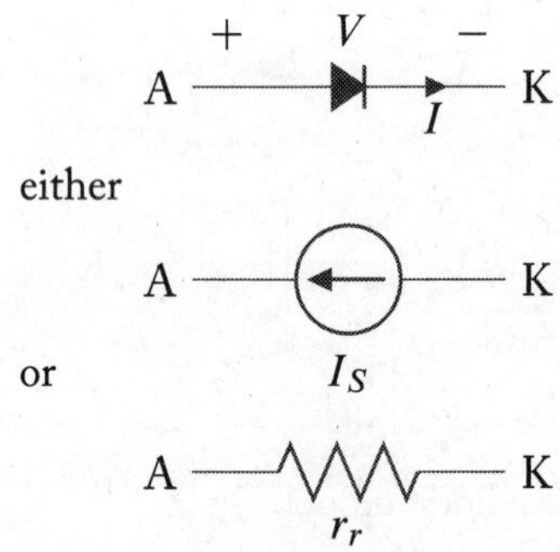

Figure 2.12: Linear reverse bias diode models.

The value of the reverse resistance, r_r, for the second model can be approximated using one of two techniques: (a) using Equation (2.9) to determine the dynamic resistance about some Q-point, or (b) assuming that the diode achieves its true reverse saturation current at the Zener breakdown voltage. While method (a) allows for an exact dynamic resistance at some point, it is often difficult to choose the proper Qpoint for a particular application. Method (b) is easier to calculate, but is less accurate at any point and underestimates the dynamic resistance for large reverse voltages.

Example 2.9

Assume the diode of Example 2.6 (with Zener breakdown occurring at a voltage of −25 V) is connected in series with a 4 V source and a resistance of 820 Ω so that the diode is *reverse biased.*

Calculate the diode current and voltage using:

(a) the current source model for a reverse biased diode

(b) the resistor model for a reverse biased diode

[6]When a large negative voltage is applied to a semiconductor diode (i.e., a voltage that exceeds some reverse threshold voltage called the Zener voltage), the diode enters a region of reverse conduction. The Zener conduction region of semiconductor diodes is discussed thoroughly in Section 2.7.

Solution:

(a) If the diode is replaced by a 1 nA current source then all circuit elements carry 1 nA of current and the diode current is −1 nA. The voltage across the resistor is given by:

$$V_r = (1\,\text{nA})(820\,\Omega) = 820\,\text{nV}.$$

Kirchhoff's Voltage Law applied to the loop gives the resulting diode voltage:

$$V_d = V_r - 4 = -3.99999918\,\text{V} \approx -4.00\,\text{V}.$$

(b) The reverse resistance can be approximated as:

$$r_r \approx |-25\,\text{V}|/(1\,\text{nA}) = 25\,\text{G}\Omega.$$

The diode voltage and current are given the following:

$$V_d = \frac{25\,\text{G}\Omega}{25\,\text{G}\Omega + 820\,\Omega}(-\ 4\,\text{V}) = -\ 3.999999869 \approx -\ 4$$
$$I_d = \frac{-4\,V}{25\,\text{G}\Omega + 820\,\Omega} = 160\,\text{pA}.$$

Comments: Each solution yields diode voltage solutions that are extremely close - essentially all the source voltage appears across the reverse biased diode. The current values vary by more than a factor of 6, but it must be remembered that these are extremely small values that are difficult to verify experimentally. Qualitatively, the two solutions are the same.

Numerical solutions of the type outlined in Example 2.3 suffer from lack of precision capability and, due to the almost zero slope of the V-I relationship, often cannot converge to a solution. In this particular problem, MathCAD is unable to effectively find solutions if the source voltage is more negative than about −1 V.

2.6 DIODE APPLICATIONS

Typical applications of diodes are considered in this section. The diode circuits studied are:

- Limiter or Clipping Circuit
- Full- and Half-Wave Rectifiers
- Peak Detector
- Clamping or DC Restoring Circuit
- Voltage Multiplier

- Diode Logic Gates
- "Superdiode"

All of the circuits analyzed in this section perform some form of wave shaping operation on the input signal to yield a desired output. The clipping circuit "truncates" the input to some desired value beyond which the signal is not to exceed. Full- and half-wave rectifiers pass only the signals of the desired polarity (positive or negative amplitude) and are commonly used in DC power supply designs. The peak detector follows only the maximum amplitudes of an incoming signal and is commonly used in amplitude modulation (AM) radio receivers in communications applications. Clamping circuits perform a level shifting operation on the input waveform, and are used to measure the duty cycle of a pulse waveform. Clamping circuits are commonly used to detect information carried on pulse-width modulated signals (i.e., the information of interest is represented by increasing or decreasing the pulse-width of a pulse waveform) by retrieving the DC component of the modulated signal. Voltage multipliers perform an integer multiplication on the input signal to yield a higher output voltage. Diode logic gates are simple circuits for performing Boolean operations. The "Superdiode" is a combination of an OpAmp and diode which eliminates the undesirable diode threshold voltage and dynamic resistance characteristics. Diodes in a circuit can, in most instances, be replaced by Superdiodes to design precision circuits.

2.6.1 LIMITER OR CLIPPING CIRCUIT

Diodes are often used in waveshaping applications. In particular, when used with a DC voltage in series with the diode, the output signal can be limited to the reference voltage level of the DC voltage source. Examples of clipping circuits are shown in Figure 2.13.

The simplified forward bias diode model of Figure 2.11 can be used to analyze clipping circuits.

The circuit of Figure 2.13a will be used as an example of this analysis. When the input voltage $v_i \leq V_d + V_{ref}$, the diode is reverse biased (or OFF).[7] Therefore, the diode can be thought of as an open circuit. The output voltage in this case follows the input voltage,

$$v_o = v_i .$$

When the voltage $v_i > V_d + V_{ref}$, the diode is forward biased (or ON). Using the piece-wise linear model of the forward biased diode, a simplified equivalent circuit of the clipping circuit of Figure 2.13a is developed in Figure 2.14.

The output voltage v_o of the clipping circuit when the diode is forward biased is found by analyzing the circuit in Figure 2.14 using superposition and voltage division,

$$v_o = \frac{r_d}{R_s + r_d} v_i + \frac{R_s}{R_s + r_d} \left(V_d + V_{ref} \right) . \tag{2.11}$$

[7]The diode resistance, r_r under reverse bias conditions is assumed to be much larger than the series resistance R_s in the derivation. ($r_r \gg R_s$.)

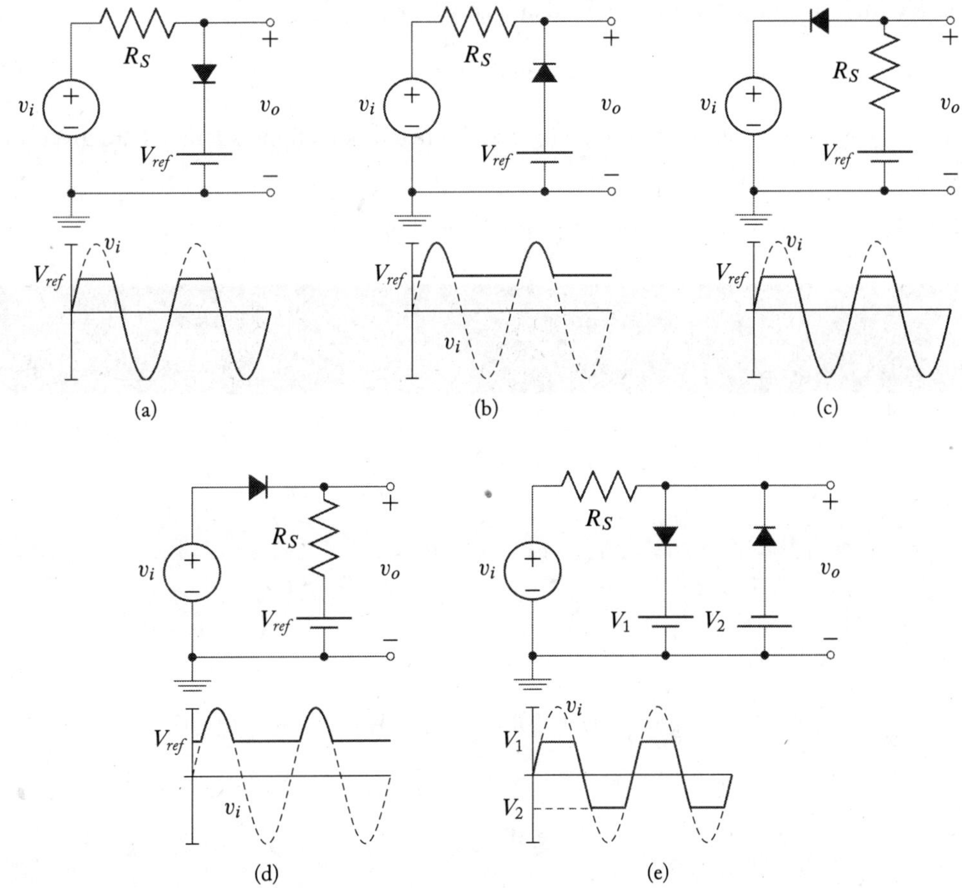

Figure 2.13: Diode clipping circuits.

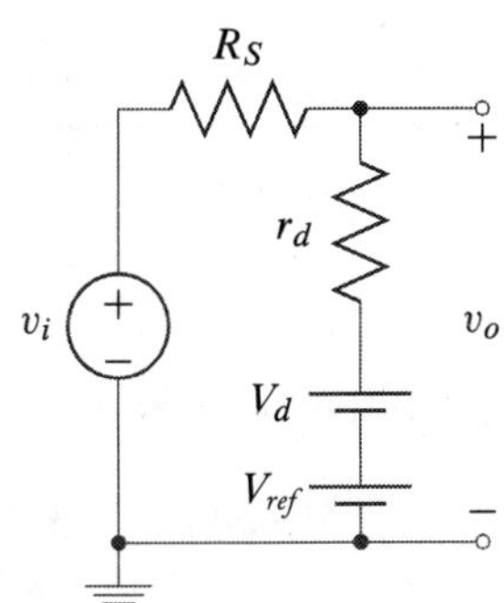

Figure 2.14: Simplified equivalent circuit of the clipping circuit of Figure 2.13a for $v_i > V_d + V_{ref}$.

If $r_d \ll R_S$ then the output voltage is held at a constant value

$$V_o = V_d + V_{ref} \tag{2.12}$$

The input-output voltage relationships for the five diode clippers circuits are given in Table 2.1.

Table 2.1: Input-output voltage relationships for diode clipping circuits

Clipping Circuits of Figure 2.13	Output Voltage, v_o		Simplified Output Voltage, v_o	
(a)	$v_o = \frac{r_d}{R_s + r_d} v_i + \frac{R_s}{R_s + r_d}(V_\gamma + V_{ref})$,	$v_i > V_\gamma + V_{ref}$	$v_o \approx V_\gamma + V_{ref}$,	$v_i > V_\gamma + V_{ref}$
	$v_o = v_i$,	$v_i \le V_\gamma + V_{ref}$	$v_o = v_i$,	$v_i \le V_\gamma + V_{ref}$
(b)	$v_o = \frac{r_d}{R_s + r_d} v_i + \frac{R_s}{R_s + r_d}(V_{ref} - V_\gamma)$,	$v_i < V_{ref} - V_\gamma$	$v_o \approx V_{ref} - V_\gamma$,	$v_i < V_{ref} - V_\gamma$
	$v_o = v_i$,	$v_i \ge V_{ref} - V_\gamma$	$v_o = v_i$,	$v_i \ge V_{ref} - V_\gamma$
(c)	$v_o = V_{ref}$,	$v_i + V_\gamma > V_{ref}$	$v_o = V_{ref}$,	$v_i + V_\gamma > V_{ref}$
	$v_o = \frac{r_d}{R_s + r_d} V_{ref} + \frac{R_s}{R_s + r_d}(v_i + V_\gamma)$,	$v_i + V_\gamma \le V_{ref}$	$v_o \approx v_i + V_\gamma$,	$v_i + V_\gamma \le V_{ref}$
(d)	$v_o = V_{ref}$,	$v_i - V_\gamma < V_{ref}$	$v_o = V_{ref}$,	$v_i - V_\gamma < V_{ref}$
	$v_o = \frac{r_d}{R_s + r_d} V_{ref} + \frac{R_s}{R_s + r_d}(v_i - V_\gamma)$,	$v_i - V_\gamma \ge V_{ref}$	$v_o \approx v_i - V_\gamma$,	$v_i - V_\gamma \ge V_{ref}$
(e)	$v_o = \frac{r_d}{R_s + r_d} v_i + \frac{R_s}{R_s + r_d}(V_1 + V_\gamma)$,	$v_i > V_\gamma + V_1$	$v_o \approx V_\gamma + V_1$,	$v_i > V_\gamma + V_1$
	$v_o = \frac{r_d}{R_s + r_d} v_i - \frac{R_s}{R_s + r_d}(V_2 + V_\gamma)$,	$v_i < -V_2 - V_\gamma$	$v_o \approx -V_2 - V_\gamma$,	$v_i < -V_2 - V_\gamma$
	$v_o = v_i$,	$-V_\gamma - V_2 \le v_i \le V_\gamma + V_1$	$v_o = v_i$,	$-V_\gamma - V_2 \le v_i \le V_\gamma + V_1$

Example 2.10

For the clipping circuit shown, find the output waveform v_o for the input voltage,

$$v_i = 5 \sin\omega_o t.$$

The diode has the following characteristics:

$$r_d = 15\,\Omega, \quad V_d = 0.7\,\text{V}, \text{ and } r_r \approx \infty$$

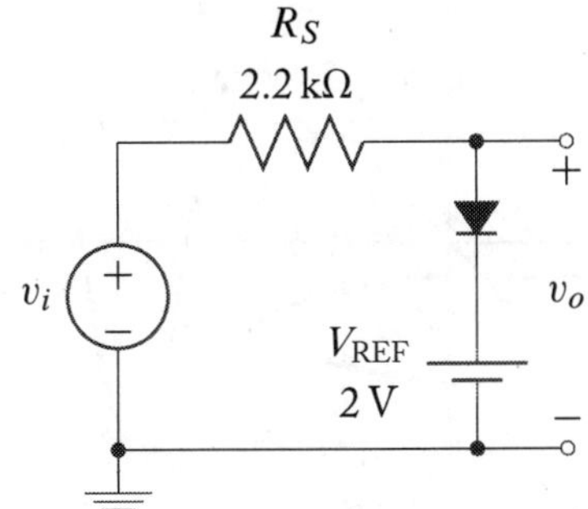

Solution #1:

Since $R_s \gg r_d$, the simplified output voltage result from Table 2.1 can be constructed. That is,

$$\begin{aligned} v_o &= V_d + V_{ref}, & v_i &> V_d + V_{ref} \\ v_o &= v_i, & v_i &\leq V_d + V_{ref}. \end{aligned}$$

Therefore, when $v_i > 2.7\,\text{V}$, $v_o = 2.7\,\text{V}$ and when $v_i < 2.7\,\text{V}$, $v_o = 5 \sin \omega_o t$. The output waveform is shown in Figure 2.15.

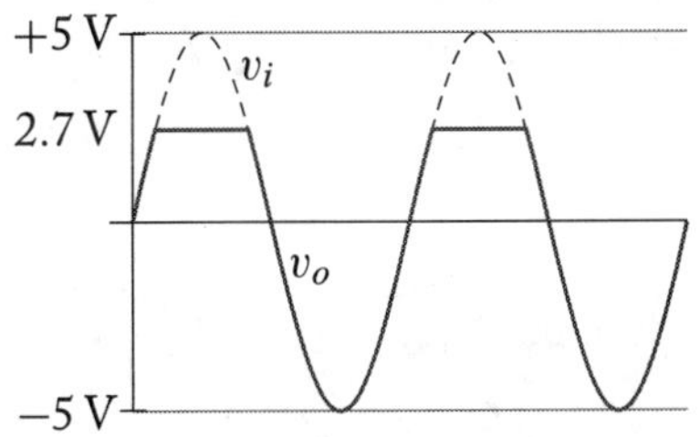

Figure 2.15: Output waveform for Example 2.10.

Solution #2: Transfer Function Analysis

A solution can be constructed from a transfer function analysis of the circuit as shown below. A transfer function defines the input/output relationship of a circuit. In this case, the transfer function is described in Table 2.1.

Solution #3: MultiSim (SPICE) Solution

Figure 2.17 is the output of a transient analysis performed on the given circuit. The SPICE parameter, RS (parasitic resistance), was changed in the virtual diode model to match the given value (15 Ω).

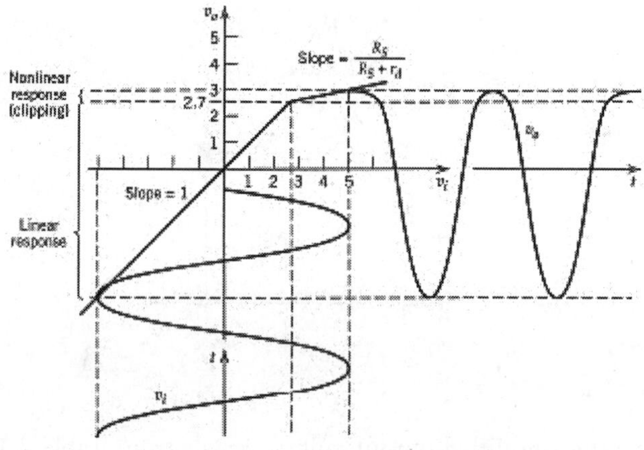

Figure 2.16: Transfer function solution to Example 2.10.

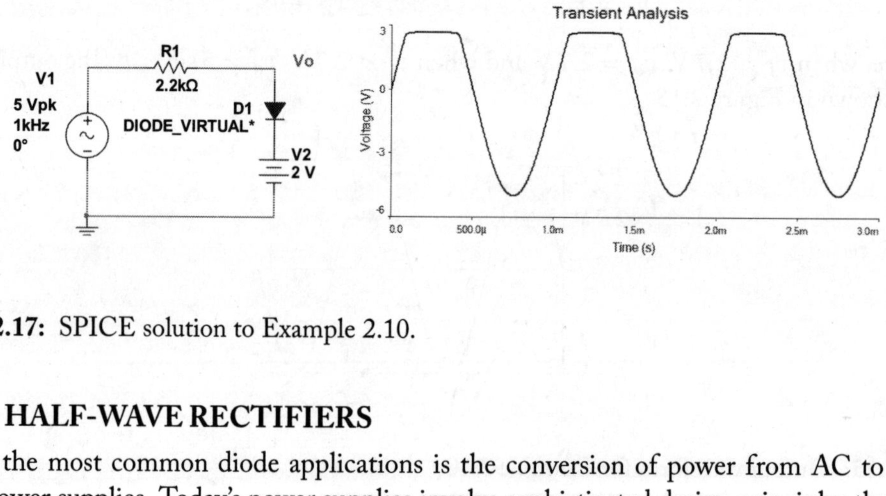

Figure 2.17: SPICE solution to Example 2.10.

2.6.2 HALF-WAVE RECTIFIERS

One of the most common diode applications is the conversion of power from AC to DC for use as power supplies. Today's power supplies involve sophisticated design principles that will be detailed in Chapter 14 (Book 4) in the fourth book of this series. However, the basic principles of converting from AC to DC power can be explored in this section.

Figure 2.18 shows a half-wave rectifier circuit. The circuit is so named because it only allows current from the positive half cycle of the input to flow through the load resistor, R. Figure 2.18 is identical to the clipping circuit of Figure 2.13d with zero reference voltage.

If v_i is a sinusoidal voltage with peak voltage V_m and radian frequency ω,

$$v_i(t) = V_m \sin \omega t$$

the average voltage V_{dc} across the load, R, is

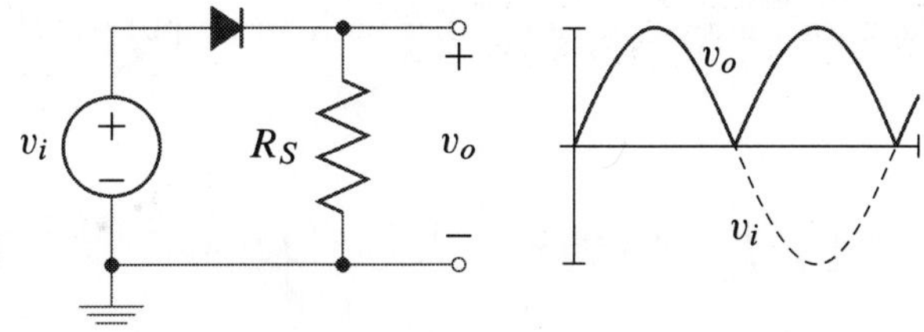

Figure 2.18: Half-wave rectifier with the output voltage waveform.

$$V_{dc} = \frac{1}{T}\int_0^T v_o(t)\,dt, \tag{2.13}$$

where T is the period of the sinusoid.

Since the diode is OFF in the interval $1/2\,T \leq t \leq T$, output voltage is[8]

$$\begin{aligned} v_o(t) &= V_m \sin\omega t - V_\gamma, & 0 \leq t < \frac{T}{2} \\ &= 0, & \frac{T}{2} \leq t \leq T. \end{aligned} \tag{2.13a}$$

Substituting Equation (2.13a) into (2.13) to solve for V_{dc},

$$\begin{aligned} V_{dc} &= \frac{1}{T}\int_0^{\frac{T}{2}} \left(V_m \sin\omega t - V_\gamma\right) dt \\ &= \frac{V_m}{\omega T}\left(-\cos\omega t\right)\Big|_0^{\frac{T}{2}} - \frac{V_\gamma}{T}\,(t)\Big|_0^{\frac{T}{2}} \\ &= \frac{-V_m}{\omega T}\left(\cos\omega t - 1\right) - \frac{V_\gamma}{2}. \end{aligned} \tag{2.14}$$

But $\omega = \frac{2\pi}{T}$ so that Equation (2.14) simplifies to

$$V_{dc} = \frac{V_m}{\pi} - \frac{V_\gamma}{2}. \tag{2.15}$$

Recall that the "effective" or root-mean-squared (RMS) voltage quantifies the amount of energy delivered to a resistor in T seconds. The use of RMS comes from the desire to compare the ability of a sinusoid to deliver energy to a resistor with the ability of a DC source.

The RMS value of any periodic waveform $v_o(t)$ is defined as

$$V_{rms} = \left[\frac{1}{T}\int_0^T v_o^2(t)\,dt\right]^{\frac{1}{2}}. \tag{2.16}$$

[8]In reality, the interval over which the diode is ON is slightly less than $1/2T \leq t \leq T$ due to the diode threshold voltage V_γ. The analysis assumes that $V_m \gg V_\gamma$. If V_m is very small, the analysis becomes more complex.

The output RMS voltage for the half-wave rectifier with an output waveform defined by Equation (2.13a) is

$$\begin{aligned} V_{rms} &= \left[\frac{1}{T}\int_0^{\frac{T}{2}} \left(V_m \sin\omega t - V_\gamma\right)^2 dt\right]^{\frac{1}{2}} \\ &= \left[\frac{1}{T}\int_0^{\frac{T}{2}} \left(V_m^2 \sin^2\omega t - 2V_\gamma V_m \sin\omega t + V_\gamma^2\right) dt\right]^{\frac{1}{2}} \\ &= \left[\frac{V_m^2}{4} - \frac{2V_\gamma V_m}{2\pi} + \frac{V_\gamma^2}{2}\right]^{\frac{1}{2}}. \end{aligned} \tag{2.17}$$

If $V_\gamma \ll V_m$ then Equation (2.17) reduces to

$$\begin{aligned} V_{rms} &= \left[\frac{1}{T}\int_0^{\frac{T}{2}} \left(V_m \sin\omega t\right)^2 dt\right]^{\frac{1}{2}} \\ &= \frac{1}{\sqrt{2}}\frac{V_m}{\sqrt{2}} = \frac{V_m}{2}. \end{aligned} \tag{2.18}$$

The efficiency of rectification is defined as

$$\eta = \frac{P_{dc}}{P_{ac}}, \tag{2.19}$$

where P_{ac} and P_{dc} are AC and DC powers respectively. For the half-wave rectifier in Figure 2.18, the efficiency is

$$\eta = \frac{P_{dc}}{P_{ac}} = \frac{\dfrac{\left(\dfrac{V_m}{\pi}\right)^2}{R}}{\dfrac{\left(\dfrac{V_m}{2}\right)^2}{R}} = \frac{4}{\pi^2} \Rightarrow 40.6\%. \tag{2.20}$$

The result of Equation (2.20) is for an ideal half-wave rectifier and represents the maximum efficiency attainable. In real systems, the efficiency will be lower due to power losses in the resistor and diode.

In order to produce a DC voltage from a half-wave rectifier, a large capacitor is placed in parallel to the load resistor. The capacitor must be large enough so that the RC time constant of the capacitor and load resistor is large compared to the period of the output waveform. This has the effect of "smoothing" the output waveform. Clearly, and efficient filter is required to eliminate any ripple in the output waveform.

In many rectifier applications, it is desirable to transformer couple the input voltage source to the rectifier circuit. This method is commonly used in the design of power supplies where there is a requirement to "step-down" the AC input voltage to a lower DC voltage. For example, a 15 V peak half-wave rectified voltage can be derived from a 120 VAC (household power is defined in RMS volts) source through the use of a transformer. Transformers also provide isolation of the circuit from the household power line, providing protection from the possibility of shock from those lines.

The turns ratio N_p/N_s (primary winding over the secondary windings) determines the "step-down" ratio. For example, if the voltage input at the primary is 120 VAC (RMS) which is approximately 170 V peak, a transformer turns ratio of 11:1 (actually 11.3 : 1) is required to yield a 15 V peak half-wave rectified signal. If the coefficient of coupling is nearly 1.0, implying no loss occurs in the transformer, the inductance ratio of the primary coil L_p and the secondary coil L_s is,

$$\frac{L_p}{L_s} = \left(\frac{N_p}{N_s}\right)^2. \tag{2.21}$$

Example 2.11

For the circuit below, determine the inductance of the secondary coil for a transformer turns ratio of 11:1. What is the peak output voltage? Assume that the coefficient of coupling is $k = 0.99$ and that the diode voltage V_d of the 1N4148 is 0.76 V . L_p.

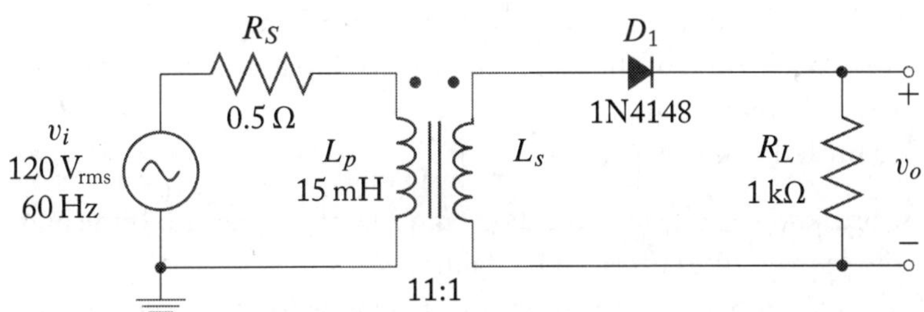

Solution:

Apply Equation (2.21) to find the secondary inductance:

$$L_s = \frac{L_p {N_s}^2}{{N_p}^2} = \frac{(15 \times 10^{-3})\,(1^2)}{11^2} = 124\,\mu\text{H}.$$

The peak output voltage is,

$$v_o = \frac{(v_{i\ \text{peak}})}{\frac{N_p}{N_s}} - V_d = \frac{120\sqrt{2}}{11} - 0.76 = 14.64\,\text{V peak}.$$

In order to simulate the circuit using SPICE, several items must be added to the circuit. The SPICE circuit is shown in Figure 2.19.

Observe that in the original circuit (Figure 2.14), the output did not reference to a common (ground) point. A large isolation resistor at the secondary facilitates the reference to ground at the secondary. The model statement for the 1N4148 is,

.model D1N4148 D(Is=0.1pA Rs=16 CJO=2p Tt=12n Bv=100 Ibv=0.1p)

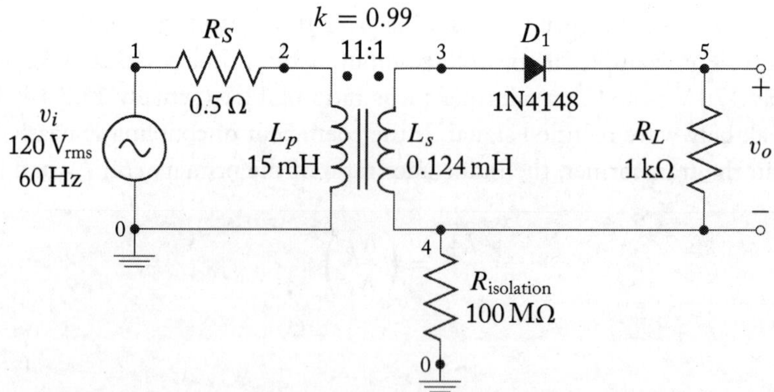

Figure 2.19: SPICE circuit topology.

Many of the SPICE model parameters are useful for frequency analysis that is used in the latter chapters. However, several of the parameters are of interest at this time. They are:

Is = reverse saturation current, and

Bv = reverse breakdown voltage.

The reverse breakdown voltage of the diode must not be exceeded. If the breakdown voltage is exceeded, the diode may suffer catastrophic failure.

The SPICE circuit model and the output voltage waveform are shown below.

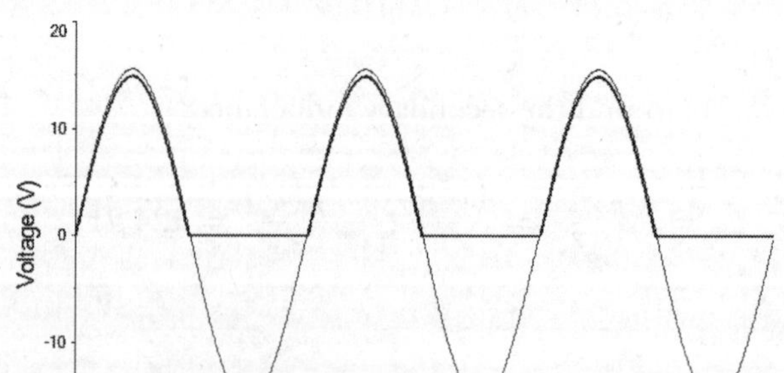

If the transformer is assumed to have a coupling coefficient of nearly 1.0, then it may be replaced by one of the equivalent circuits of the ideal transformer shown in Figure 2.20.

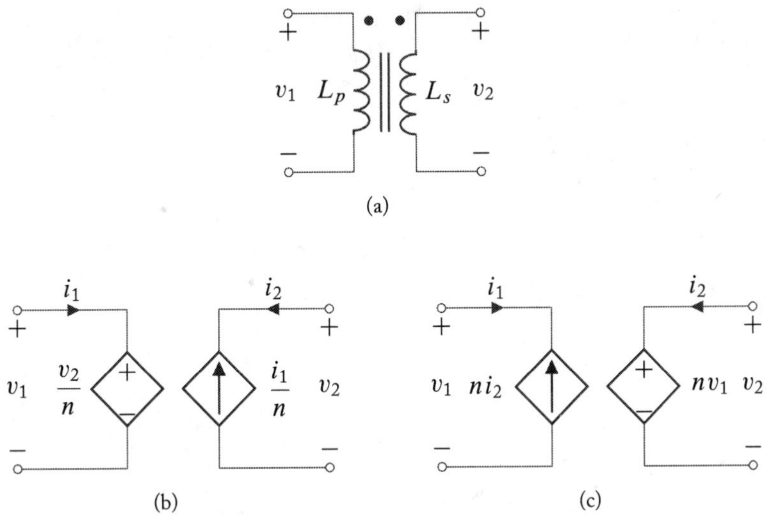

Figure 2.20: (a) Ideal transformer; (b) and (c) equivalent circuits.

2.6.3 FULL-WAVE RECTIFIERS

To remove the ripple from the output of a half-wave rectifier may require a very large capacitance. In many instances, the capacitor required to reduce the ripple on the half-wave rectified output voltage to the desired design specification may be prohibitively large.

A full-wave rectifier circuit can be used as a more efficient way to reduce ripple on the output voltage. A center-tapped input transformer-coupled full-wave rectifier is shown in Figure 2.21. Each half of the transformer with the associated diode acts as a half-wave rectifier. The diode D1 conducts when the input $v_i > V_\gamma$ and D2 conducts when the input $v_i < V_\gamma$. Note that the secondary winding is capable of providing twice the voltage drop across the load resistor. Additionally, the input to the diodes and the output share a common ground between the load resistor and the center-tap.

An isolation transformer is not required to design a full-wave rectifier. If ground isolation is not required, only a center-tapped well coupled coil is required as shown in Figure 2.22.

An alternate configuration for a full-wave rectifier exists with an addition of two diodes. In the alternate configuration, called the *bridge rectifier* shown in Figure 2.23, the source and load do not share an essential common terminal. Additionally, the secondary transformer does not

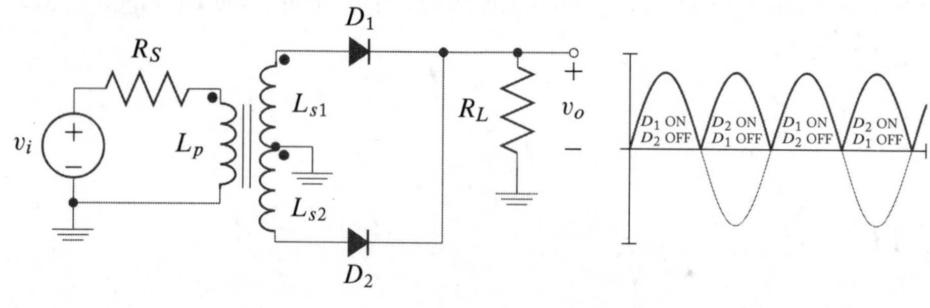

Figure 2.21: Full-wave rectifier with center-tapped transformer.

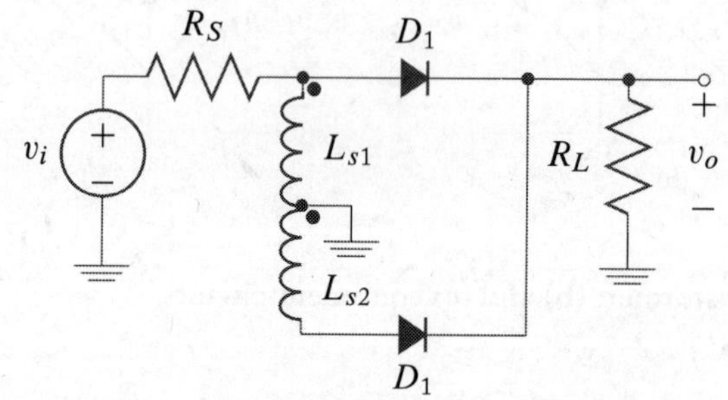

Figure 2.22: Full-wave rectifier without an isolation transformer.

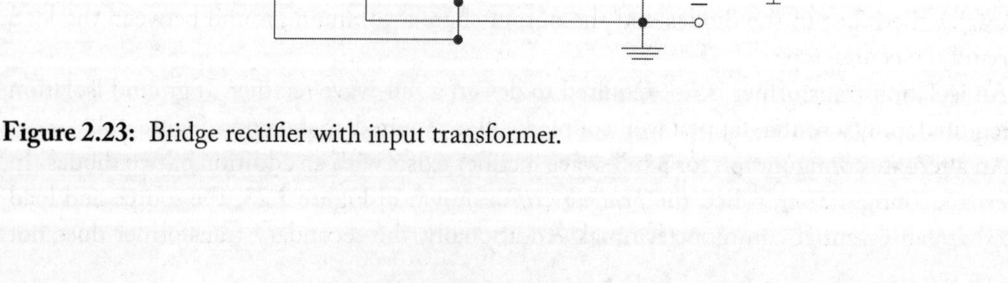

Figure 2.23: Bridge rectifier with input transformer.

require a center tap and provides a voltage only slightly greater than half that of the secondary in Figure 2.21.

In the bridge rectifier circuit, diodes D_2 and D_4 are ON for the positive half cycle of the voltage across the secondary of the transformer. Diodes D_1 and D_3 are off in the positive half cycle since their anode voltages are less than the cathode voltages. This is due to the voltage drop across the ON diodes and the load resistor. In the negative half cycle of the voltage across the secondary of the transformer, diodes D_1 and D_3 are ON, with D_2 and D_4 OFF. In both half cycles, the current through the load resistor is in the same direction. Therefore, each half cycle, the output voltage appears in the same polarity.

From Equation (2.13), the output DC voltage of a full-wave rectifier circuit is twice that of the half-wave rectifier since its period is half that of the half-wave rectifier circuit,

$$
\begin{aligned}
[V_{dc}]_{\text{full-wave}} &= \frac{1}{T}\int_0^T v_o(t)\,dt \\
&= 2[V_{dc}]_{\text{half-wave}} \\
&= \frac{2V_m}{\pi}.
\end{aligned}
\tag{2.22}
$$

Similarly, the RMS output voltage of a full-wave rectifier is found by applying Equation (2.16),

$$
\begin{aligned}
V_{rms} &= \left[\frac{1}{T}\int_0^T v_o^2(t)\,dt\right]^{\frac{1}{2}} \\
&= \frac{V_m}{\sqrt{2}}.
\end{aligned}
\tag{2.23}
$$

The maximum possible efficiency of the full-wave rectifier is significantly greater than that of the half-wave rectifier since power from both positive and negative cycles are available to produce a DC voltage,

$$
\eta_{\text{full-wave}} = \frac{P_{dc}}{P_{ac}} = \frac{\dfrac{\left({}^{2V_m}/_{\pi}\right)^2}{R_L}}{\dfrac{\left({}^{V_m}/_{\sqrt{2}}\right)^2}{R_L}} = \frac{8}{\pi^2} = 2\eta_{\text{half-wave}} \Rightarrow 81.2\%. \tag{2.24}
$$

In order to produce a DC source from the output of a full-wave rectifier, a capacitor is placed in parallel to the load resistor as shown in Figure (2.24). The RC time constant must be long with respect to $1/2T$ to "smooth out" the output waveform.

Let t_1 and t_2 be the time between two adjacent peaks of the filtered rectified voltage as shown in Figure 2.25. Then the output voltage between t_1 and t_2 is,

$$
v_o = V_m e^{\frac{-(t-t_1)}{R_L C}}, \qquad t_1 \le t \le t_2. \tag{2.25}
$$

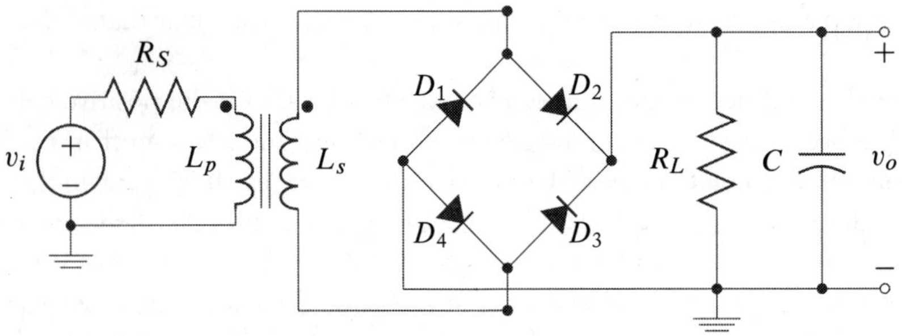

Figure 2.24: Filtered full-wave rectifier circuit.

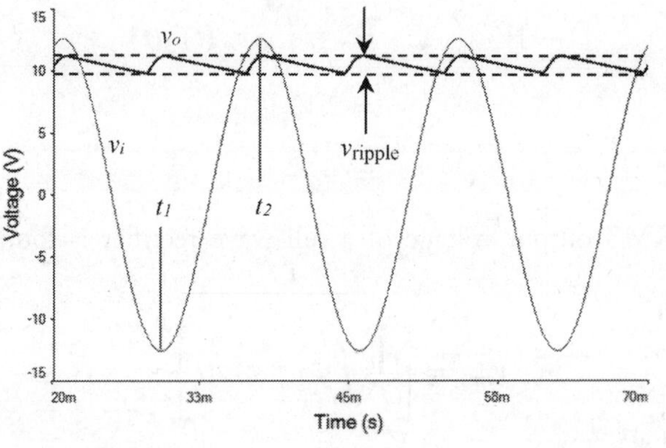

Figure 2.25: Full-wave rectified voltage with ripple.

The peak-to-peak ripple is defined as,

$$\begin{aligned} v_r &= v_o(t_1) - v_o(t_2) \\ &= V_m \left(1 - e^{\frac{-(t_2 - t_1)}{R_L C}} \right). \end{aligned} \tag{2.26}$$

If $R_L \gg (t_2 - t_1)$ then the exponential approximation can be used,

$$e^{-x} \approx 1 - x, \qquad \text{for } |x| \ll 1. \tag{2.27}$$

Applying Equation (2.27) to (2.25), a good approximation for the peak-to-peak ripple voltage can be derived,

$$v_r\,(\text{peak-to-peak}) \approx V_m \frac{t_2 - t_1}{R_L C}. \tag{2.28}$$

Since $t_2 - t_1 \approx \frac{T}{2} = \frac{1}{2f_0}$, where f_o is the frequency of the input signal and T is the period of that signal, the peak-to-peak ripple voltage for a full-wave rectifier circuit is,

$$v_r\,(\text{peak-to-peak})_{\text{full-wave}} \approx \frac{V_m}{2f_oR_LC}. \tag{2.29}$$

The DC component of the output signal is,

$$\begin{aligned} V_{O,dc} = V_{dc} &= V_m - \frac{1}{2}v_r \\ &= V_m\left(1 - \frac{1}{4f_oR_LC}\right). \end{aligned} \tag{2.30}$$

For a half-wave rectifier, $t_2 - t_1 \approx T = 1/f_o$, since only half the cycle of the input signal is passed. Therefore, the peak-to-peak ripple voltage of a half-wave rectifier is,

$$v_r\,(\text{peak-to-peak})_{\text{half-wave}} \approx \frac{V_m}{f_oR_LC}. \tag{2.31}$$

In both the half- and full-wave rectifiers, the DC voltage is less than the peak rectified voltage.

Example 2.12
Consider the full-wave rectifier circuit of Figure 2.24 with $C = 47\,\mu$F and transformer winding ratio of 14:1. If the input voltage is 120 VAC (RMS) at 60 Hz, what is the load resistor value for a peak-to-peak ripple less than 0.5 V? What is the output DC voltage?

Solution:

Since the transformer turns ratio is 14:1, the voltage across the secondary is,

$$V_m = \frac{120\sqrt{2}}{14} = 12.1\text{ V}.$$

From Equation (2.29) for peak-to-peak ripple,

$$R \leq \frac{V_m}{2f_oC\,v_r} = \frac{12.1}{2\,(60)\left(47\times10^{-6}\right)(0.5)} \leq 4.29\text{ k}\Omega.$$

The DC voltage at the output is found by using Equation (2.30),

$$V_{dc} \leq V_m - \frac{1}{2}v_r \leq 12.1 - \frac{1}{2}(0.5) \leq 11.9\ V.$$

2.6.4 PEAK DETECTOR

One of the first applications of the diode was a "detector" in radio receivers that retrieved information from "amplitude modulated" (AM) radio signals. The AM signal consists of a radio-frequency "carrier" wave which is at a high frequency and varies in amplitude at an audible frequency. The detector circuit, shown in Figure 2.26, is similar to a half-wave rectifier. The RC time constant is approximately the same as the period of the carrier so that the output voltage can follow the variation in amplitude of the input.

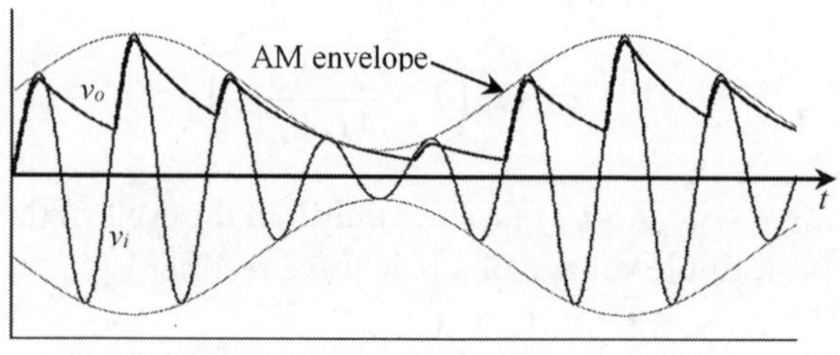

Figure 2.26: Peak detector and associated waveforms.

2.6.5 CLAMPING OR DC RESTORING CIRCUITS

Diode circuits can be designed clamp a voltage so that the output voltage is shifted to never exceed (or fall below) a desired voltage. A clamping circuit is shown in Figure 2.27.

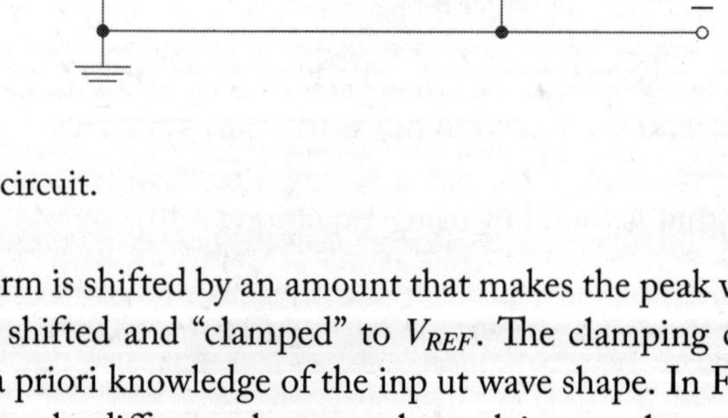

Figure 2.27: Clamping circuit.

The input waveform is shifted by an amount that makes the peak voltage equal to the value V_{REF}. The waveform is shifted and "clamped" to V_{REF}. The clamping circuit allows shifting of the waveform without a priori knowledge of the inp ut wave shape. In Figure 2.27, the capacitor charges to a value equal to the difference between the peak input voltage and the reference voltage of the clamping circuit, V_{REF}. The capacitor then acts like a series battery whose value is the voltage across the capacitor, shifting the waveform to the value shown in Figure 2.28.

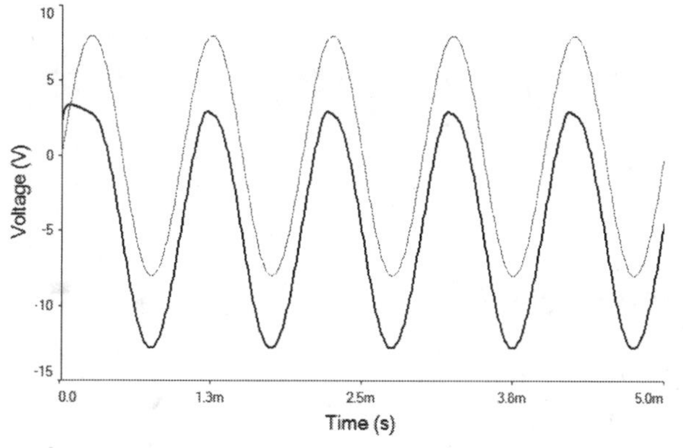

Figure 2.28: Input and output wave forms for the clamping circuit.

The clamping circuit configuration in Figure 2.27 clamps the *maximum* to the reference voltage. If the diode is reversed, the circuit will clamp the minimum voltage of the signal to the reference voltage.

2.6.6 VOLTAGE MULTIPLIER

Diode circuits may be used as voltage doublers as shown in Figure 2.29. The circuit is a clamper formed by C_1 and D_1, and a peak rectifier formed by D_1 and C_2. With a peak input signal V_m, the clamping section yields the waveforms shown in Figure 2.30. The positive voltage is clamped to zero volts. Across diode D_1, the negative peak reaches $-V_m$ due to the charge stored in capacitor C_1. The voltage stored in C_1 is $V_{C1} = V_m$ corresponding to the maximum negative input voltage. Therefore, the voltage across diode D_1 is,

$$v_{D1} = V_m \sin\omega t - V_{C1}. \tag{2.32}$$

The section consisting of D_2 and C_2 is a peak rectifier. Therefore, the output voltage v_o across C_2 is, after some time, a DC voltage shown in Figure 2.30,

$$v_o = -(V_m + V_{C1}). \tag{2.33}$$

By adding more capacitor and diode sections, higher multiples of the input voltage are achievable.

2.6.7 DIODE LOGIC GATES

Diodes together with resistors can be used to perform logic functions. Figure 2.31 shows diode AND and OR gates.

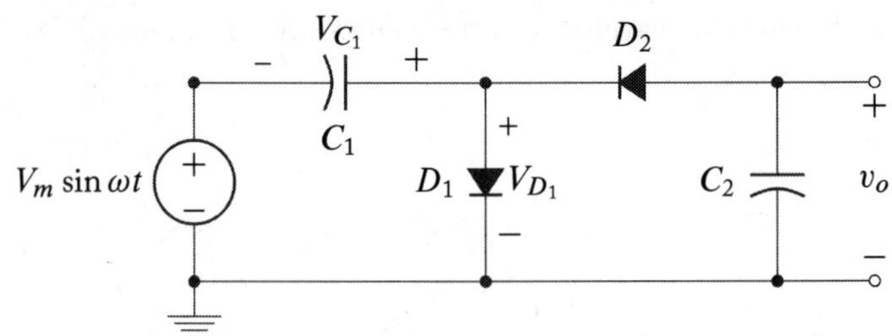

Figure 2.29: Diode voltage doubler circuit.

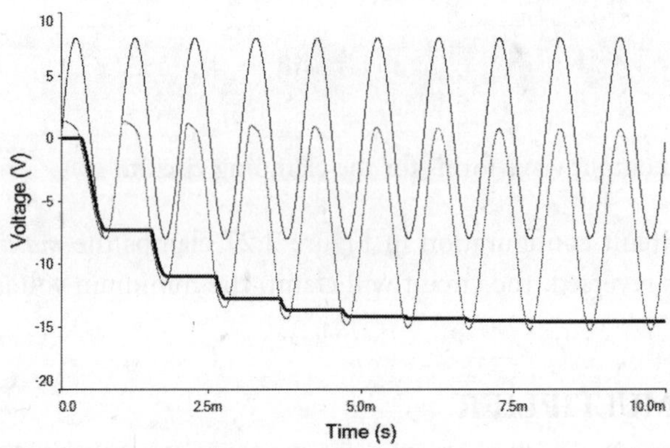

Figure 2.30: Output signal from a voltage doubler circuit.

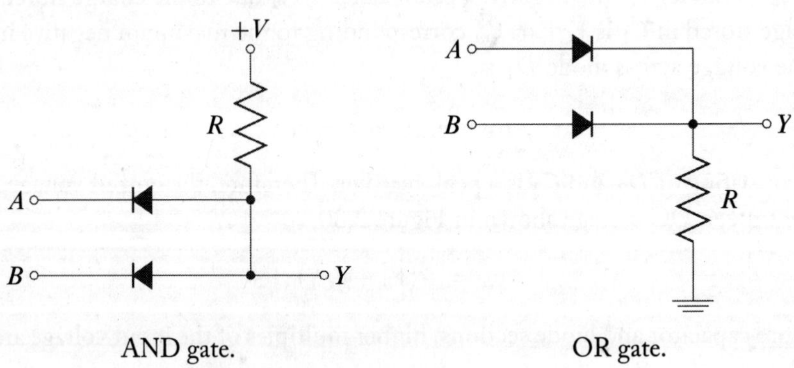

AND gate. OR gate.

Figure 2.31: Diode logic AND and OR gates.

In the AND gate, when either input is connected to ground, the diode in series with that input is forward biased. The output is then equal to one forward biased diode voltage drop above ground which is interpreted as logic "0." When both inputs are connected to $+V$, both diodes are zero biased, yielding an output voltage of $+V$ which is interpreted as logic "1." The Boolean notation for the circuit is,

$$Y = A \bullet B.$$

In the OR gate, if one or both of the diodes is connected to $+V$, that (those) diode(s) will conduct, clamping the output voltage to a value equal to $+V - V_d$, or logic "1." Therefore, the Boolean notation for the circuit is,

$$Y = A + B.$$

2.6.8 THE SUPERDIODE

Figure 2.32 shows a precision half-wave rectifier using a "superdiode." The superdiode consists of an OpAmp and diode. The operation of the circuit is as follows: For positive v_i, the output of the OpAmp is will go positive causing the diode to conduct. This in turn closes the negative feedback path creating an OpAmp voltage follower. Therefore,

$$v_o = v_i, \qquad v_i \geq 0. \tag{2.34}$$

The slope of the voltage follower transfer function is unity.

For $v_i < 0$, the output voltage of the OpAmp follows the input and goes negative. The diode will not conduct since it is reverse biased. Therefore, no current flows through the resistor R, and

$$v_o = 0, \qquad v_i < 0. \tag{2.35}$$

The advantage of the superdiode is the very small turn-on voltage exhibited and ideal transfer function for positive v_i.

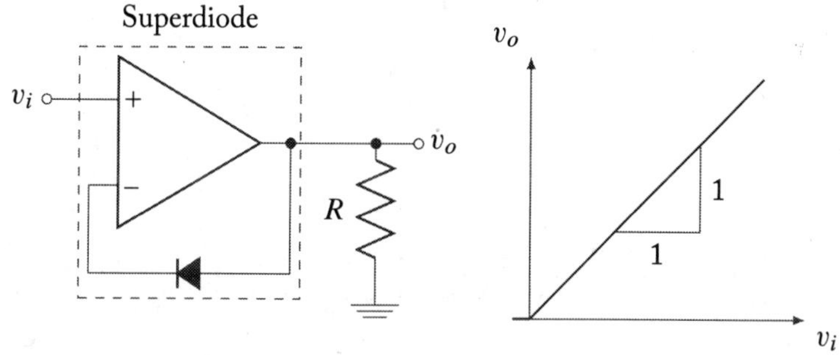

Figure 2.32: Precision rectifier using a Superdiode.

Precision circuits (clamper, peak detector, etc.) can be designed using the Superdiode in place of regular diodes to eliminate dynamic resistance and diode threshold voltage effects. The superdiode is commonly used in small-signal applications and not used in power circuits.

2.7 ZENER DIODES AND APPLICATIONS

Diodes that at designed with adequate power dissipation capabilities to operate in breakdown are called Zener diodes and are commonly used as voltage reference or constant voltage devices.

The two mechanisms responsible for the breakdown characteristics of a diode are avalanche breakdown and Zener breakdown. Avalanche breakdown occurs at high voltages (≥ 10 V) where the charge carriers acquire enough energy to create secondary hole-electron pairs which act as secondary carriers. This chain reaction causes and avalanche breakdown of the diode junction and a rapid increase in current at the breakdown voltage. Zener breakdown occurs in the heavily doped p- and n-regions on both sides of the diode junction and occurs when the externally applied potential is large enough to create a large electric field across the junction to force bound electrons from the p-type material to *tunnel* across to the n-type region. A sudden increase in current is observed when sufficient external potential is applied to produce the required ionization energy for tunneling.

Regardless of the mechanism for breakdown, the breakdown diodes are usually called Zener diodes. The symbol and characteristic curve of a low voltage (referring to the breakdown voltage) Zener diode are shown in Figure 2.33. The forward bias characteristic is similar to conventional p-n junction diodes. The reverse bias region depicts the breakdown occurring at V_Z which is nearly independent of diode current. A wide range of Zener diodes are commercially available over a wide range of breakdown voltages and power ratings to 100 W.

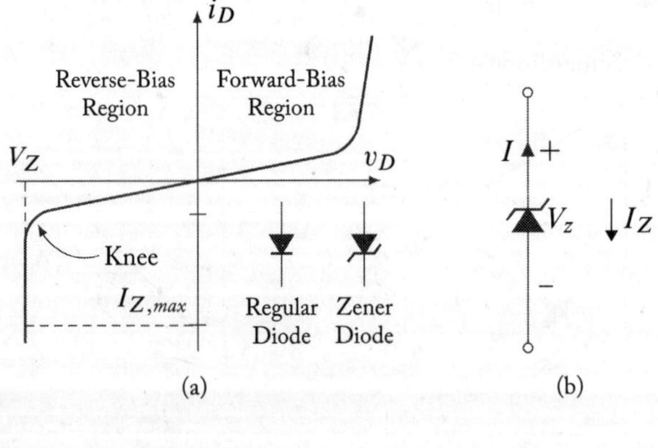

Figure 2.33: (a) Characteristic curve of a Zener diode; (b) The Zener diode symbol.

Changes in temperature generally cause a shift in the breakdown voltage. The temperature coefficient is approximately $+2\,\text{mV}/°\text{C}$ for Zener breakdown. For avalanche breakdown, the temperature coefficient is negative.

The simplified SPICE model of a Zener diode is identical to that of the conventional diode with the addition of the reverse breakdown "knee" voltage B_V and the corresponding reverse breakdown "knee" current I_{BV}. The relationship between B_V and I_{BV} is shown in the reverse-bias portion of the Zener diode characteristic curve of the Zener diode in Figure 2.34. To obtain a steeper reverse breakdown characteristic, a higher breakdown current I_{BV} may, in general, be used without incurring significant errors. The Zener diode model statement for Figure 2.34 uses $B_V = 5$ and $I_{BV} = 10\,\text{m}$ for a Zener voltage of 5 V at 10 mA. Both B_V and I_{BV} are positive quantities. If I_{BV} is large, the reverse breakdown curve is steeper.

The dynamic resistance of the Zener diode in the reverse breakdown region, r_Z , is the slope of the diode curve at the operating reverse bias current. Since the reverse current increases rapidly with small changes in the diode voltage drop, r_Z is small (typically 1 to 15 Ω). The Zener diode piece-wise linear model and its simplified version are shown in Figure 2.35.

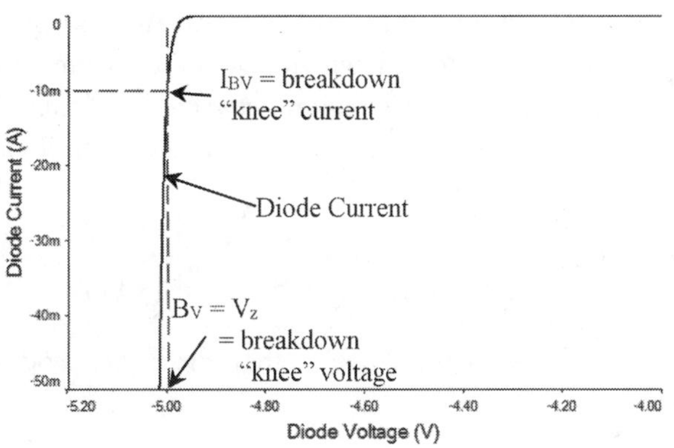

Figure 2.34: The reverse breakdown "knee" voltage B_V and the corresponding current I_{BV}.

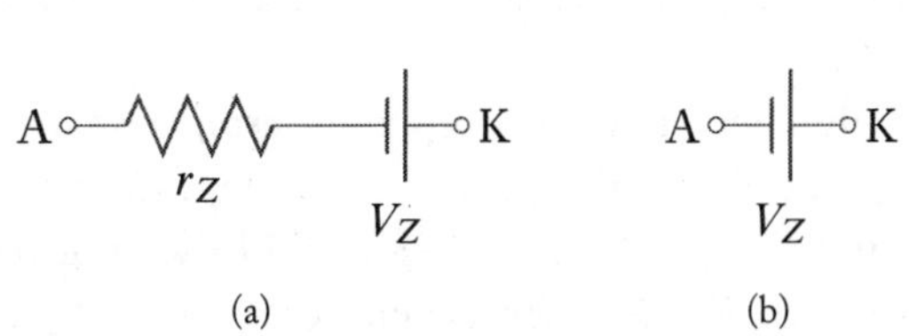

Figure 2.35: (a) The Zener diode piece-wise linear model; (b) The simplified Zener diode model.

A typical application of Zener diodes is in the design of the voltage reference circuit shown in Figure 2.36.

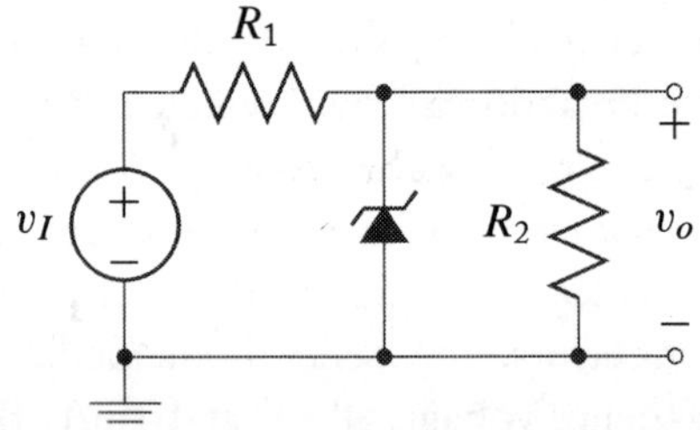

Figure 2.36: Zener diode voltage reference circuit.

To simplify the analysis, first consider the circuit without the Zener diode. The output voltage is simply that of a voltage divider,

$$v_{O,\,\text{no diode}} = \frac{R_2}{R_1 + R_2} v_I \tag{2.36}$$

where the input voltage v_I and output voltage v_o have both AC and DC components. Replace the Zener diode in the circuit. If the breakdown voltage V_Z is greater than $v_{o,\,\text{no diode}}$, then the Zener diode is operating in the reverse-bias region between the Zener knee and 0 V or in the forward-bias region. Therefore, the Zener diode operates as a conventional diode[9] and,

$$v_O = \begin{cases} -V_\gamma, & v_I \le \frac{R_1+R_2}{R_2}(-V_\gamma) \\ v_{O,\,\text{no diode}}, & v_I > \frac{R_1+R_2}{R_2}(-V_\gamma) . \end{cases} \tag{2.37}$$

If the breakdown voltage V_Z is less than $v_{O,\,\text{no diode}}$, then the Zener diode operates in the breakdown region beyond V_Z and forces

$$v_O = V_Z, \qquad \text{when } v_I \ge \frac{R_1 + R_2}{R_2}(V_Z) . \tag{2.38}$$

Figure 2.37 shows the input, output, and transfer characteristics of a Zener voltage reference circuit.

The Zener voltage reference circuit clamps the output voltage to the Zener breakdown voltage. The current through the Zener diode forces the voltage drop across resistor R_1 to $v_O - v_I$ since V_Z is invariant over a wide range of currents I_Z and the dynamic resistance in the breakdown region is negligible compared to R_2. R_1 must be chosen to limit I_Z to safe operating values as specified by the diode manufacturer.

[9]A piece-wise simplified linear model of a diode is used in this example.

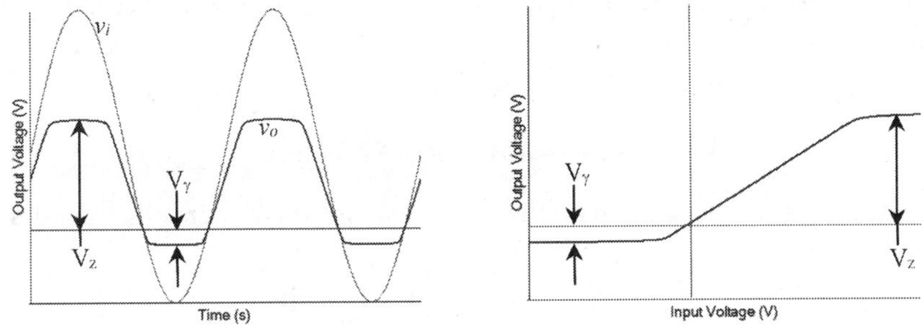

(a)

Figure 2.37: (a) The input and output waveforms and (b) transfer characteristic of the Zener voltage reference shown in Figure 2.36.

Example 2.13

The Zener diode in the circuit shown has a working range of current for proper regulation:

$$5\,\text{mA} \leq I_z \leq 50\,\text{mA}$$

and a Zener voltage

$$V_z = 50\,\text{V}.$$

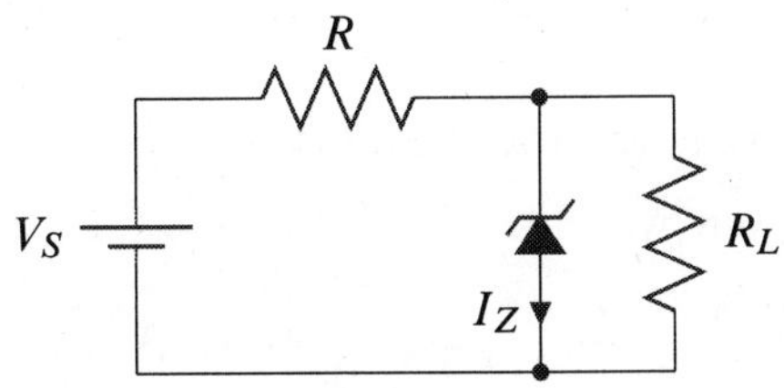

(a) If the input voltage, V_s, varies from 150 and 250 V and $R_L = 2.2\,\text{k}\Omega$, determine the range of values for the resistor, R, to maintain regulation.

(b) If R is chosen as the midpoint of the range determined in part (a), how much variation in the load resistance, R_L, is now possible without losing regulation?

Solution:

(a) The current through the load is given by:

$$I_L = \frac{50\ V}{2.2\ \text{k}\Omega} = 22.74\,\text{mA}.$$

The current through the resistor R must lie in the range of the load current plus the diode current:

$$5\,\text{mA} + 22.74\,\text{mA} \leq I \leq 50\,\text{mA} + 22.74\,\text{mA}$$
$$27.74\,\text{mA} \leq I \leq 72.74\,\text{mA}$$

but this current is also dependent on the source voltage, the Zener voltage, and the resistance R:

$$I = \frac{V_s - V_Z}{R} = \frac{V_s - 50}{R}$$

or

$$R = \frac{V_s - 50}{I}.$$

Since regulation must occur for both extremes of the source voltage, R must be the intersection of the limits determined by the above equation:

$$\frac{V_{s(\text{min})} - 50}{I_{(\text{max})}} \leq R \leq \frac{V_{s(\text{min})} - 50}{I_{(\text{min})}} \quad \& \quad \frac{V_{s(\text{max})} - 50}{I_{(\text{max})}} \leq R \leq \frac{V_{s(\text{max})} - 50}{I_{(\text{min})}}$$

or, after applying the intersection of the ranges:

$$\frac{V_{s(\text{max})} - 50}{I_{(\text{max})}} \leq R \leq \frac{V_{s(\text{min})} - 50}{I_{(\text{min})}}$$
$$\frac{250 - 50}{72.74} \leq R \leq \frac{150 - 50}{27.74} \quad \Rightarrow \quad 2.75\,\text{k}\Omega \leq R \leq 3.60\,\text{k}\Omega.$$

(b) The midpoint of the above range is $R = 3.175\,\text{k}\Omega$.

The resistor current is given by:

$$I = \frac{V_s - 50}{R} \quad \Rightarrow \quad 31.5\,\text{mA} \leq I \leq 63\,\text{mA}.$$

Since regulation must hold for all values of R, the load current must lie in the intersection of the possible ranges of the resistor current minus the diode current:

$$31.5 - 50\,\text{mA} \leq I_L \leq 31.5 - 5\,\text{mA} \quad \& \quad 63 - 50\,\text{mA} \leq I_L \leq 63 - 5\,\text{mA}$$

thus,

$$13\,\text{mA} \leq I_L \leq 26.5\,\text{mA}.$$

Since

$$R_L = \frac{V_Z}{I_L},$$

the range of R_L is found to be

$$3.85\,\text{k}\Omega \geq R_L \geq 1.93\,\text{k}\Omega.$$

Another typical application of Zener diodes occurs in AC-DC conversion. Recall that the output of a filtered rectifier has residual voltage ripple. By adding a Zener diode voltage reference circuit at the output of a rectifier and filter circuit, as shown in Figure 2.38, the residual voltage ripple at the output can be eliminated. The resistors R_1 and R_L must be carefully selected to yield the desired voltage output. If R_1 is chosen to be much less than R_L, and if the voltage across the capacitor is greater than V_Z, then the output voltage will be clamped at V_Z.

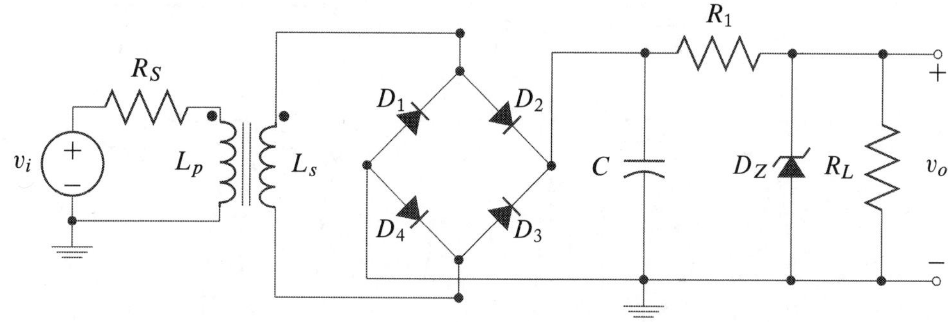

Figure 2.38: Full-wave rectifier circuit with output voltage reference.

Example 2.14

Consider the full-wave rectifier circuit with a Zener clamp shown in Figure 2.38. If the input voltage is 120 VAC (RMS) at 60 Hz, transformer winding ratio is 14:1, $C = 47\,\mu\text{F}$, $R_L = 3.8\,\text{k}\Omega$, $R_1 = 220\,\Omega$, show that the output is clamped to 5 V if a Zener diode with $V_Z = 5\,\text{V}$ is used.

Solution:

This is the same problem as Example 2.12. The resistance for the circuit without R_1 and the Zener diode was found to be less than $4.29\,\text{k}\Omega$. The DC output voltage was determined to be less than or equal to 11.9 V. Since $V_Z < 11.9\,\text{V}$, the resulting output voltage is clamped to 5 V. To confirm the result, a SPICE simulation of the circuit is performed. The SPICE output file and the input and outputs at various points on the circuit follows.

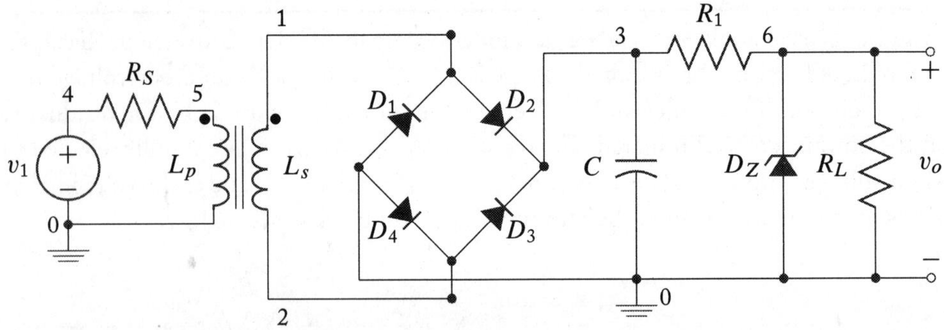

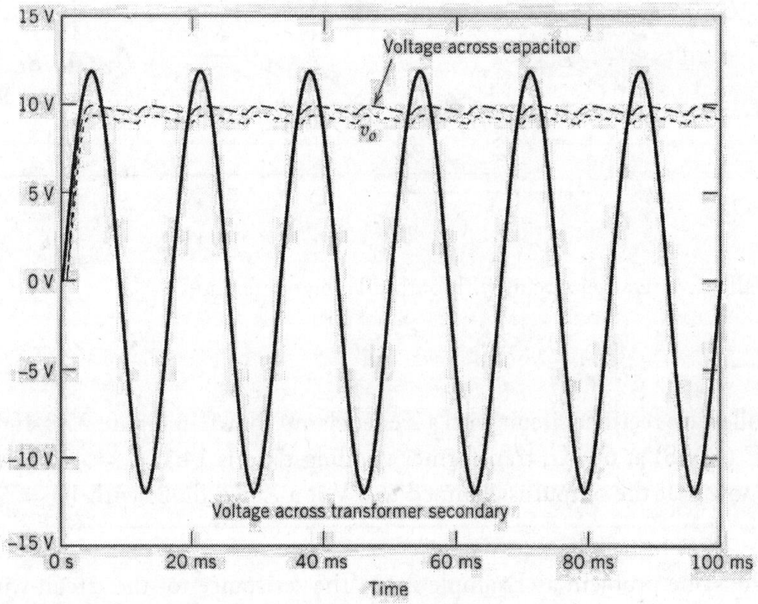

A design trade-off exists between using a large capacitor filter to remove the output voltage ripple or a smaller capacitor and a Zener diode clamp. Large capacitors require large "real-estate" or space on a printed circuit board and height above the circuit board. However, capacitor filters only dissipate energy when in the charging or discharging cycles.

Although the Zener diode is small in size and requires substantially less circuit volume than large capacitors, the diode must be able to dissipate from zero to the maximum current delivered to the load. If the Zener diode is to carry large currents over much of the operating cycle, the power dissipation is high and a large capacitor filter may be preferred. The Zener diode may be preferred if the power dissipation in the diode can be limited.

2.8 OTHER COMMON DIODES AND APPLICATIONS

In the previous sections of this chapter, the conventional *p-n* junction diode and the Zener diode and their applications were introduced. Although the conventional *p-n* junction diode and the Zener diode are the most common diode types used in electronic design, other types of diodes are designed into certain electronic applications. Some of the different types of diode include the tunnel diode, backward diode, Schottky barrier diode, Varactor Diode, *p-i-n* diode, IMPATT diode, TRAPATT diode, BARRITT diode, solar cell, photodiode, light-emitting diode, and semiconductor laser diode.

Unfortunately, it is far beyond the scope of this book to discuss all of the different types listed above. However, four common types of diodes from the above list are presented in this section for discussion. They are the

- Tunnel diode
- Schottky barrier diode
- Photodiode
- Light-Emitting Diode

2.8.1 TUNNEL DIODE

The tunnel diode (also called the Esaki diode after the L. Esaki who announced the new diode in 1958) voltage-current characteristic is shown in Figure 2.39. The figure shows that the tunnel diode is an excellent conductor in the reverse direction. Figure 2.40 is the circuit symbol for the tunnel diode.

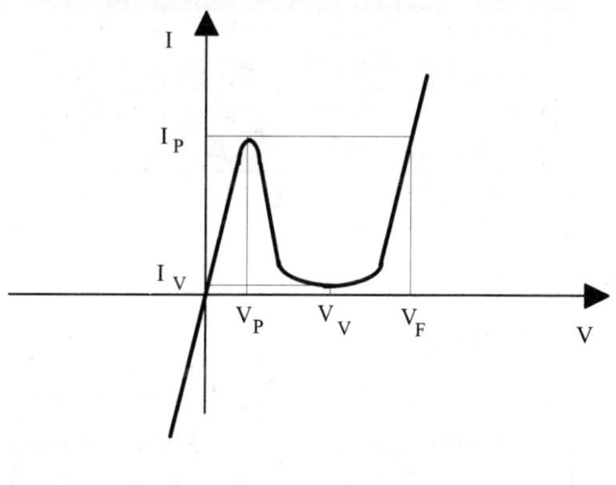

Figure 2.39: Voltage-current characteristic of a tunnel diode.

For small forward voltages (in the order of 50 mV in Ge), the resistance is in the order of 5 Ω. At the peak current I_P corresponding to the voltage V_P, the slope of the characteristic

Figure 2.40: Tunnel diode symbol.

curve is zero. As the voltage increases beyond V_P, the current also decreases. The tunnel diode characteristic curve in this region exhibits a *negative dynamic resistance* between the peak current I_P and the minimum or valley current I_P. At the valley voltage, V_V, corresponding to the valley current I_V, the slope of the characteristic curve is again zero. Beyond V_V, the curve remains positive. At the peak forward voltage, V_F, the current again reaches the value of I_P.

Since it is difficult to manufacture silicon tunnel diodes with a high ratio of peak-to-valley current I_P/I_V, most commercially available tunnel diodes are made from germanium (Ge) or gallium arsenide (GaAs). Table 2.2 is a summary of some of the static characteristics of these devices.

The operating characteristics of the tunnel diode are highly dependent on the load line of the circuit in which the diode is operating. Some load lines may intersect the tunnel diode characteristic curve in three places: the region between 0 and V_P, between V_P and V_V, and beyond V_V. This multi-valued feature makes the tunnel diode useful in high speed pulse circuit design. High frequency (microwave) oscillators are often designed so that the tunnel diode is biased in its negative dynamic resistance region.

Table 2.2: Typical tunnel diode parameters

Parameters	Ge	GaAs	Si
I_P/I_V	8	15	3.5
V_P (V)	0.055	0.15	0.065
V_V (V)	0.35	0.5	0.42
V_F (V)	0.5	1.1	0.7

2.8.2 SCHOTTKY BARRIER DIODE

The Schottky barrier diode (or simply Schottky diode) is a metal-semiconductor diode. The circuit symbol of the Schottky diode is shown in Figure 2.41. Metal-semiconductor diodes are formed by bonding a metal (usually aluminum or platinum) to n- or p-type silicon. Metal-semiconductor diode voltage-current characteristics are very similar to conventional p-n junction diodes and can be described by the diode equation with the exception that the threshold voltage V_γ is in the range from 0.3 V to 0.6 V. The physical mechanisms of operation of the conventional p-n junction diode and the metal-semiconductor diode are not the same.

Figure 2.41: Schottky diode symbol.

The primary difference between metal-semiconductor and *p*-*n* junction diodes is in the charge storage mechanism. In the Schottky diode, the current through the diode is the result of the drift of majority carriers. The Schottky diode switching time from forward to reverse bias is very short compared to the *p*-*n* junction diode.

Therefore, Schottky diodes are often used in integrated circuits for high speed switching applications. The Schottky diode is easy to fabricate on integrated circuits because of its construction. The low noise characteristics of the Schottky diode is ideal for the detection of low-level signals like those encountered in radio frequency electronics and radar detection applications.

2.8.3 PHOTODIODE

The photodiode converts optical energy to electric current. The circuit symbol of the photodiode is shown in Figure 2.42.

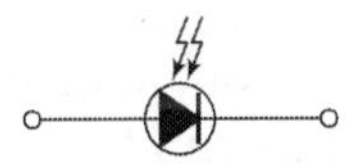

Figure 2.42: Photodiode symbol.

In order to make this energy conversion, the photodiode is reverse biased. Intensifying the light on the photodiode induces hole-electron pairs that increase the magnitude of the diode reverse saturation current. The useful output of the photodiode its *photocurrent* which, for all practical purposes, is proportional to the light intensity (in Watts) on the device. The proportionality constant is called the Responsivity, R, which is usually given in amperes per watt and is dependent on the wavelength of the light. Figure 2.43 shows a photodiode characteristic curve.

If the intensity of the light on the photodiode is constant, the photodiode can be modeled as a constant current source so long as the voltage does not exceed the avalanche voltage. Naturally, the photocurrent will vary with varying input light intensity. Since the photocurrent can be very small, an electronic amplifier is used in many applications to both boost the signal level and to convert from a current to a voltage output. For example, in optical fiber communication receivers, the average intensity of a time varying infrared light on the photodiode can be significantly less than 1 μW. Taking a typical photodiode responsivity for fiber optic application of 0.7 A/W, 1 μW of light will produce 0.7 μA of average current. This low level signal must be amplified by electronic amplifiers for processing by other electronic circuits to retrieve the transmitted information.

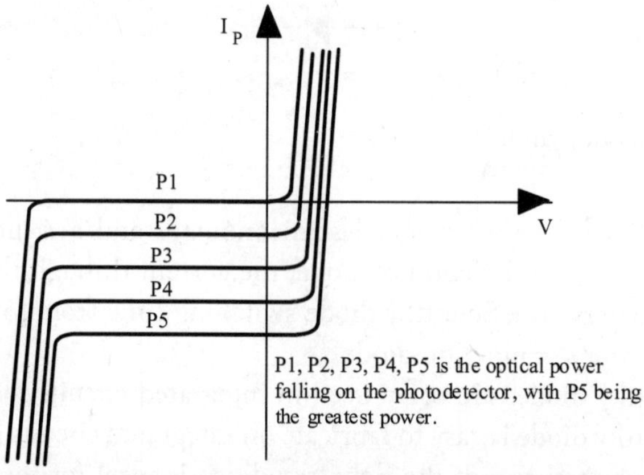

Figure 2.43: Characteristic curves of a photodiode.

2.8.4 LIGHT-EMITTING DIODE

The light-emitting diode (LED) converts electric energy to optical energy (light). LEDs are used for displays and are used as the light source for low cost fiber optic communication transmitters. By appropriate doping, the emission wavelength of the LED can be varied from the near-infrared ($< 2\,\mu$m) to the visible (400 nm to 780 nm). The symbol for the LED, shown in Figure 2.44, is similar to that of the photodiode except for the direction of the arrows representing light being emitted.

Figure 2.44: LED symbol.

When the LED is conducting, its diode voltage drop can be about 1.7 V although like small-signal diodes, they can vary due to materials used to fabricate the LED. The intensity of the light emitted from the LED is proportional to the current through the diode and is characterized by the so called light intensity-current (L-I) curve shown in Figure 2.45a. The LED also has a current-voltage relationship depicted in Figure 2.45b.

When using the LED in a circuit, a series current-limiting resistor is used to prevent destruction should large currents flow through the LED. The magnitude of the current-limiting resistor is calculated by limiting the current though the LED to a desired level I_{OP}, less than the maximum operating current with a diode threshold voltage of V_γ. For example, in Figure 2.46, if the diode threshold voltage is 1.7 V and $I_{OP} = 10$ mA provides a satisfactory optical output, the

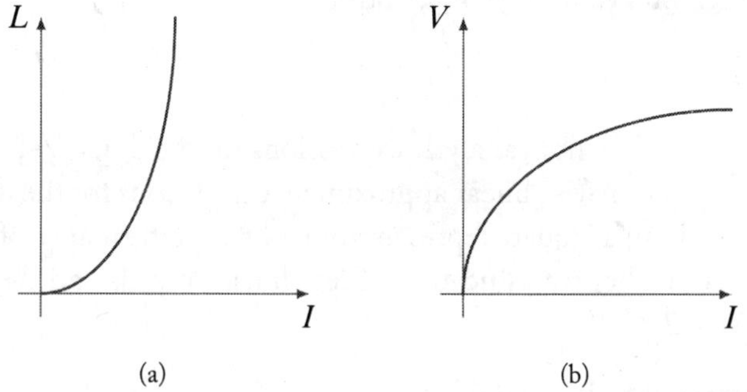

Figure 2.45: (a) L-I characteristic curve of the LED; (b) LED I-V characteristic curve at low bias current.

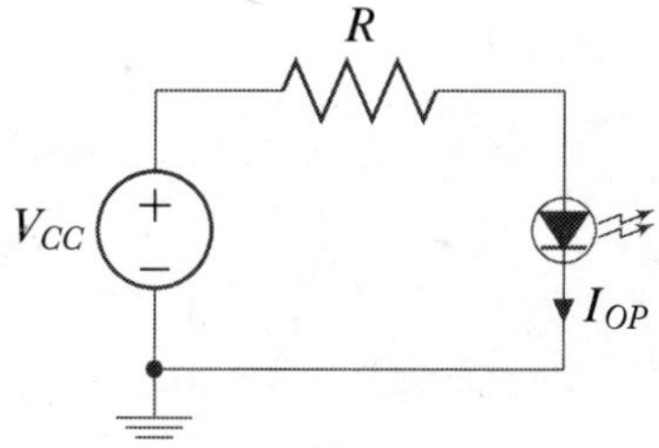

Figure 2.46: Simple LED driving circuit.

current-limiting resistor is

$$\begin{aligned} R &= \frac{V_{CC} - V_\gamma}{I_{OP}} \\ &= \frac{5 - 1.7}{10 \times 10^{-3}} = 330\,\Omega. \end{aligned}$$

2.9 CONCLUDING REMARKS

The semiconductor diode has been described in this chapter as the most basic non-linear electronic device. It is a two terminal device that provides small resistance when currents flow through the diode from the anode to the cathode and extremely large resistance to currents in the reverse direction. Large reverse voltages force the diode into breakdown and the dynamic resistance again becomes small. Diode applications utilize the characteristic properties of the diode in one or more of these three regions of operation:

- the forward bias region ($V \geq V_\gamma$),

- the reverse bias region ($V_\gamma > V \geq V_Z$), and
- the zener region ($V < V_Z$).

While nearly-exact, all-region, analytic expressions for the diode V-I relationship were presented, it was shown that piece-wise linear approximate expressions for the diode V-I relationship can, in many cases, provide an adequate representation of the performance of the diode. Other applications allow for even further simplification. These linear models and their simplified versions are summarized in Table 2.3.

Table 2.3: Regional linear models of a diode

Region	Linear Model	Simplified Model
Forward bias	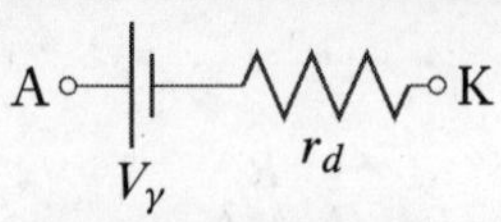	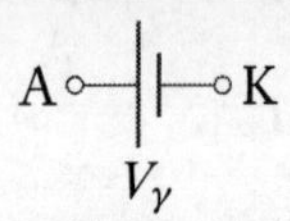
Reverse bias	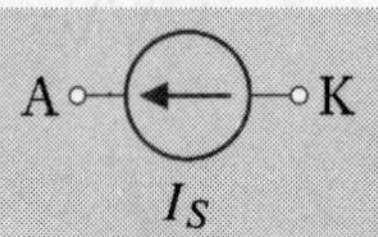 or 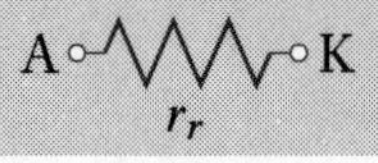	
Zener	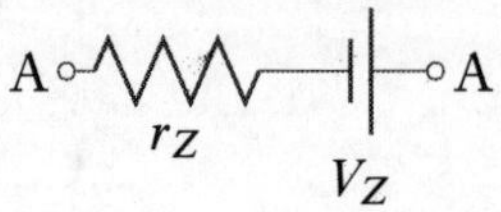	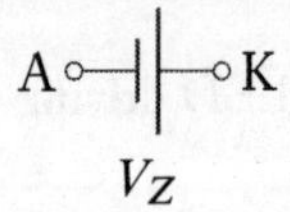

The diode was shown to be useful in a number of applications determined by the surrounding circuitry. The Zener diode, tunnel diode, Schottky diode, photodiode, and the LED were introduced as diodes with properties that necessary for specialized electronic applications. In later chapters, additional linear and non-linear diode applications will be examined.

Summary Design Example

Many electronic applications have a need for backup electrical power when there is a primary power failure. This backup may come in the form of very large-value capacitors (1 F or greater), battery systems, or generators powered by an internal combustion engine. One such electronic application operates with the following power requirements:

- Allowable input voltage — $5\,\text{V} < V_{CC} < 9\,\text{V}$
- Current draw — $100\,\text{mA} < I < 500\,\text{mA}$.

The primary power source, V_{CC}, provides a nominal voltage of $8\,\mathrm{V} \pm 1\,\mathrm{V}$. Design a backup power system that will provide adequate auxiliary power when the primary power source fails (i.e., $V_{CC} = 0$). The maximum duration of primary power failure is 2 hours.

Solution:

The total power requirement for this system lies in the range:

$$0.5\,\mathrm{W} < P_T < 4.5\,\mathrm{W}.$$

Capacitive backup is a poor choice for this system: even if the largest available capacitors are used, it will only last a few seconds. Conversely, motorized generators are inappropriately large for this system: a 0.04 Hp motor would be more than adequate to power a 5 W generator. Of the three given choices, the best for this application is a rechargeable battery backup.

The design goals for the backup system include:

- the battery backup powers the system only during primary power failure
- the battery backup is recharged at a controlled rate during normal operation
- the battery backup is protected against excessive recharging .

All design goals can be met with a network composed of diodes and a resistor. The proposed design is as shown. During normal operation the system state is:

- D_1 - ON, D_2 - OFF, & D_z either OFF or in the Zener region. Resistor, R, controls the rate of charging of the battery and D_z prevents overcharging the battery.

During backup operation the system state is:

- D_1 - OFF, D_2 - ON, & D_z OFF. D_1 blocks current discharge to the primary power source and D_2 provides a low-impedance path to the load.

The properties of the system components must be specified.

1. The battery voltage must lie within acceptable ranges for both providing auxiliary power and receiving current for recharge. Auxiliary power voltage must be greater than the minimum voltage required by the load *plus* V_γ and smaller than the minimum voltage provided by the primary power source *minus* V_γ:

$$5 + V_\gamma < \text{battery voltage} < 7 - V_\gamma.$$

A 6 V battery is a good choice for this system. The battery must also have a capacity so that backup will be provided during the entire time of primary power failure. Battery capacity is measured in the product of current and time (i.e., ampere-hours). This system requires a battery with a capacity greater than one ampere-hour.

$$\text{Battery capacity} > (\text{max. current})(\text{time}) = (500\,\mathrm{mA})(2\text{ hours}) = 1\text{ ampere-hour}.$$

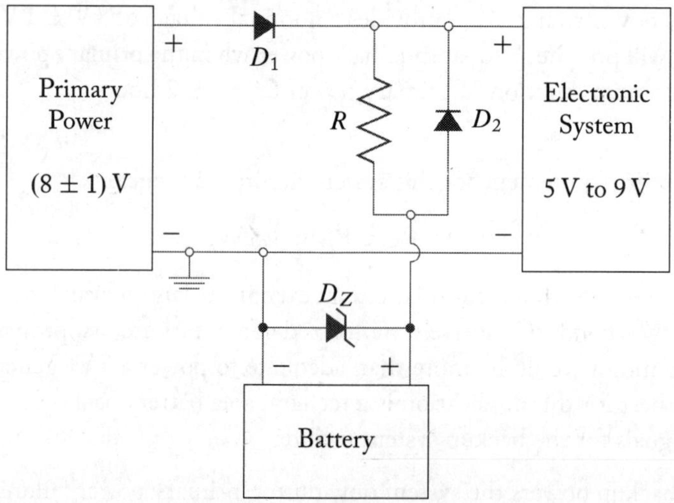

Figure 2.47: Proposed backup power system.

2. D_1 and D_2 must have sufficiently large current ratings. In each case the diode must be capable of carrying a minimum of 500 mA continuously.

3. Battery recharging current must be limited. Every battery type has a recommended rate of charge in order to maintain proper operation for a long life. A typical value for a small battery such as the one specified here is approximately 20 mA. The resistor, R, can then be determined from the nominal values for the primary power and battery voltages.

$$R = \frac{8\,\text{V} - 6\,\text{V}}{20\,\text{mA}} = 100\,\Omega\,.$$

The power rating is given by the maximum voltage squared divided by the resistance:

$$P_R > \frac{(9-6)^2}{100} = 0.03\,\text{W} \quad \Rightarrow \quad \text{a 1/4 Watt resistor will suffice.}$$

4. The Zener diode provides protection against over charging the battery. As batteries are overcharged, the battery voltage increases. Depending on the type of battery chosen, a maximum voltage will determine the Zener voltage. Maximum recharging current determines the capacity of the Zener diode. A typical overvoltage for small batteries is approximately 0.2 V, which leads to:

$$V_Z = 6.2\,\text{V}.$$

The maximum recharging current is 30 mA. This corresponds to a maximum power dissipation in the Zener diode of 0.19 W. A 1/4 W diode will suffice.

Summary component list

- One 6 V, rechargable battery with at least 1 ampere-hour capacity.
- Two power diodes with at least a 500 mA rating.
- One 1/4 W, 100 Ω resistor.
- One 6.2 V, 1/4 W Zener diode

2.10 PROBLEMS

2.1. A Silicon diode has a reverse saturation current of 0.1 nA and an empirical scaling constant, $\eta = 2$. Assume operation at room temperature.

(a) At what diode voltage will the reverse current attain 99% of the saturation value?

(b) At what diode voltage will the forward current attain the same magnitude?

(c) Calculate the forward currents at diode voltages of 0.5 V to 0.8 V in 0.1 V increments.

2.2. A Silicon diode ($\eta = 2$) at room temperature conducts 1 mA when 0.6 V is applied across its terminals.

(a) Determine the diode reverse saturation current.

(b) What will the diode current be if 0.7 V is applied across it?

2.3. Experimental data at 25°C indicates that the forward biased current, I_D, flowing through a diode is 2.5 μA with a voltage drop across the diode, V_D, of 0.53 V and $I_D = 1.5$ mA at $V_D = 0.65$ V. Determine:

(a) η

(b) I_S

(c) I_D at $V_D = 0.60$ V

2.4. For the diode in the above problem, what is the diode voltage drop, V_D, at

(a) $I_D = 1.0$ mA at 50°C

(b) $I_D = 1.0$ mA at 0°C

2.5. A Silicon diode has a reverse saturation current of 1 nA and an empirical scaling constant, $\eta = 2$. Assume operation at room temperature.

(a) A positive voltage of 0.6 V is applied across the diode, determine the diode current.

(b) What voltage must be applied across the diode to increase the diode current by a factor of ten?

(c) What voltage must be applied across the diode to increase the diode current by a factor of one hundred?

2.6. A diode is operating in a circuit in series with a constant current source of value I. What change in the voltage across the diode will occur if an identical diode is placed in parallel with the first? Assume $I \gg I_S$. What if two identical diodes are placed in parallel with the first?

2.7. A Silicon diode has a reverse saturation current of 1 nA and an empirical scaling constant, $\eta = 1.95$. Determine the percentage change in diode current for a change of temperature from 27°C to 43°C for diode voltages of:

(a) -1 V

(b) 0.5 V

(c) 0.8 V

2.8. A diode at room temperature has 0.4 V across its terminals when the current through it is 5 mA and 0.42 V when the current is twice as large. What values of the reverse saturation current and empirical scaling constant allow this diode to be modeled by the diode equation?

2.9. At room temperature, a diode with $\eta = 2$ has 2.5 mA flowing through it with 0.6 V across its terminals. (a) Find V_d when $I_D = 10$ mA. (b) Determine the reverse saturation current. (c) The diode is connected in series with a 3 V DC source and a resistance of 200 Ω. Find I_D if the diode is operating in the forward-bias region.

2.10. Let the reverse saturation current of a diode equal 15 nA and $\eta V_t = 25.6$ mV. If $I_D = 5$ mA, find

(a) V_d,

(b) V_d/I_D,

(c) r_d.

If I_D varies over the range $4.8\text{ mA} \leq I_D \leq 5.2\text{ mA}$, what is the range of

(d) V_d and

(e) r_d?

2.11. When a 20 A current is initially applied to a Silicon diode @ 300K of a particular characteristic, the voltage across the diode is $V_D = 0.69\,\text{V}$. With so much current flowing through the diode, the power dissipation raises the operating temperature of the device. The increased temperature eventually causes the diode voltage to stabilize at $V_D = 0.58\,\text{V}$.

(a) What is the temperature rise in the diode?

(b) How much power is dissipated in the diode at the two operating conditions mentioned?

(c) What is the temperature rise per watt of power dissipation?

2.12. A Silicon diode with parameters

$$I_S = 5\,\text{nA} \qquad \eta = 2$$

is placed in the given circuit.

(a) Determine the diode current and voltage.

(b) How much power is dissipated in each circuit element?

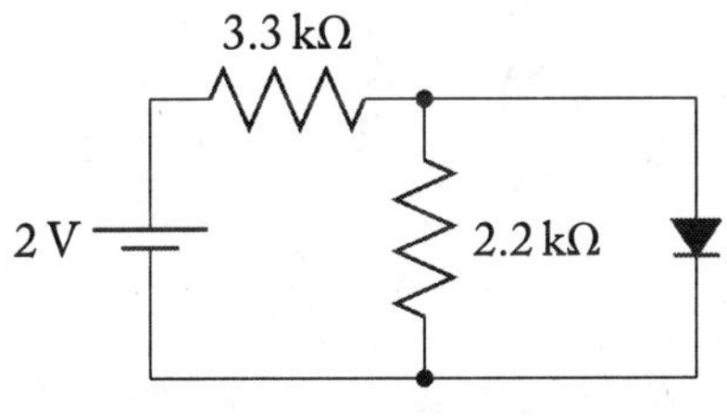

2.13. Determine the current I in the given circuit if the diode is described by:

$$I_S = 3\,\text{nA} \qquad \& \qquad \eta = 1.9$$

Hint: Find the Thévenin equivalent of the total circuit connected to the diode terminals.

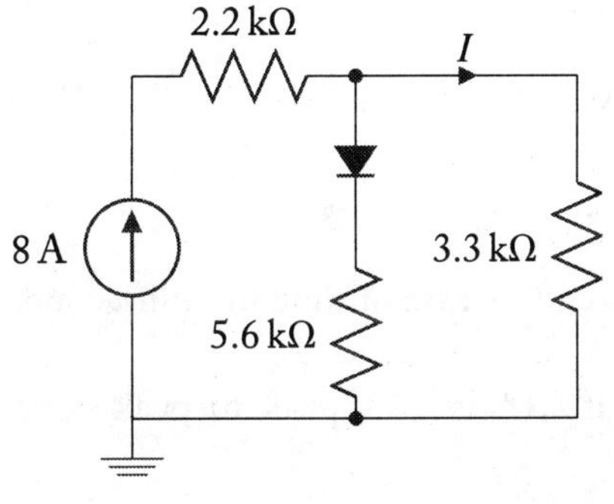

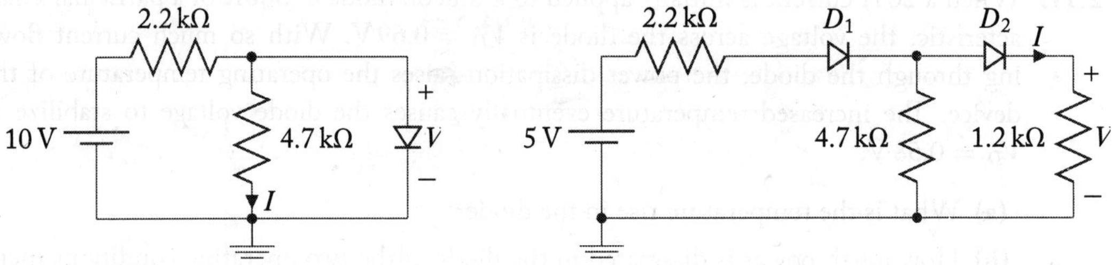

(a) (b)

2.14. Find the values of I and V for the circuits shown below. Assume that the diodes are ideal.

2.15. Find the values of all currents, I, and voltage, V, for the circuits shown below. Assume that the diodes are ideal.

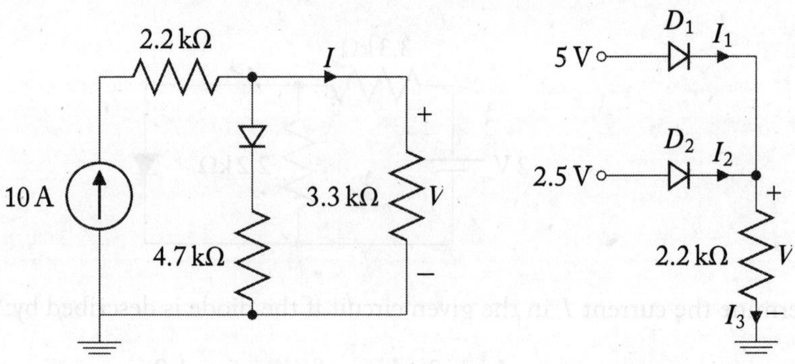

(a) (b)

2.16. The reverse saturation current for the silicon diode operating at room temperature in the circuit below is $I_S = 15\,\text{pA}$.

(a) Sketch v_o as a function of time in milliseconds for v_i.

(b) Repeat (a) if $v_i(t)$ is a 2 V peak-to-peak signal with the same period as the square wave in (a).

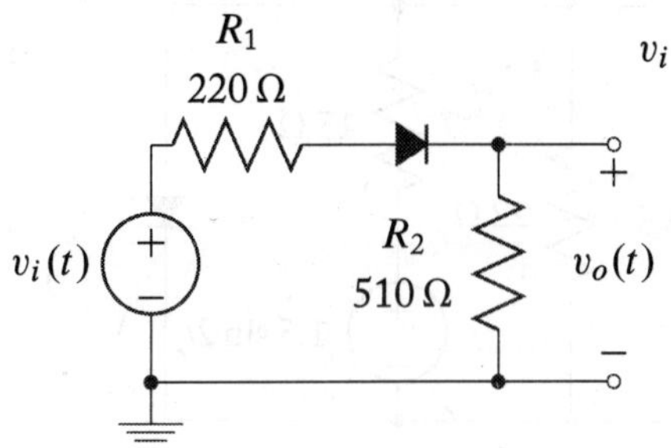

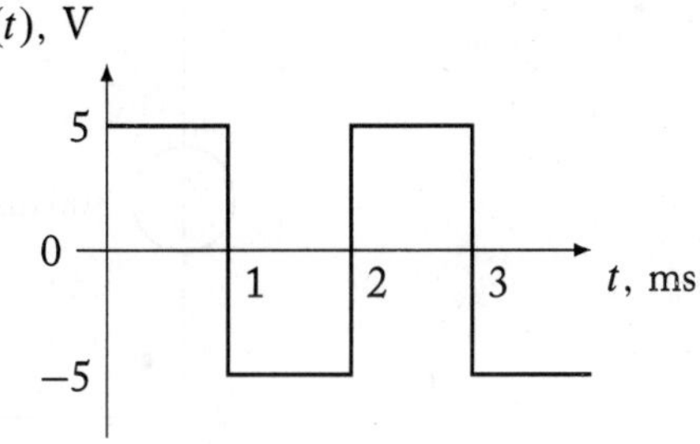

2.17. The circuit shown includes a diode with the following specifications:

$$\eta = 2 \qquad \text{and} \qquad I_S = 15\,\text{pA}$$

operating at room temperature. Let

$$v_i\,(t) = 0.1 \sin \omega_o t \text{ V}.$$

(a) Determine the quiescent ($v_i(t) = 0$) current in the diode.

(b) Find the dynamic resistance, r_d, of the diode.

(c) Find the output voltage, v_o.

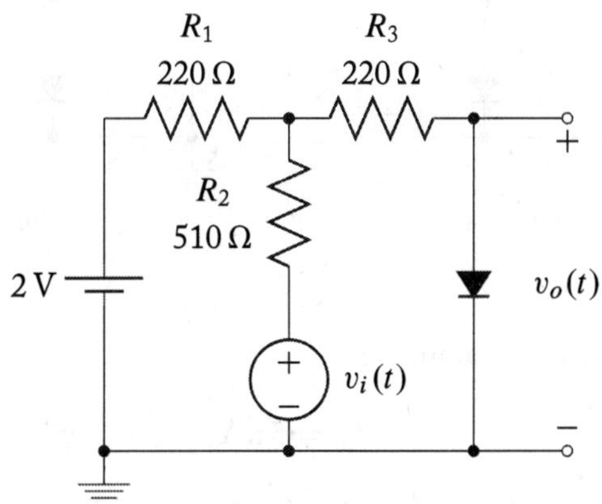

2.18. A Silicon diode has a reverse saturation current of 8 nA and an empirical scaling constant, $\eta = 2$. Find the diode current in the given circuit as a function of time.

(a) Using load line techniques

(b) Using SPICE

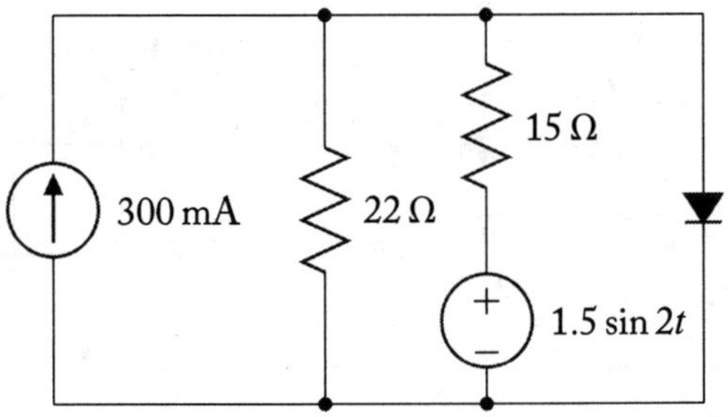

2.19. The four-diode circuits below uses identical diodes described by $\eta = 2$ and $I_S = 1.5\,\text{pA}$.

(a) For the circuit shown in Figure (a), determine the value of the current source to obtain an output voltage $V_o = 2.6\,\text{V}$.

(b) Find the change in output voltage for the circuit in Figure (b). How much current is drawn away by the load?

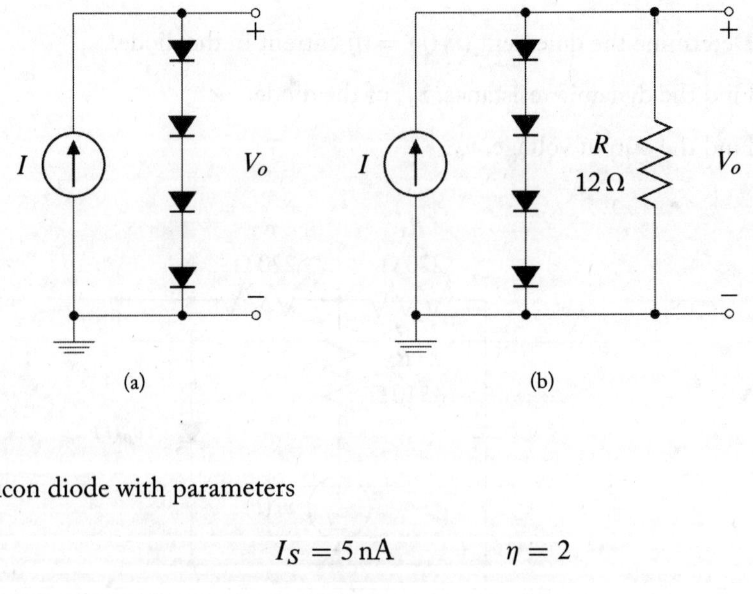

(a) (b)

2.20. A Silicon diode with parameters

$$I_S = 5\,\text{nA} \qquad \eta = 2$$

is carrying a forward current of 1 mA. Find the following:

(a) The diode forward dynamic resistance

(b) The diode threshold voltage

(c) A linear model of the diode at that operating point.

2.21. For the logarithmic amplifier shown (let $v_D \gg \eta V_t$, assume ideal OpAmps):

(a) Find the expression for the output voltage in terms of the input voltage.

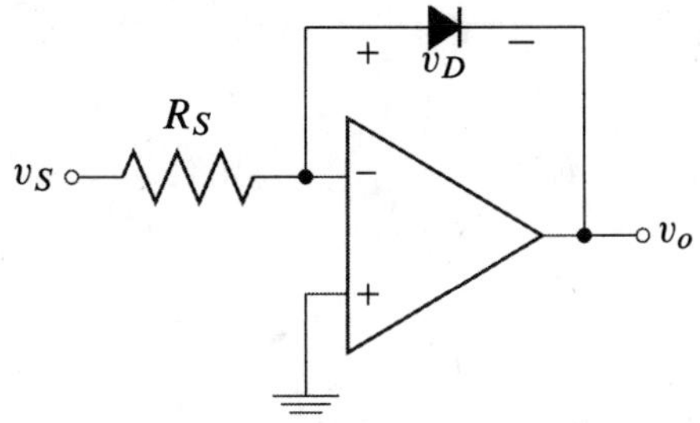

(b) Find the expression for the output voltage in terms of the input voltage for the anti-logarithmic amplifier.

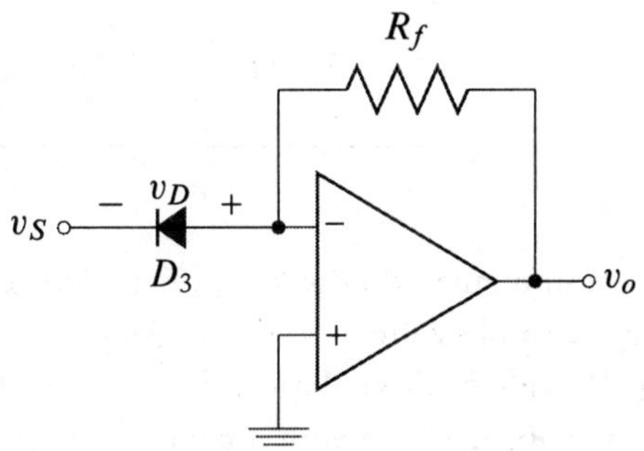

2.22. The diodes in the circuit are modeled by a simple model:

$$
\begin{aligned}
V_\gamma &= 0.7 \\
r_d &= 0 \\
r_r &= \infty
\end{aligned}
$$

Sketch the transfer characteristic for $-25 \leq v_i \leq 25\,\text{V}$. In each region indicate which diode(s) are ON. Also indicate all slopes and voltage levels on the sketch.

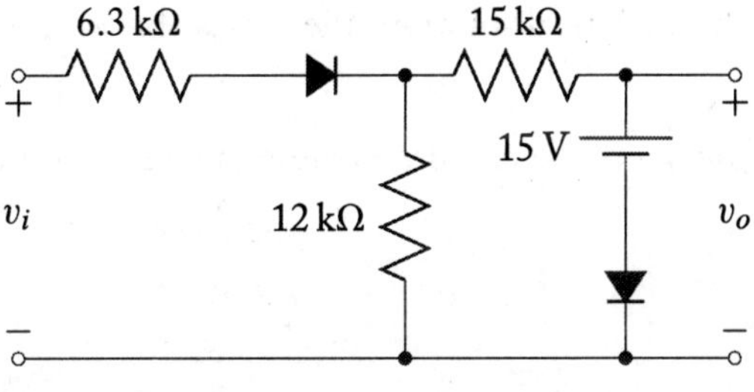

2.23. For the circuits below, plot the output voltage against the input voltage. Assume Silicon diodes.

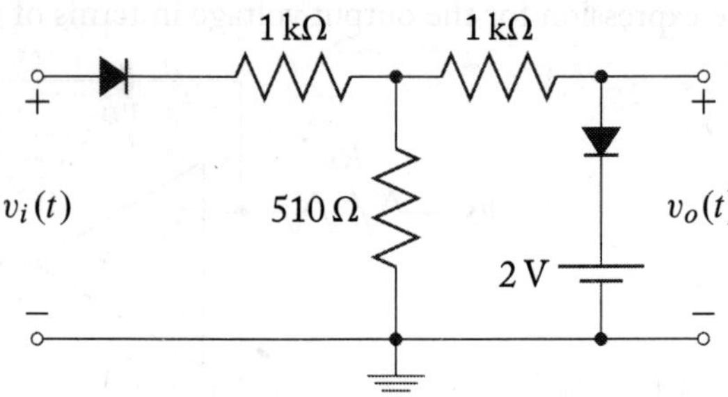

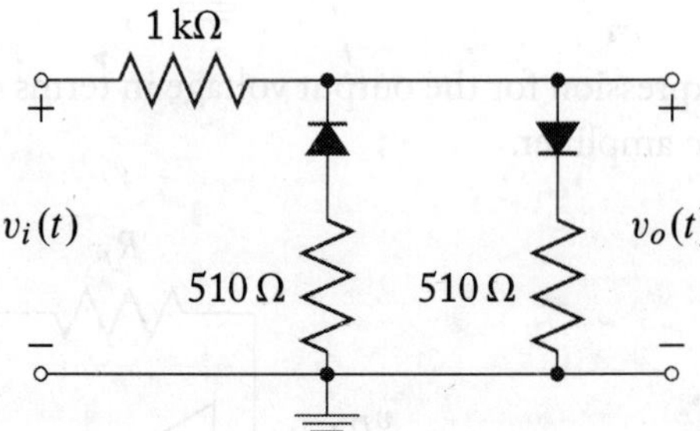

2.24. Design a half-wave unregulated power supply to provide an output DC voltage of 10 V with a peak-to-peak ripple voltage of 0.1 V. Assume a 120 V, 60 Hz supply. Use transformer coupling. Provide both analytical design and SPICE outputs. Piece-wise linear models of diodes may be used for your analytical design.

2.25. Design an unregulated half-wave rectifier power supply with transformer input coupling that has an input of 120 V_{RMS} at 60 Hz and requires a maximum DC output voltage of 17 V and a minimum of 12 V. The power supply will provide power to an electronic circuit that requires a constant current of 1 A. Determine the circuit configuration, transformer winding ratio, and capacitor size. Assume no losses by the transformer and a diode $V_\gamma =$ 0.7 V. Use SPICE to confirm the operation of the circuit.

2.26. Design a full-wave bridge rectifier to provide an output DC voltage of 10 V with a peak-to-peak ripple voltage of 0.1 V. Assume a 120 V, 60 Hz supply. Use transformer coupling. Provide only the analytical solution. Piece-wise linear models of diodes may be used for your design.

2.27. Given the following diode circuit. Assume diodes with the following properties:

$$V_\gamma = 0.7\text{ V}$$
$$r_d = 0$$
$$r_r = \infty$$

(a) What range of values for V_{CC} will produce ***both*** the following design goals?

- if $V_1 = 25$ V and $V_2 = 0$ V, then $V_o = 3$ V

- if $V_1 = V_2 = 25\,\text{V}$, then $V_o \geq 10\,\text{V}$

(b) Choose V_{CC} to be the midpoint of the range calculated in part (a). Calculate the diode currents for $V_1 = V_2 = 25\,\text{V}$.

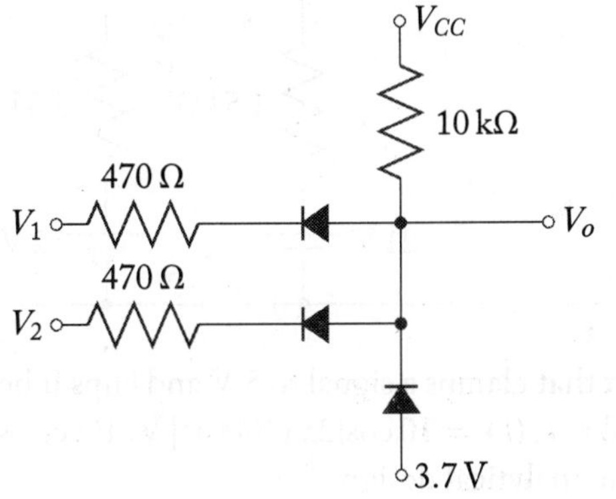

2.28. The diodes in the circuit are modeled by a simple model:

$$V_\gamma = 0.7\,\text{V}$$
$$r_d = 0$$
$$r_r = \infty$$

Sketch the transfer characteristic for $0 \leq v_i \leq 30\,\text{V}$. In each region indicate which diode(s) are ON. Also indicate all slopes and voltage levels on the sketch.

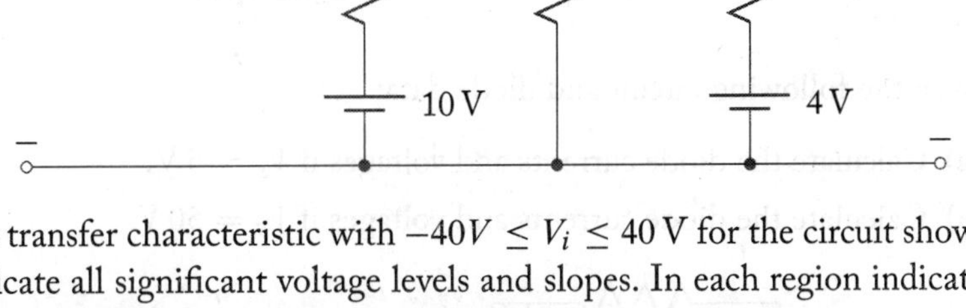

2.29. Sketch the transfer characteristic with $-40V \leq V_i \leq 40\,\text{V}$ for the circuit shown. On the sketch indicate all significant voltage levels and slopes. In each region indicate the state of each diode. The diodes have the following approximate properties:

$$r_d \approx 0$$
$$V_\gamma \approx 0.7$$
$$r_r \approx \infty$$

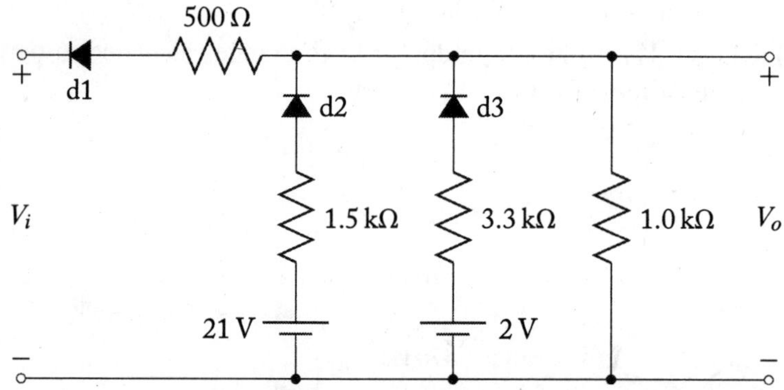

2.30. Design a circuit that clamps a signal to 5 V and clips it below −5 V. Plot the output when the input signal is $v_i(t) = 10\cos[2\pi(1000)t]$ V. Piece-wise linear models of diodes may be used for your analytical design.

2.31. Analyze the voltage tripler-quadrupler shown. Assume $v_i(t)$ is sinusoidal.

(a) Calculate the maximum voltage across each capacitor.

(b) Calculate the peak inverse voltage of each diode.

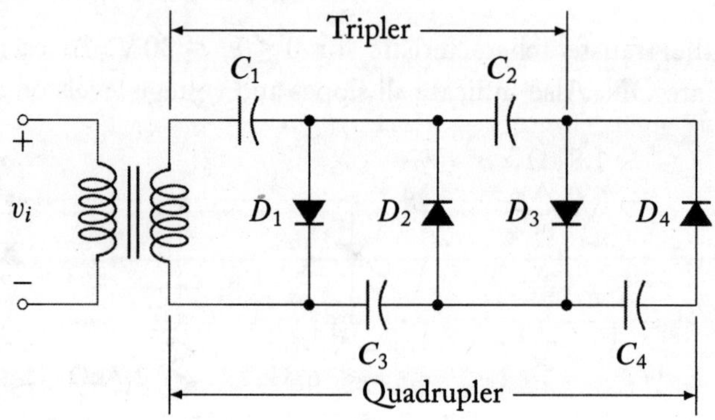

2.32. Given the following circuit and diode data:

(a) Calculate the diode currents and voltages if $V_s = 5$ V.

(b) Calculate the diode currents and voltages if $V_s = 50$ V.

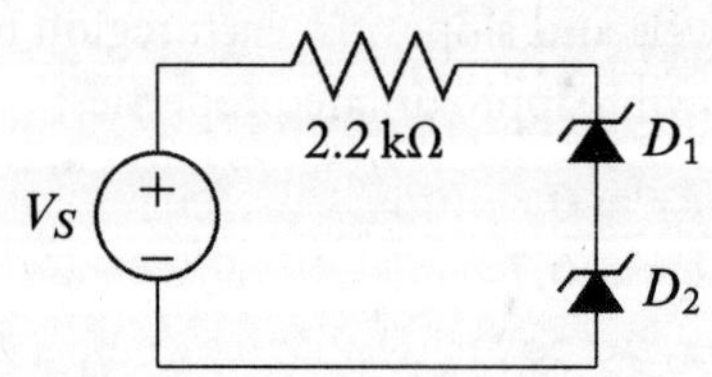

	I_S (μA)	η (−)	V_Z (V)
D_1	2	1.8	30
D_2	3	2.0	25

2.33. A simple Zener diode voltage regulator is under design. The design constraints are:

Supply voltage	—	$V_s = 150\,\text{V}$
Zener voltage	—	$V_z = 50\,\text{V}$
Zener current range	—	$5\,\text{mA} \leq I_Z \leq 50\,\text{mA}$

(a) Determine the value of the regulator resistor, R, so that voltage regulation is maintained for a load current in the range: $1\,\text{mA} \leq I_L \leq I_{L(\text{max})}$. What is the maximum load current, $I_{L(\text{max})}$

(b) What power rating is necessary for the resistor, *R*?

(c) If the variable load of part a is replaced by a $2\,\text{k}\Omega$ resistor and the regulator is constructed using the resistor *R* calculated in part (a), over what range of supply voltages will regulation be maintained?

2.34. A load draws between 20 mA and 40 mA at a voltage of 20 VDC. The available DC power varies between 100 and 140 VDC.

(a) Design a voltage regulator using the following components:

- Resistors – any size: any number
- Zener Diode – regulates for a Zener current of 5 mA to 50 mA.

(b) The value engineering department has found an "equivalent" Zener diode that can be purchased for 70% of the cost of your original diode. This diode regulates for a Zener current of 3 mA to 42 mA.

What effect on your design will the substitution have? Is a redesign necessary?

2.35. A load draws between 100 mA and 400 mA at 5 V. It is receiving power from an unregulated power supply that can vary between 7.5 V and 10 V. Design a Zener diode voltage regulator using a 5 V Zener diode that regulates for diode currents between 50 mA and 1.1 A. Show the circuit diagram for the completed design. Be sure to specify the power rating of any necessary resistors.

2.36. Design a wall socket power converter that plugs into a standard electrical wall socket and provides an unregulated DC voltage of 6 V to a portable Compact Disk player. For maximum conversion efficiency, a transformer-coupled full-wave bridge rectifier is required. The output ripple of the converter is specified to be less than 10% and must be able to provide 0.25 A of constant current. Assume no losses by the transformer and diode $V_\gamma = 0.7\,\text{V}$.

(a) Determine the circuit configuration, transformer winding ratio, and capacitor size. Use SPICE to confirm the operation of the circuit.

(b) A 6 V Zener diode is used to clamp the output of the wall-mounted converter. Adjust the circuit parameters to insure stable 6 V output from the converter. Use SPICE to confirm the operation of the circuit.

2.37. Each diode is the given circuit is described by a linearized volt-ampere characteristic with dynamic forward resistance, r_d, and threshold voltage V_γ. The values for these parameters are listed in the table. The voltage source has value:

$$V = 200\,\text{V}.$$

(a) Find the diode currents if $R = 20\,\text{k}\Omega$.

(b) Find the diode currents if $R = 4\,\text{k}\Omega$.

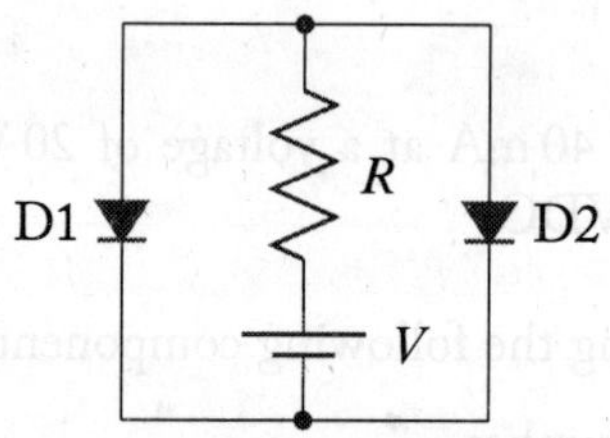

Diode	D1	D2
r_d	20	10
V_γ	0.2	0.6
r_r	∞	∞

2.38. For the circuit shown below:

(a) find the equation for the DC load line and plot on the diode characteristic curve provided.

(b) Find the AC load line equation and plot on the diode characteristic curve.

(c) Find v_o for $\omega_o = 2\pi(1000)$ rad/s.

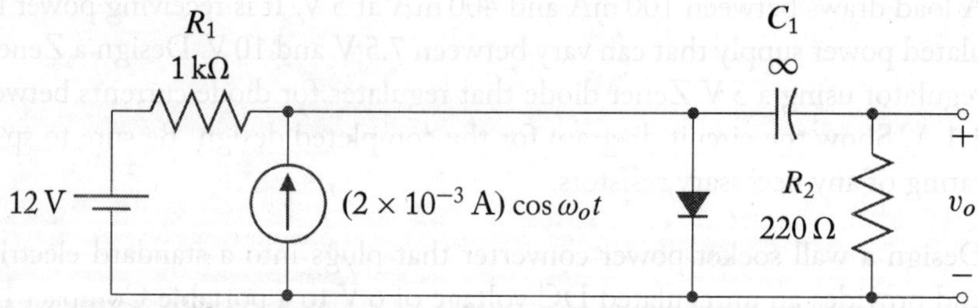

2.39. For the circuit shown below:

(a) Find the equation for the DC load line and plot on the diode characteristic curve provided.

(b) Find the AC load line equation and plot on the diode characteristic curve.

(c) Find v_o for $\omega_o = 2\pi(1000)$ rad/s.

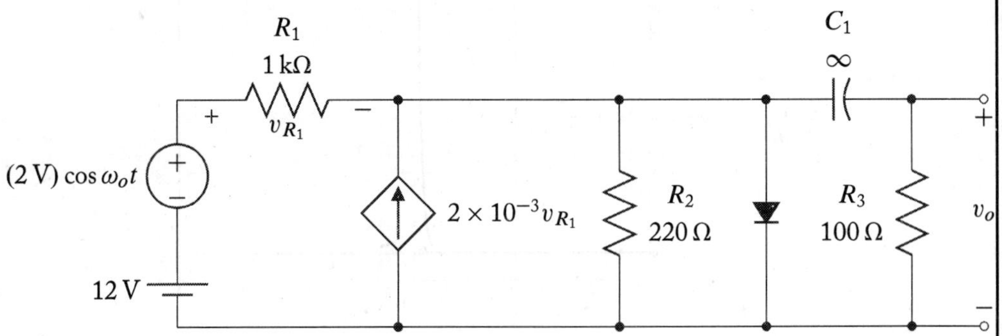

2.40. For the circuit shown below:

(a) Find the equation for the DC load line and plot on the diode characteristic curve provided.

(b) Find v_o for $\omega_o = 2\pi(1000)$ rad/s.

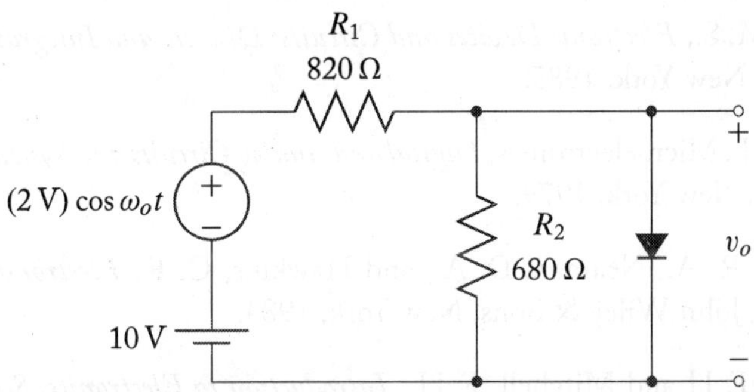

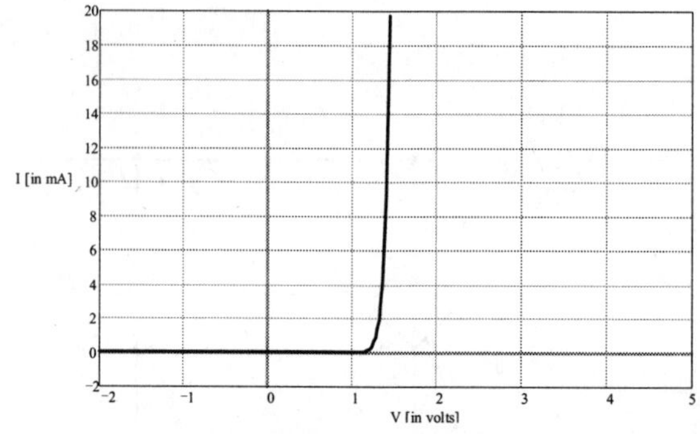

Figure 2.48: Diode curve for Problems 2.38 and 2.39.

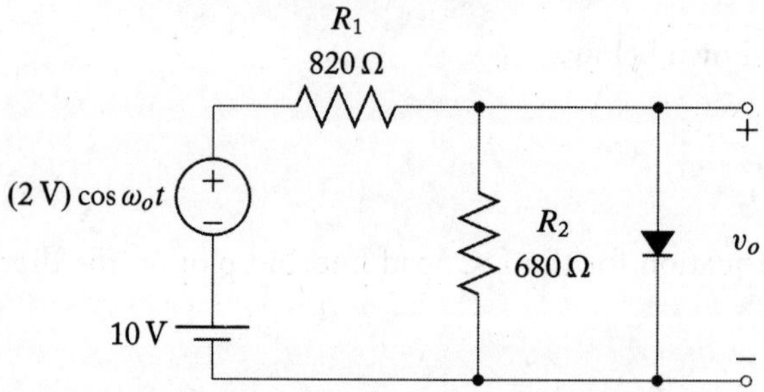

Figure 2.49: Diode curve for Problem 2.40.

2.11 REFERENCES

[1] Ghausi, M. S., *Electronic Devices and Circuits: Discrete and Integrated*, Holt, Rinehart and Winston, New York, 1985.

[2] Millman, J., Microelectronics, *Digital and Analog Circuits and Systems*, McGraw-Hill Book Company, New York, 1979.

[3] Colclaser, R. A., Neaman, D. A., and Hawkins, C. F., *Electronic Circuit Analysis: Basic Principles*, John Wiley & Sons, New York, 1984.

[4] Mitchell, F. H and Mitchell, F. H., *Introduction to Electronics, Second Edition*, Prentice-Hall, Englewood Cliffs, 1992.

[5] Savant, C. J., Roden, M. S., and Carpenter, G. L., *Electronic Design: Circuit and Systems, Benjamin/Cummings Publishing Company*, Redwood City, 1991.

[6] Sedra A. S. and Smith, K. C., *Microelectronic Circuits, Second Edition*, Holt, Rinehart and Winston, New York, 1987.

[7] Tuinenga, P. W., *SPICE: A Guide to Circuit Simulation & Analysis Using PSpice, Second Edition*, Prentice-Hall, Englewood Cliffs, 1992.

[8] Thorpe, T. W., *Computerized Circuit Analysis with SPICE*, John Wiley & Sons, New York, 1992.

[9] Colclaser, R. A. and Diehl-Nagle, S., *Materials and Devices for Electrical Engineers and Physicists*, McGraw-Hill, New York, 1985.

[10] Streetman, B. G., *Solid State Electronic Devices, Second Edition*, Prentice-Hall, Englewood Cliffs, 1990.

[11] Navon, D. H., *Semiconductor Micro-devices and Materials*, Holt, Rinehart & Winston, New York, 1986.

[12] Grebene, A. B., *Bipolar and MOS Analog Integrated Circuit Design*, John Wiley & Sons, New York, 1984.

[13] Mayer, W. and Lau, *S. S., Electronic Materials Science: For Integrated Circuits in Si and GaAs,* Macmillan, New York, 1990.

CHAPTER 3

Bipolar Junction Transistor Characteristic

The Bipolar Junction Transistor (BJT) is perhaps the most basic of three-terminal semiconductor devices. It can be found, for example, as a vital component in digital and analog integrated circuits, audio and other frequency range amplifiers, radio electronics, and electronic control devices with a wide range of applications. The BJT is an active device that is highly non-linear, and, along with applications in non-linear circuitry, plays an important part in many linear electronic applications. The apparent contradiction of a non-linear device being useful in linear applications is placated by a region of BJT operation that is nearly linear. Non-linear BJT operation typically involves a transition between BJT operating regions.

BJTs are constructed with two *p-n* junctions sharing a common region, identified as the base region. This common region, lying between two regions of the complementary doping, causes the two diode-like *p-n* junctions to become coupled.[1] The base region may be doped as either a *p*-region or an *n*-region: the two types of BJT formed are identified as *npn* or *pnp* respectively.

Before proceeding with technical descriptions of the operation of a BJT, it is necessary to define appropriate descriptive conventions. The two circuit symbols for BJTs are shown in Figure 3.1.

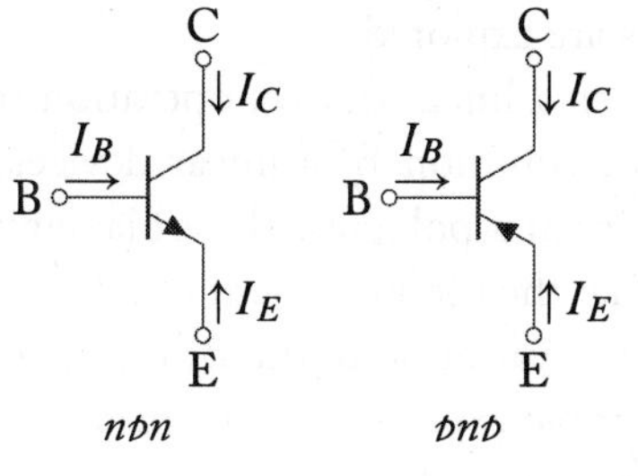

Figure 3.1: BJT circuit symbols.

[1]Extensive discussions of the semiconductor physics that lead to coupled *p-n* junctions forming a bipolar junction transistor are not within the scope of this electronics text. The authors suggest several texts in semiconductor physics and electronic engineering materials at the end of this chapter for those readers interested in these aspects of physical electronics.

The three terminals of a BJT are uniquely defined by the circuit symbols and are identified as:

- C—the Collector
- B—the Base
- E—the Emitter

The ordering of the characterizing letters *npn* or *pnp* indicates the doping type of the Collector, Base, and Emitter regions, respectively. The two types of BJT have unique symbolic representation characterized by the direction of the arrow on the Emitter terminal, which indicates the direction current would flow if the base-emitter junction were forward biased. The current entering each terminal is identified with the subscript of the terminal: the positive direction for all currents is *into* the device (i.e., I_B is the current entering the base of the transistor as is drawn in Figure 3.1). The BJT terminal voltage differences will be identified with standard double-subscript notation: voltage at first subscript with the second subscript as reference. For example, V_{BE} is the voltage at the base terminal with the emitter terminal taken as reference ($V_{BE} = V_B - V_E$).

As is true of all chapters in this book, the focus of this chapter is on quasistatic (low-frequency), large-signal analysis. Book 2 of the series will focus on small-signal linear applications (amplifiers, etc.) of BJTs. Book 3 will explore the higher frequency ranges.

This chapter begins with discussion of the principles of BJT operation: the non-linear characteristics of BJT operation are explored through the non-linear Ebers-Moll equations. Quantitative results are obtained through graphic techniques and analytic characterization using SPICE and MathCAD. In order to simplify the analysis of BJT operation, four simple, linear models for the BJT, one for each of its four regions of operation, are derived. Digital logic gates provide a good example of circuitry effectively using BJT regional transitions: the operational characteristics of three BJT logic gate families are explored.

In order to use the BJT as a linear device, operation must be restricted to single region operation. Amplifiers, the most common BJT linear devices, operate with BJTs biased into a linear region using a variety of circuit topologies. These bias circuits are explored with close interest on two significant design criteria: the establishment of a fixed quiescent operating point and the stability of that operating point to variation in the BJT characteristic parameters.

The summary design example explores a non-linear use of a BJT as a controlled switch in a Zener diode voltage regulator circuit. Such usage can greatly increase circuit efficiency while reducing component cost.

3.1 BJT *V-I* RELATIONSHIPS

Much like the semiconductor diode, the Bipolar Junction Transistor can be described empirically by a set of experimentally derived curves or theoretically by a set of equations. Because there are three terminals and the action of the two *p–n* junctions are coupled, a single V-I relationship (as was possible for the diode) is not applicable to the BJT: a *set* of curves or equations is necessary.

A set of empirical curves for a typical *npn* BJT is shown in Figure 3.2(a–d). While only the first (a & b) or last (c & d) pair of curves are necessary for complete description of the BJT, both pairs are shown for completeness. These characteristic curves are grouped into two categories:

Common base characteristics (Figure 3.2a–b):

Input characteristics: In common base configurations, the emitter terminal is the input and the base-emitter junction is the primary control. Figure 3.2a is a plot of the input (emitter) current as a function of the control (base-emitter) voltage with the output (base-collector) voltage as a parameter.

Output characteristics: The output for this configuration is at the collector terminal. Figure 3.2b is a plot of the output (collector) current as a function of the output (collector-base) voltage with the input (emitter) current as a parameter.

Common emitter characteristics (Figure 3.2c–d):

Input characteristics: In common emitter configurations, the base terminal is the input and the base-emitter junction is the primary control. Figure 3.2c is a plot of the input (base) current as a function of the control (base-emitter) voltage with the output (collector-emitter) voltage as a parameter.

Output characteristics: The output for this configuration is also at the collector terminal. Figure 3.2d is a plot of the output (collector) current as a function of the output (collector-emitter) voltage with the input (base) current as a parameter.

One of the most accurate and simple theoretical characterizations of the BJT are the Ebers-Moll equations, initially derived by J. J. Ebers and J. L. Moll.[2] These equations relate the collector and emitter currents to the base-collector junction and base-emitter junction voltages. The third BJT current, base current, may then be calculated using Kirchhoff's Current Law applied to the BJT as a whole:

$$I_B = -(I_C + I_E). \tag{3.1}$$

The collector-emitter voltage may be derived from the base-collector and base-emitter voltages by applying Kirchhoff's Voltage Law around the BJT terminals:

$$V_{CE} = V_{BE} - V_{BC}. \tag{3.2}$$

[2] J. J. Ebers and J. L. Moll, "Large Signal Behavior of Junction Transistors," *Proceedings of the IRE*, Vol. 42, No. 12, December 1954, pp. 1761–1772.

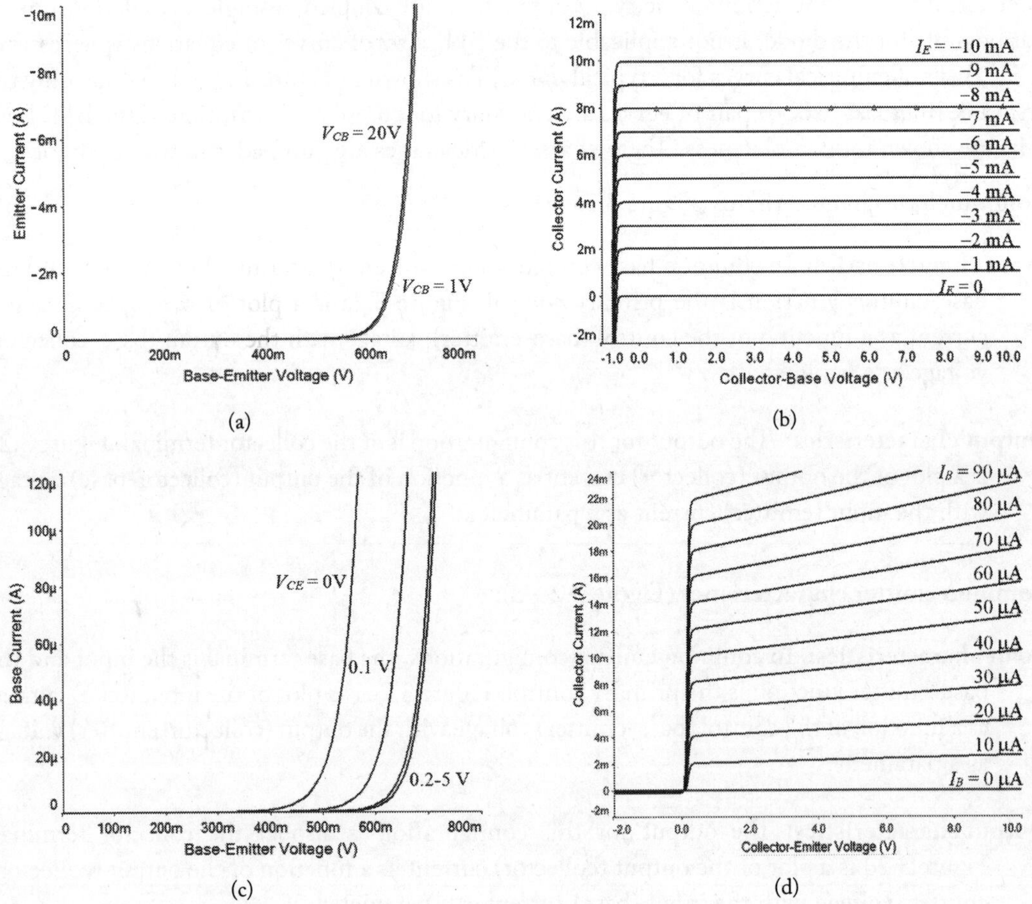

Figure 3.2: 2N2222A input and output characteristics.

The Ebers-Moll equations for an *npn* BJT[3] are given by:

$$I_E = -I_{ES}\left(e^{\frac{V_{BE}}{\eta V_t}} - 1\right) + \alpha_R I_{CS}\left(e^{\frac{V_{BC}}{\eta V_t}} - 1\right) \tag{3.3a}$$

$$I_C = -I_{CS}\left(e^{\frac{V_{BC}}{\eta V_t}} - 1\right) + \alpha_F I_{ES}\left(e^{\frac{V_{BE}}{\eta V_t}} - 1\right) \tag{3.3b}$$

[3] *pnp* BJTs are constructed with the *p-n* junctions in the opposite direction from *npn* BJTs. Therefore, the polarities the junction voltages and currents must be reversed. The Ebers-Moll equations for *pnp* BJTs are:

$$I_E = I_{ES}\left(e^{\frac{V_{EB}}{\eta V_t}} - 1\right) - \alpha_R I_{CS}\left(e^{\frac{V_{CB}}{\eta V_t}} - 1\right)$$

$$I_C = I_{CS}\left(e^{\frac{V_{CB}}{\eta V_t}} - 1\right) + \alpha_F I_{ES}\left(e^{\frac{V_{EB}}{\eta V_t}} - 1\right)$$

where V_t is the voltage equivalent temperature defined in Chapter 2, η is an empirical scaling constant that depends on geometry, material, and doping levels,[4] and I_{ES} and I_{CS} are the emitter and collector *p-n* junction saturation currents for a specified temperature. As with diode leakage currents, I_{ES} and I_{CS} have temperature variation similar to *p-n* junction temperature variation (described in Section 2.1): the saturation currents roughly double with every 6°K increase in temperature.

The quantities α_F and α_R have particular significance. In order to understand this significance, it is best to mentally perform two mental experiments upon the equations. As a first test, set $V_{BD} \gg \eta V_t$ and $V_{BC} \ll -\eta V_t$, that is strongly forward bias the base-emitter junction while the base-collector junction is strongly reverse biased. The terms with I_{ES} in Equations (3.3) become negligible and I_E and I_C are related by α_F:

$$I_C - \alpha_F I_E \tag{3.4}$$

α_F is therefore identified as the DC collector-emitter current gain.

Figure 3.2b & d graphically demonstrate this relationship for two different circuit connections of a BJT. It can be seen from the curves that the quantity α_F is very close to, but slightly smaller than, unity. Equation (3.1) can be substituted into Equation (3.4) to get the ratio of the collector current to the base current.

$$I_C = \frac{\alpha_F}{1-\alpha_F} I_B = \beta_F I_B, \qquad \text{where } \beta_F = \frac{\alpha_F}{1-\alpha_F}. \tag{3.5}$$

Figure 3.2d graphically demonstrates this relationship. It can be seen that these bias conditions lead to a region where an approximate *linear relationship* exists among the BJT currents, this linear region is identified as the *forward-active* region and is defined by forward biased base-emitter and reverse biased base-collector junctions.

If the opposite biasing scheme is used: set $V_{BC} \gg \eta V_t$ and $V_{BE} \ll -\eta V_t$, that is strongly forward bias the base-collector junction while the base-emitter junction is strongly reverse biased. The terms with I_{CS} in Equations (3.3) become negligible and I_E and I_C are now related by α_R:

$$I_E \approx -\alpha_R I_C \tag{3.6}$$

α_R is identified as the DC collector-emitter reverse current gain.

A relationship similar to Equation (3.5) can be determined relating the emitter and base currents.

$$I_E = \frac{\alpha_R}{1-\alpha_R} I_B = \beta_R I_B, \quad \text{where } \beta_R = \frac{\alpha_R}{1-\alpha_R}. \tag{3.7}$$

[4]The Ebers-Moll equations are written here with a single value of η. In fact, it is possible for each of the *p-n* junctions to have an individual value of this scaling constant. In practice, the values are nearly identical: hence the use of a single value here. SPICE and most other circuit emulators allow for the possibility of individual junction values of η and require the user to input both values should a change from the default value of unity be desired.

Figure 3.2d also demonstrates this relationship ($V_{CE} < 0$). Here the bias conditions again lead to an approximate linear relationship among the BJT currents. This linear region is identified as the *inverse-active* region and is defined by forward biased base-collector and reverse biased base-emitter junctions. Notice that β_R is typically much smaller than β_F. While it is *possible* for a BJT to have the forward and reverse values of β to be nearly identical, bipolar junction transistors are typically designed for optimal performance in the forward-active region: this design process leads to the significantly larger value for β_F than for β_R. Manufacturing conditions lead to a relationship between the forward and reverse current gain expressions:

$$\alpha_R I_{CS} = \alpha_F I_{ES} = I_S. \tag{3.8}$$

Where I_S, the *transistor saturation current*, is a constant for any particular BJT. It should be noted that Equations (3.4) & (3.5) and (3.5) & (3.7) apply to *different biasing conditions* and cannot be valid simultaneously: each set of bias conditions has its own application.

There are two other regions of particular interest in the operation of a BJT. The first of these occurs when both base-collector and base-emitter junctions are reverse-biased. Under these bias conditions the emitter and collector currents become (simplifying Equation (3.3))

$$I_E \approx I_{ES} - \alpha_R I_{CS} \tag{3.9}$$

and

$$I_C \approx I_{CS} - \alpha_F I_{ES}. \tag{3.10}$$

The currents become basically junction leakage currents. This region of BJT operation corresponds to two *p-n* junctions that are reverse biased or turned off. The region is called the *cutoff* region.

The final region of interest occurs when both *p-n* junctions are forward-biased. This region is highly non-linear and is shown in Figure 3.2b & d by the convergence of curves near the origin of the horizontal axis. In this region the relationships between the BJT currents is not clear, however the terminal base-emitter and base-collector voltages correspond to voltage across forward biased *p-n* junctions and therefore lie in the range of approximately 0.6 V to 0.9 V. This region is called the *saturation* region.

The Ebers-Moll equations can be easily converted into a circuit model consisting of a pair of diodes and a pair of dependent current-controlled current sources. This highly useful model is shown for an *npn* BJT[5] in Figure 3.3.

Notice that Kirchhoff's current law applied to the emitter and collector nodes produces the following equations:

$$I_E = -I_F + \alpha_R I_R \tag{3.11}$$

$$I_C = -I_R + \alpha_F I_F \tag{3.12}$$

[5]The Ebers-Moll model for a pnp Bipolar Junction Transistor takes exactly the same form as that of an *npn* BJT except ***the two diodes are reversed in direction***. The *dependent current sources and their controlling currents keep the same polarity* even though forward-biased and reverse-biased now imply currents in the opposite direction.

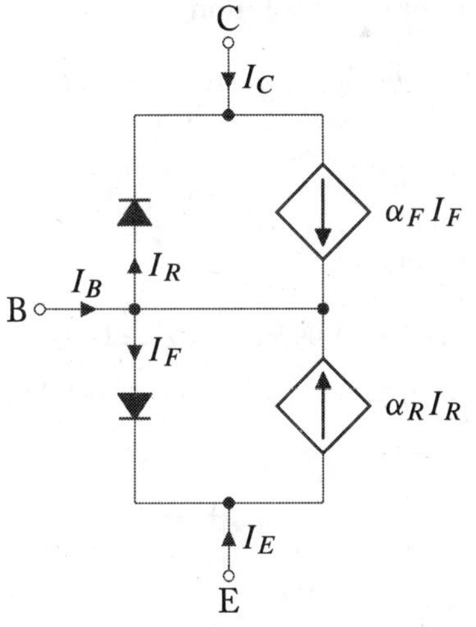

Figure 3.3: The Ebers-Moll model of an *npn* BJT.

where the currents in the diodes are given by:

$$I_F = I_{ES}\left\{e^{\frac{V_{BE}}{nV_t}} - 1\right\} \tag{3.13}$$

$$I_R = I_{CS}\left\{e^{\frac{V_{BC}}{nV_t}} - 1\right\}. \tag{3.14}$$

Substitution of these diode equations, as derived in Chapter 2, into Kirchhoff's equations at the nodes produces the usual form of the Ebers-Moll equations as seen in Equation (3.3) (exercise left to the reader).

The Ebers-Moll model and the related Ebers-Moll equations accurately predict the behavior of a BJT throughout its regions of operation at low frequencies. The model does not include power dissipation restrictions or the possibility of the reverse breakdown of either junction due to excessive reverse voltages being applied. Addition of capacitors across the junctions as described in Chapter 10 (Book 3), of the series, expands the model to include high frequency effects.

3.2 THE BJT AS A CIRCUIT ELEMENT

The operation of a BJT in a simple circuit can be derived in much the same manner as the operation of a diode in a simple circuit. There are several choices, among the most obvious are:

- Use the Ebers-Moll set of equations and obtain a numerical solution
- Use the empirical V-I curves and obtain a graphical solution

- Use computer simulation to obtain a solution

The choice of solution technique depends strongly on the complexity of the circuit in which the BJT is operating.

Example 3.1
A BJT with the following parameters:

$$\alpha_F = 0.995 \qquad \alpha_R = 0.95$$
$$\eta = 1 \qquad I_S = 0.1\,\text{fA}$$

is connected as shown.

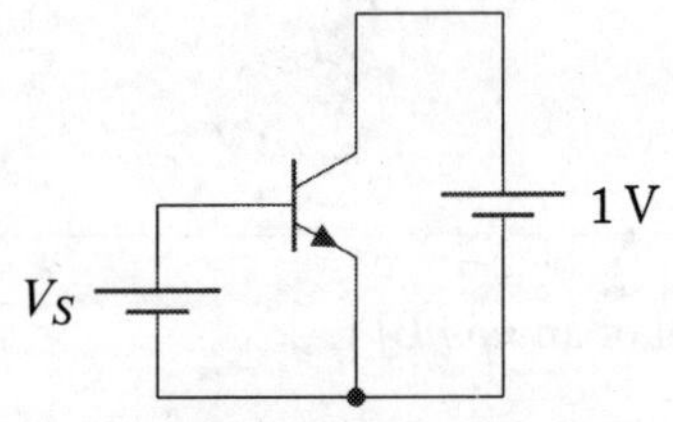

Determine the collector, base, and emitter currents for the following values of the source voltage, V_S: 0.4, 0.6, 0.7, 0.8, 0.9, and 1.0 V.

Solution:

The simple connection of the voltage sources leads directly to the determination of the base-emitter and base-collector junction voltages: thus, using the Ebers-Moll Equations seems the most direct method of solution. The Ebers-Moll equations are:

$$I_E = -I_{ES}\left(e^{\frac{V_{BE}}{\eta V_t}} - 1\right) + \alpha_R I_{CS}\left(e^{\frac{V_{BC}}{\eta V_t}} - 1\right)$$
$$I_C = -I_{CS}\left(e^{\frac{V_{BC}}{\eta V_t}} - 1\right) + \alpha_F I_{ES}\left(e^{\frac{V_{BE}}{\eta V_t}} - 1\right).$$

Here, the following substitutions due to the circuit connections can be made:

$$V_{BE} = V_S \qquad V_{CE} = V_S - 1.0$$
$$I_{ES} = I_S/\alpha_F = 0.101\,\text{fA} \qquad I_{CS} = I_S/\alpha_R = 0.105\,\text{fA}$$

I_B can be calculated by applying Kirchhoff's Current Law to the BJT:

$$I_B = -(I_C + I_E).$$

Direct substitution of these values into the Ebers-Moll Equations yields the values shown in Table 3.1.

Table 3.1:

V_S (V)	I_C	I_E	I_B
0.4	480.2 pA	−482.7 pA	2.413 pA
0.6	1.052 mA	−1.058 mA	5.289 nA
0.7	49.27 mA	−49.51 mA	247.6 nA
0.8	2.306 mA	−2.318 mA	11.59 μA
0.9	108.0 mA	−108.5 mA	542.5 mA
1.0	5.054 A	−5.079 A	25.40 mA

Notice that when V_S is small (i.e., $V_S \approx 0.4\,\text{V}$) the collector current is very small. When V_S increases by 50% to 0.6 V, the current jumps by a factor of more than 2000: the base-emitter junction of the BJT has become forward biased while the base-collector junction has remained reverse-biased: the BJT has transitioned from the cut-off region to the forward-active region. For $0.6\,\text{V} \le V_S \le 1.2\,\text{V}$, the BJT remains in the forward-active region and the base and collector currents are related by:

$$I_C = \frac{\alpha_F}{1-\alpha_F} I_B = \frac{0.995}{1-0.995} I_B = 199 I_B .$$

Notice that in the forward-active region, small changes in the base current produce significantly larger changes in the collector (and consequently the emitter) current.

Example 3.2
A BJT with the following parameters:

$$\alpha_F = 0.995 \qquad \alpha_R = 0.95$$
$$\eta = 1 \qquad I_S = 0.1\,\text{fA}$$

is connected as shown, with the following resistor values:

$$R_b = 3.3\,\text{k}\Omega \qquad R_c = 220\,\Omega .$$

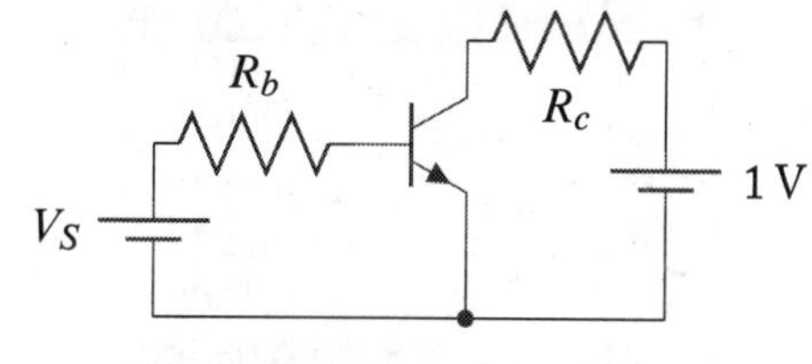

Determine the collector, base, and emitter currents for the following values of the source voltage, V_S: 0.4 V, 0.6 V, 0.7 V, 0.8 V, 0.9 V, 1.0 V, and 1.2 V.

Solution #1 (Ebers-Moll Equations):

Here the Ebers-Moll Equations are not sufficient to determine the currents. Two additional equations, dependent on the circuit topology and parameters, are needed. Loop equations around base-emitter and base-collector loops yield:

$$V_{BE} = V_S - I_B R_b$$
$$V_{BC} = V_S - I_B R_b - (1.0 - I_C R_C).$$

Combining these two equations with the Ebers-Moll Equations and searching for solutions is, by hand, quite complex. Realistically, a computer search for solutions is the only practical method of solution. A Computer search solution using MathCAD for the above circuit can be performed as shown in Figure 3.4:

Repeated use of this MathCAD program (changing the value of V_S) yields a set of results for the various input voltage values (Table 3.2).

Table 3.2:

V_S (V)	I_C	I_E	I_B
0.4	480.2 pA	−482.6 pA	2.413 pA
0.6	1.052 μA	−1.057 μA	5.285 nA
0.7	47.78 μA	−48.03 μA	240.1 nA
0.8	1.125 mA	−1.131 mA	5.654 μA
0.9	4.088 mA	−4.114 mA	25.62 μA
1.0	4.292 mA	−4.346 mA	54.68 mA
1.2	4.405 mA	−4.764 mA	113.6 μA

Notice that the addition of a resistor on the collector of the BJT creates a transition from the forward-active region to the saturation region. In the saturation region of a BJT the base and collector currents no longer have a constant linear relationship. For example, in this circuit the collector-to-base current relationships are shown in Table 3.3.

Table 3.3:

V_S (V)	I_C/I_B	V_{CE} (V)
0.4	199	1.000
0.6	199	1.000
0.7	199	0.989
0.8	199	0.7525
0.9	160	0.1005
1.0	78.5	0.0558
1.2	38.6	0.0349

Solution of Simple *npn* BJT with 2 Resistors—Example 3.2-2—T.F. Schubert & E.M. Kim

Defining BJT and circuit parameters

$I_s := 0.1 \times 10^{-15}$ $\alpha_F := 0.995$ $\alpha_R := 0.95$ $V_t := 0.026$ $R_b := 3300$

$I_{es} := \frac{I_s}{\alpha_F}$ $I_{cs} := \frac{I_s}{\alpha_R}$ $\eta := 1$ $V_S := 1.0$ $R_c := 220$

Guess Values

$I_c := 0.002$ $I_e := -0.0021$ $V_{be} := 0.8$ $V_{bc} := -0.2$

GIVEN (Solve Block)

$$V_{be} = V_S + (I_e + I_c)R_b \qquad V_{bc} = V_{be} - (1 - I_c R_c)$$

$$I_e = -I_{es}\left[\exp\left(\frac{V_{be}}{\eta V_t}\right) - 1\right] + \alpha_R I_{cs}\left[\exp\left(\frac{V_{bc}}{\eta V_t}\right) - 1\right]$$

$$I_x = -I_{cs}\left[\exp\left(\frac{V_{bc}}{\eta V_t}\right) - 1\right] + \alpha_F I_{es}\left[\exp\left(\frac{V_{be}}{\eta V_t}\right) - 1\right]$$

$x :=$ Find$(I_c, I_e, V_{be}, V_{bc})$

$$x = \begin{bmatrix} 4.29171 \times 10^{-3} \\ -4.34639 \times 10^{-3} \\ 0.81956 \\ 0.76374 \end{bmatrix} \qquad I_B := -(x_0 + x_1) \qquad I_B = 5.468 \times 10^{-5}$$

Figure 3.4: MathCAD solution to Example 3.2.

The transition to the saturation region is signaled by a change in the ratio of the collector current to the base current. Saturation occurs when

$$\frac{I_C}{I_B} < \beta_F,$$

which appears to occur in this circuit application when V_S increases to a value larger than something slightly less than 0.9 V.

The last column in the table reports the voltage at the collector of the BJT. Notice that for small values of the input voltage ($V_S \le 0.6$), the collector voltage is essentially the collector supply voltage (1.0 V): for large values of the input voltage ($V_S > 0.9$), the collector voltage nears zero. This property of a near-constant output voltage value for a range of input values has special significance in digital applications circuitry. These applications are discussed at length in Section 3.5.

Solution #2 (Graphical Techniques):

The empirical curves for the BJT coupled with load line techniques provide a more direct form of solution. The two supplemental equations derived in the first solution are actually the equations for load lines: one in the base-emitter loop and one in the collector-emitter loop.

$$V_{BE} = V_S - I_B R_b \tag{3.15}$$

$$V_{BC} = V_S - I_B R_B - (1.0 - I_C R_C). \tag{3.16}$$

Solution takes the form of plotting the load lines on the transistor curves: Equation (3.15) is plotted on the output curves (Figure 3.5b), and Equation (3.16) is plotted as a series of parallel lines on the input transistor curves (Figure 3.5a), one line for each of the specified values of V_S. Unfortunately, the load lines cross several of the transistor parameter curves. It is the task of the circuit analyst/designer to obtain a solution that is consistent with all constraints. The V-I relationships for this BJT are shown in Figure 3.5a & b.

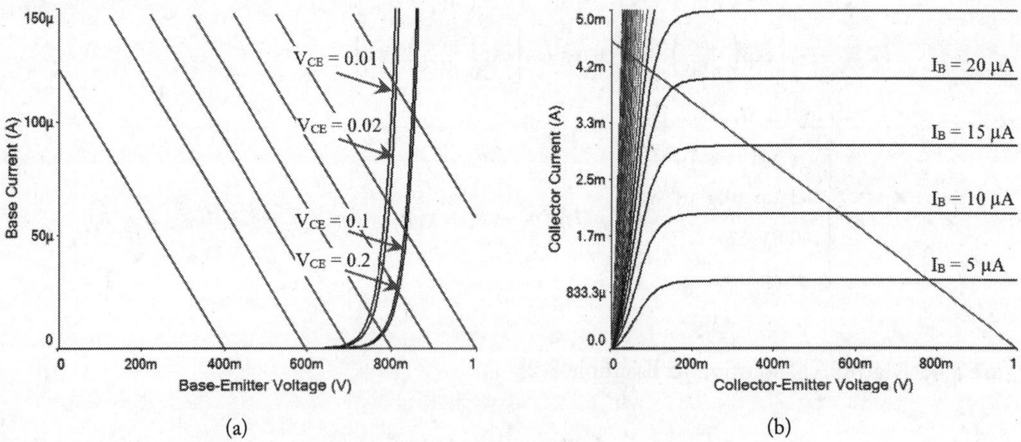

Figure 3.5: Transistor curve traces for Example 3.2.

For example, take the case where $V_S = 0.9\,\text{V}$. Looking at the input curves, one concludes that the base current must approximately lie in the range:

$$20\,\mu\text{A} \le I_B \le 40\,\mu\text{A}.$$

From the output curves, using those values of I_B, one then concludes that V_{CE} must lie in the range:

$$0.08\,\text{V} \le V_{CE} \le 0.15\,\text{V}.$$

This set of restrictions reduces the range of allowable base currents to:

$$23\,\mu\text{A} \le I_B \le 27\,\mu\text{A},$$

which, in turn, reduces the range of V_{CE} to:

$$0.09\,\text{V} \le V_{CE} \le 0.11\,\text{V}.$$

Continuing along this path, one progresses to a final solution for $V_S = 0.9$. This solution takes the form of the following voltage and current values:

$$V_{CE} \approx 0.1\,\text{V}, \qquad I_B \approx 25\,\mu\text{A},$$
$$I_C = \frac{1.0 - V_{CE}}{220} \approx \frac{1.0 - 0.1}{220} = 4.1\,\text{mA}.$$

While reading values off graphs will introduce some margin of error, this result is very close to that of Solution #1: other values of V_S will similarly produce results equivalent to those previously obtained.

Solution #3 (SPICE Simulation):

This problem is particularly suited for DC analysis using SPICE. Of particular importance is the modeling of the BJT to suit the Ebers-Moll parameters given. SPICE uses a model for BJTs that in its most simple formulation becomes the Ebers-Moll model. The parameters necessary for input for this problem are shown in Table 3.4.

Table 3.4:

Parameter	SPICE variable	Value
I_s	IS	0.1×10^{-15}
η	NF	1
	NR	1
β_F	BF	$\frac{\alpha_F}{1-\alpha_F} = \frac{0.995}{1-0.995} = 199$
β_R	BR	$\frac{\alpha_R}{1-\alpha_R} = \frac{0.95}{1-0.95} = 19$

When comparing simulation techniques, it is also necessary to make sure all other equation parameters are identical. The MathCAD simulation used the common approximation: $V_t = 0.026\,\text{V}$. This approximation is quite valid for room temperature $\approx 300°\text{K}$ – it corresponds to $\approx 28.6°\text{C} \approx 301.7°\text{K}$. The output values for the SPICE simulation are shown in Figure 3.6. Notice the results are within $\pm 0.11\%$ of those calculated using the Ebers-Moll equations solved directly with MathCAD.

3.3 REGIONS OF OPERATION IN BJTS

BJT operation has been seen to fall into four basic regions of useful operation. The regions are described by the state of bias of the two p-n junctions within the transistor. The four possible combinations and the corresponding region names are shown in Figure 3.7. Briefly, the four regions of operation are:

1. The cutoff region is defined by both base-emitter and base-collector junctions being reverse biased. Reverse biasing both junctions reduces all currents in a BJT to small leakage values in the picoampere to nanoampere range: the BJT essentially looks like an open circuit. Applications for this region are primarily in the switching and digital logic areas.

2. The saturation region is defined by both junctions being forward biased. Here the possibility exists for large current flow between the collector and emitter terminals with minimal dynamic resistance ($V_{CE} \approx 0$). Applications again fall in the switching and digital logic areas.

3. The forward-active region. This region is defined as a forward biased base-emitter junction and a reverse biases base-collector junction. BJTs operating in this region are characterized by a relatively constant collector current to base current ratio. The region is most commonly used for amplification with the parameters α_F and β_F describing the amplification.

V1	I(Q1[IC])	I(Q1[IE])	I(Q1[IB])	V(vb)	V(vb) – V(vc)
0.4	4.807E-10	-4.831E-10	2.381E-12	4.000E-01	-6.000E-01
0.6	1.050E-06	-1.056E-06	5.279d-9	6.000E-01	-2.998E-01
0.7	4.769E-05	-4.793E-05	2.396E-07	6.992E-01	-2.903E-01
0.8	1.124E-03	1.000E+01	5.646E-06	7.814E-01	2.857E-03
0.9	4.088E-03	-4.114E-03	2.560E-05	8.155E-01	7.149E-01
1.0	4.292E-03	-4.346E-03	5.466E-05	8.196E-01	7.638E-01
1.2	4.387E-03	-4.500E-03	1.136E-04	8.251E-01	7.902E-01

Figure 3.6: SPICE simulation output values.

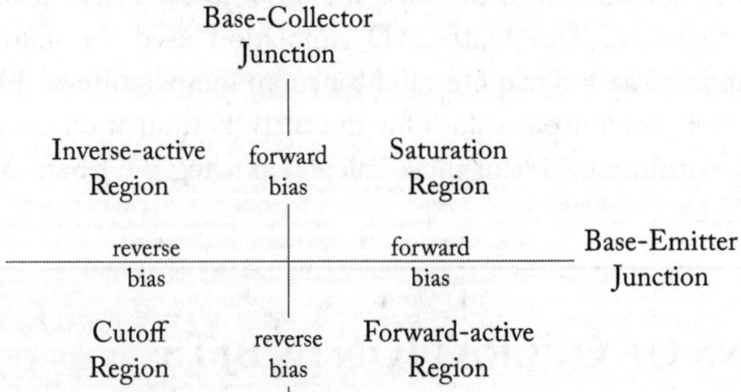

Figure 3.7: The four basic regions of BJT operation.

4. The inverse-active region. This region is the direct opposite of the forward-active region: the base-collector junction is forward biased and the base-emitter junction is reverse biased. Here emitter current is a multiple of base current with α_R and β_R describing the amplification. While it is certainly possible to manufacture a BJT with amplification in the inverse-active region as large as that of the forward-active region, most BJTs are optimized for forward amplification resulting in much smaller values for α and β in this region. The region is rarely used with the notable exception of the input stages of the transistor-transistor-logic (TTL) family of digital logic gates.

In addition to these four regions, there is an additional region of severe consequences: the breakdown region. A BJT enters the breakdown region when one or both of the *p-n* junctions are sufficiently reverse biased so that a Zener-like breakdown occurs. Transistors are not manufactured to withstand extended use in the breakdown region and typically will exhibit catastrophic thermal run-away and destruction. Manufacturers list maximum voltages that can be safely applied to the junctions and maximum power limitations in order to ensure that the devices are operated safely.

3.4 MODELING THE BJT IN ITS REGIONS OF OPERATION

The Ebers-Moll model for Bipolar Junction Transistors is a flexible, but rather complex, model that can be used in all four useful regions of operation. It can, however, be simplified within each region to provide a set of elementary BJT models: one for each region. The derivation of each of the individual models is described below and a summary table listing each model and conditions to test the validity of each model appears in Figure 3.8.[6]

1. The *cutoff region* is defined by both base-emitter an base-collector junctions being reverse biased. Reverse biased p-n junctions provide extremely high impedance and low leakage currents; the low currents in the junctions reduce the dependent current sources of the Ebers-Moll model to near-zero value. Therefore, a simple model of a BJT in cutoff is three terminals with open circuits between. The typical turn-on voltage (the voltage at which a junction becomes forward biased) for a Silicon BJT *p-n* junction is:

$$V_{BE(on)} = V_{BC(on)} = 0.6\,\mathrm{V}.$$

2. The *saturation region* is defined by both *p-n* junctions being forward biased. Forward biased junctions can be modeled by a voltage source in series with a small resistance. The model most commonly used ignores this small resistance and models the BJT with two constant

[6]In Figure 3.8 the voltage source polarities are correct for *npn* BJTs. The circuit diagrams for *pnp* BJTs are identical with the polarity of the voltage sources reversed. One common error in making this change involves the polarity of the dependent current sources: the polarity (or direction) of the dependent current sources *stays the same* for both types of BJT. The currents flow in opposite directions, but the *ratio*, as indicated by the direction of the dependent current sources, does not change polarity.

Single Region Model Equivalent Circuits (*npn* voltage polarities shown) (*pnp voltage* polarities are reversed)	Characteristics of the Models *npn*	*pnp*
1. Cutoff Region Model 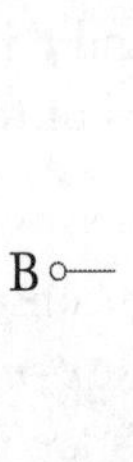	$I_C = I_B = I_E = 0$ $V_{BE} < V_{BE(\text{on})}$ $V_{BC} < V_{BC(\text{on})}$	$I_C = I_B = I_E = 0$ $V_{BE} > -V_{BE(\text{on})}$ $V_{BC} > -V_{BC(\text{on})}$
2. Saturation Region Model 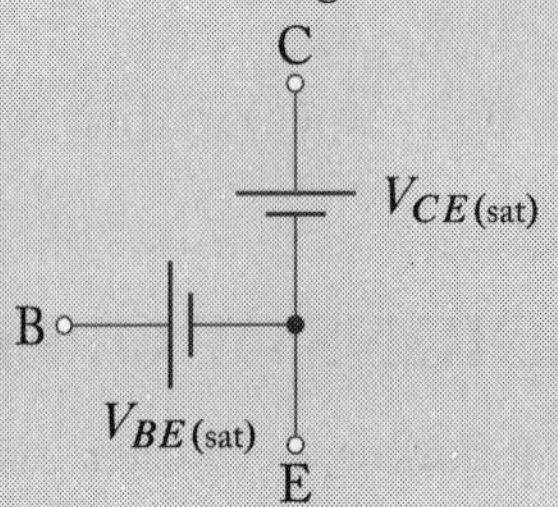	$I_B > 0$ $I_C > 0$ $I_C < \beta_F I_B$ $V_{BE} > V_{BE(\text{on})}$	$I_B < 0$ $I_C < 0$ $\lvert I_C \rvert < \lvert \beta_F I_B \rvert$ $V_{BE} < -V_{BE(\text{on})}$
3. Forward-active Model 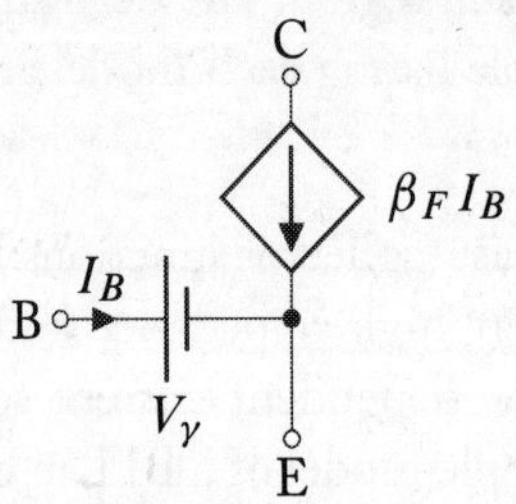	$I_C = \beta_F I_B$ $I_B > 0$ $V_{BE} > V_{BE(\text{on})}$ $V_{CE} > V_{CE(\text{sat})}$	$I_C = \beta_F I_B$ $I_B < 0$ $V_{BE} < -V_{BE(\text{on})}$ $V_{CE} < -V_{CE(\text{sat})}$
4. Inverse-active Model 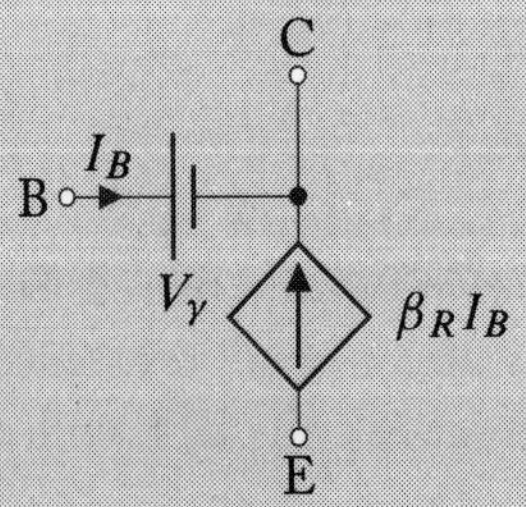	$I_E = \beta_F I_B$ $I_B > 0$ $V_{BC} > V_{BE(\text{on})}$ $V_{EC} > V_{CE(\text{sat})}$	$I_E = \beta_F I_B$ $I_B < 0$ $V_{BC} < -V_{BE(\text{on})}$ $V_{EC} < -V_{CE(\text{sat})}$

Figure 3.8: Bipolar junction transistors: linear model.

voltage sources. Rather than show a voltage source connected from the base to each of the other terminals, it is most common to model the BJT with a voltage source from the base to the emitter and another from the emitter to the collector: this second voltage source will be a small value source since it models the difference in forward bias voltage for the two junctions. Typically, one junction will tend to be more forward biased than the other (i.e., the base-emitter junction is often more forward biased than the base-collector junction) due to the different currents passing through each junction. For Silicon transistors typical model values are:

$$V_{BE(sat)} = 0.8\,\text{V} \quad \text{and} \quad V_{CE(sat)} = 0.2\,\text{V}.$$

3. The *forward-active region* is defined as a forward biased base-emitter junction and a reverse biased base-collector junction. Here the forward biased base-emitter junction is modeled by a voltage source and the reverse biased junction by an open circuit. The current through the forward biased base-emitter junction activates the dependent current source between the collector and base in the Ebers-Moll model while the other current source is inactive (its controlling current is near zero value). The typical value for Silicon BJTs is:

$$V_\gamma = 0.7\,\text{V}.$$

4. The *inverse-active region* is the opposite of the forward-active region. The base-collector junction is modeled by a voltage source and a dependent current source is connected between the emitter and base.

When one complex model is separated into a group of simple models, several difficulties can occur. The most prevalent of these difficulties is the choice of the proper model for the circumstances in question. Whenever a circuit element is replaced by a model that is only correct for one region of operation of the circuit element, the assumptions upon which the model must be tested in order to verify the validity of the replacement. Experience leads to the proper choice of correct model on the first guess: incorrect guesses, when tested for validity, give clues as to which is the correct model to choose next.

Example 3.3
Given the circuit shown with element values

$$V_{bb} = 2\,\text{V} \qquad R_b = 22\,\text{k}\Omega \qquad R_c = 2\,\text{k}\Omega$$
$$V_{cc} = 10\,\text{V} \qquad R_e = 100\,\Omega$$

and a Silicon BJT with $\alpha_F = 0.99$.

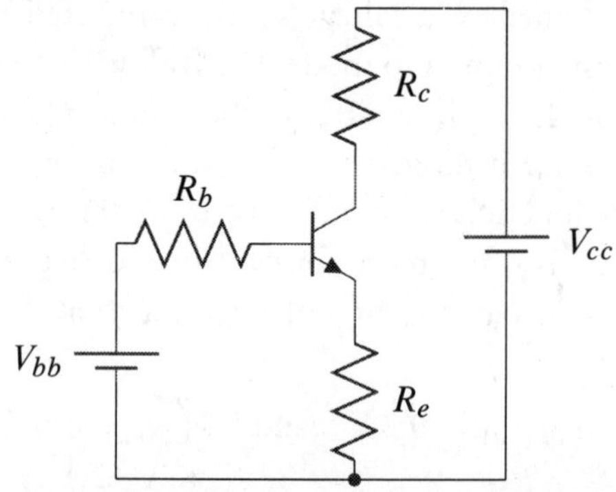

Determine the region of operation for the BJT and the base and collector currents.

Solution:

First a region of operation must be chosen. It seems clear that the inverse-active region can clearly be eliminated — the base must be at higher voltage than the collector with positive base and negative collector currents. Also easily eliminated is the cutoff region — with no current flowing in the BJT, the base-emitter and base-collector junctions would clearly be forward biased (a violation of the assumptions for that region). The choice clearly lies between the forward-active region and the saturation region. The correct choice is not obvious.

As a first try, assume the BJT is operating in the saturation region.

Attempt #1 (saturation)

Replace the BJT with its saturation region model (indicated in Figure 3.8) and then calculate the terminal currents for the BJT.

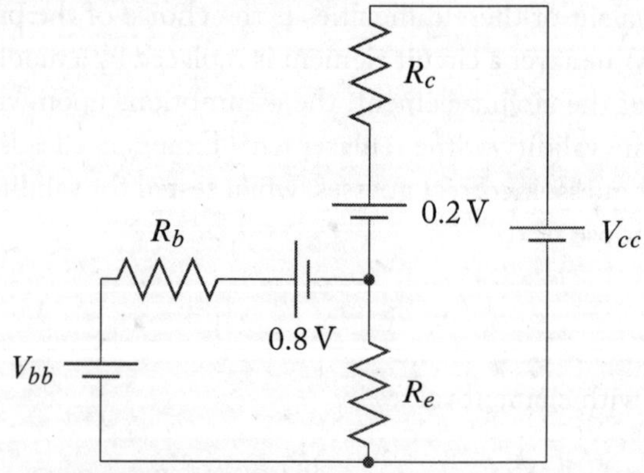

Around the left mesh, Kirchhoff's voltage law gives:

$$V_{bb} - R_b I_B - 0.8 - R_e(I_B + I_C) = 0.$$

Around the right mesh, the equation is:

$$V_{CC} - R_C I_C - 0.2 - R_e(I_B + I_C) = 0.$$

Inserting the circuit values yields two equation in the two unknowns, I_B and I_C:

$$\begin{aligned} 22,100 I_B + 100 I_C &= 1.2 \\ 100 I_B + 2,100 I_C &= 9.8. \end{aligned}$$

The solutions are:

$$I_B = 33.19\,\text{mA} \qquad \text{and} \qquad I_C = 4.665\,\text{mA}.$$

These solutions must now be checked to see if they are consistent with the saturation region model of the BJT. From Figure 3.8, the *npn* saturation region characteristics of interest are:

$$I_B > 0, \quad I_C > 0, \quad I_C < \beta_F I_B.$$

The base and collector currents are positive but ratio of these currents is:

$$\frac{I_C}{I_B} = \frac{4.665\text{ mA}}{33.19\ \mu\text{A}} = 140.6 > \beta_F = \frac{\alpha_F}{1-\alpha_F} = \frac{.99}{1-.99} = 99.$$

Thus, the basic assumptions of the saturation region model have been violated. The verification of assumptions has indicated that, in all likelihood, the collector current to base current ratio is not smaller than β_F: a sign that the BJT must be in the forward-active region.

Attempt #2 (forward-active)

Replace the BJT with its forward-active region model (indicated in Figure 3.8) and then calculate the terminal currents for the BJT.

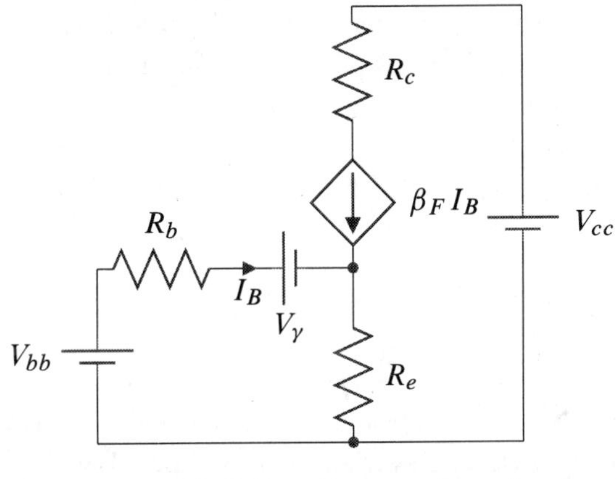

Around the left mesh, Kirchhoff's Voltage Law produces (notice that the current out of the emitter is the base current + the collector current):

$$V_{bb} - I_B R_b - V_\gamma - (1+\beta_F) I_B R_e = 0.$$

With the circuit element values and the value for $V_\gamma(0.7)$ inserted , the only unknown is I_B. The solution of the equation is:

$$I_B = 40.6\,\text{mA}$$

I_C can then be calculated as:

$$I_C = \beta_F I_B = 4.02\,\text{mA}.$$

The appropriate region verification checks are:

$$I_B > 0 \qquad \text{and} \qquad V_{CE} > V_{CE(sat)} = 0.2.$$

The base current has the correct sign: V_{CE} can be calculated by using Kirchhoff's Voltage around the right mesh:

$$V_{CC} - I_C R_c - V_{CE} - (I_C + I_B) R_e = 0.$$

Which yields:

$$V_{CE} = 1.54\,\text{V}.$$

The forward-active region verifications have been shown to be valid: thus the currents calculated for the forward-active region are the correct values.

3.5 DIGITAL ELECTRONICS APPLICATIONS

Many digital electronics applications are based upon a BJT changing from one region of operation to another. The simple regional models of BJT operation are sufficient to understand the basic operation of several logic gate types.[7]

In digital systems there are only a few basic logic operations which must typically be performed. The most common of these operations are: NOT, AND, and OR. A common property of these operations is that a variety of inputs produce an output in binary form: that is, an output that exists in one of two possible states. Typical electronic circuitry assigns to each logic value (i.e., "1" or "0," "HIGH" or "LOW," "ON" or "OFF") a specific range of voltage values. As an example, the transistor-transistor logic (TTL) circuit family, one of the basic logic circuit families that will be discussed in this section, assigns two logic voltage ranges:

$$\begin{aligned} &\text{low voltage range} && — && 0\,\text{V} \le V_L \le 0.8\,\text{V} \\ &\text{high voltage range} && — && 2.0\,\text{V} \le V_H \le 5.5\,\text{V}. \end{aligned}$$

One of the major goals of an electronic circuit that will perform a logic operation is to provide a constant value output that is invariant to this form of voltage variation on the input. The simple circuit of Example 3.2 showed this property of virtually invariant output voltage to a range of input voltage values. A circuit with similar properties is shown in Figure 3.9.

[7]Gate speed and some of the more advanced digital circuitry topics will be discussed in Chapter 16 (Book 4).

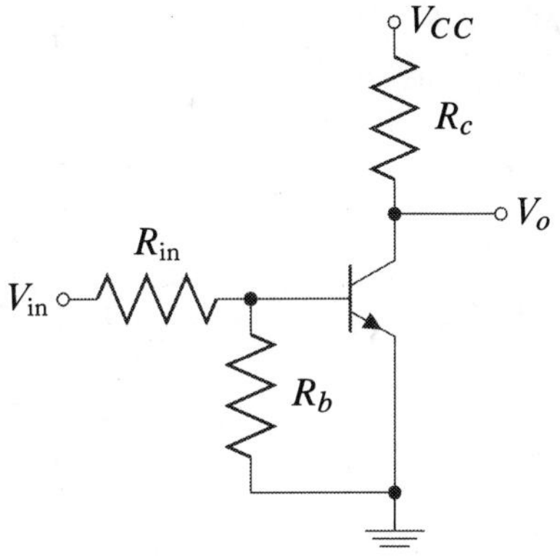

Figure 3.9: A simple logic inverter.

3.5.1 A LOGIC INVERTER CIRCUIT

If one assumes that V_{CC} is a positive voltage, the operation of the circuit of Figure 3.9 can be described as:

1. For small values of V_{in}, the BJT is in the cut-off region. The base and collector currents are near zero, and the output voltage, V_o, is basically the same value as V_{CC}.

2. As V_{in} increases, the base-emitter voltage on the BJT will increase until the BJT turns on and the BJT enters the forward-active region. This transition of regions will occur at an input voltage,
$$V_{in}=\frac{R_b+R_{in}}{R_b}V_{BE(on)}. \tag{3.17}$$
As the input voltage continues to increase, the output voltage will decrease steadily until it approaches $V_{ce(sat)}$:
$$V_o = V_{CC} - \beta_F I_B R_c$$
where
$$I_B = \frac{V_{in} - V_{BE(on)}}{R_{in}} - \frac{V_{BE(on)}}{R_b}.$$

3. When the output voltage reaches $V_{ce(sat)}$, the BJT will enter the saturation region, further increases in the input voltage will result in negligible changes in the output voltage.

This simple circuit forms the basis of a "NOT" gate (also known as a logic inverter): low value input voltages become high output voltages and vice versa. The BJT switches between the cut-off and the saturation regions to form the two logic levels. The high output region (BJT in cut-off) is described as #1 above, and the low region (BJT in saturation) is described as #3. A transition region (BJT in forward-active region) is described as #2, and serves as a buffer to isolate the two regions.

Example 3.4

The circuit of Figure 3.9 has the following circuit element values

$$V_{CC} = 5\,\text{V} \quad R_{in} = 5.6\,\text{k}\Omega \quad R_b = 15\,\text{k}\Omega \quad R_c = 2.2\,\text{k}\Omega$$

and a Silicon BJT with $\beta_F = 50$.

Determine the voltage transfer relationship for $0\,\text{V} \leq V_{in} \leq 5\,\text{V}$.

Solution:

The voltage transfer relationship is best shown with a plot V_o as a function of V_{in} as the desired result.

It is best to begin at one end of the input range, for example begin at $V_{in} = 0$. Clearly, both BJT p-n junctions are reverse biased and the transistor is in the cut-off region. With the transistor off, $V_o = 5\,\text{V}$. As the input voltage increases, the BJT will remain in the cut-off region until the base-emitter voltage becomes sufficiently large to turn the BJT on. Equation (3.17) yields the transition input voltage:

$$V_{in(1)} = \frac{R_b + R_{in}}{R_b} V_{BE(on)} = \frac{15\,\text{k} + 5.6\,\text{k}}{15\,\text{k}}(0.6) = 0.824\,\text{V}.$$

Section "1" of the transfer relationship can be drawn (seen in Figure 3.10) with the above derived result.

When $V_{in} > V_{in(1)}$ the BJT is in the forward-active region. The circuit can then be redrawn using the forward-active model of the BJT.

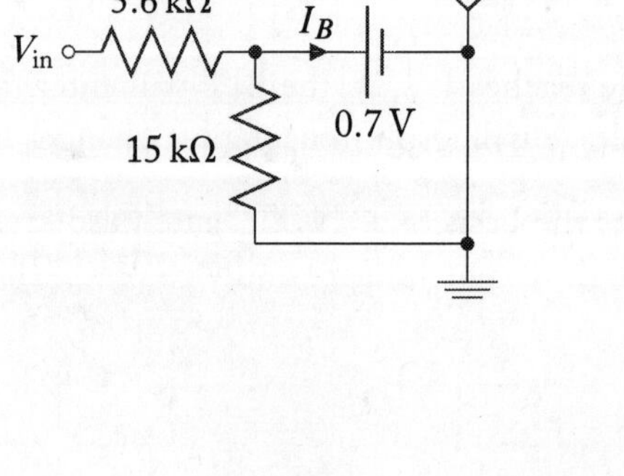

The circuit defining equations are:

$$I_B = \frac{V_{in} - 0.7}{5.6\,\mathrm{k}} - \frac{0.7}{15\,\mathrm{k}}$$
$$I_B = 178.6 V_{in} - 171.7\,\mu\mathrm{A}$$

and

$$V_o = 5 - 2.2\,\mathrm{k}(50 I_B)$$

or

$$V_o = -19.65 V_{in} + 23.88.$$

The input-output transfer relationship for this region of operation is of particular interest. Notice that *the output voltage is multiple of the input voltage* (in this case, -19.65) with a DC offset (in this case, 23.88 V). Small variations of the input voltage create larger variations at the output! This important property, known as amplification, often associated with the forward-active region of a BJT is discussed thoroughly in Chapter 5 (Book 2), of the series.

The BJT forward-active portion of the transfer relationship is valid until the BJT enters the saturation region: that is, when $V_o = V_{CE(sat)} = 0.2\,\mathrm{V}$. Solving for the input voltage that corresponds to the transition to the BJT saturation region yields:

$$0.2 = -19.65 V_{in(2)} + 23.88 \qquad \Rightarrow \qquad V_{in(2)} = 1.205\,\mathrm{V}.$$

Section "2" of the transfer relationship can be plotted (shown in Figure 3.10) using the above relationships.

For $V_{in} > V_{in(2)} = 1.205\,\mathrm{V}$, the BJT is in saturation and $V_o \approx 0.2\,\mathrm{V}$. That result is shown in section "3" of the transfer relationship. It should be noted that the simple linearized models for BJTs have greatest error near the transition between regions. In this example, the errors are greatest near $V_{in(1)} \approx 0.83\,\mathrm{V}$. Here, because of the extreme steepness of the curve in region "2," region "1" ends at $V_{in(1)} = 0.824\,\mathrm{V}$ and region "2" begins (the output voltage cannot exceed V_{CC}, therefore when $V_o = 5$) at $V_{in} = 0.960\,\mathrm{V}$. Experience tells us to fill in the gap with a smooth curve.

Notice in the transfer relationship, shown in Figure 3.10, the two constant voltage levels are quite wide compared to the very rapid transition region. This form of transfer characteristic is quite desirable in digital logic circuitry in that it allows for variation in the input (often caused by noise) without change in the output. *Noise Margins* are quantitative measures of the allowable variation in inputs. Each is defined as the variation in input from the nominal input (voltage levels that a similar gate would deliver if attached to the input):

$$\mathrm{NM}(0) \equiv \text{allowable LOW input variation}$$
$$\mathrm{NM}(1) \equiv \text{allowable HIGH input variation.}$$

For this logic inverter circuit:

$$\begin{aligned} \text{NM}(0) &= V_{in(1)} - V_{o(low)} = 0.824 - 0.2 = 0.624\,\text{V} \\ \text{NM}(1) &= V_{in(2)} - V_{o(high)} = 1.205 - 5 = -3.795\,\text{V}. \end{aligned}$$

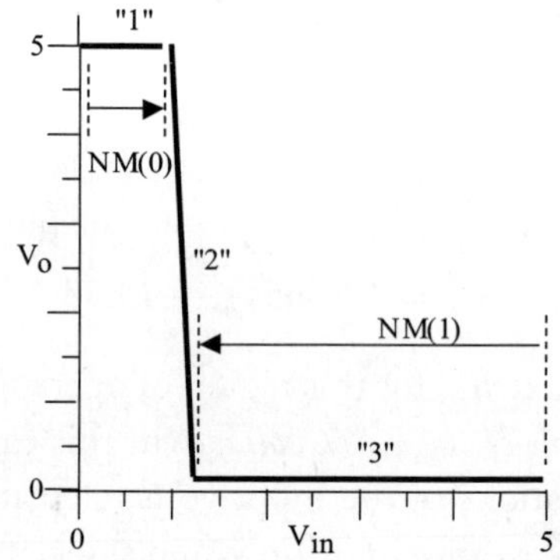

Figure 3.10: Voltage transfer relationship for Example 3.4. (Obtained using three linearized models of BJT operation): (1) Cut-off; (2) Forward active; (3) Saturation.

The logic inverter circuit serves a useful purpose in digital circuitry: it performs the NOT operation. In order to perform the more complex operations such as: AND and OR it is necessary to increase the complexity of the circuitry. As a first example, the addition of diodes to the input of a BJT leads to the common Diode-Transistor Logic (DTL) family of gates. The most basic circuit of the DTL family is the NAND (NOTAND) gate. It is formed by a connection of a diode logic AND and a transistor logic inverter. In Figure 3.11, this gate is shown with two input diodes, D_{1a} and D_{1b}: additional inputs could be implemented with additional similarly-connected diodes.

3.5.2 DIODE-TRANSISTOR LOGIC GATE

The operation of this circuit can be described as:[8]

1. For small values of *either* input voltage (or *both* input voltages), the corresponding input diode, D_1, will turn on. The voltage at the anode of the input diodes then becomes:

   $$V_{anode} = V_{in} + V_{\gamma}.$$

   If V_{anode} is sufficiently small, that is if

   $$V_{anode} < V_{BE(on)} + 2V_{\gamma}$$

[8] V_{CC} is once again assumed to be a positive voltage. Often V_{CC} is chosen to be $\approx +5$ V, but there are several varieties of DTL circuitry that use other positive values.

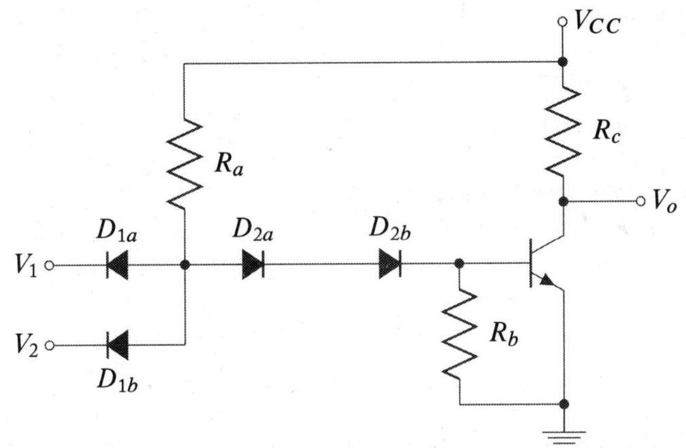

Figure 3.11: A DTL NAND gate.

or equivalently,

$$V_{in} < V_\gamma + V_{BE(on)}.$$

the BJT will be in the cut-off region and the output voltage,

$$V_o = V_{CC}.$$

2. If the minimum value of *both* inputs is V_{in}, and as V_{in} increases beyond the constraints of the above region of operation, i.e., when

$$V_{anode} > 2V_\gamma + V_{BE(on)}$$

or equivalently,

$$V_{in} > V_\gamma + V_{BE(on)}.$$

The BJT enters the forward active region. The output voltage will steadily decrease until the BJT enters the saturation region. The forward-active region of the BJT will end when the input diodes *both* turn off; that is, when:

$$V_{in} > V_{BE(sat)} + 2V_\gamma - V_\gamma = V_{BE(sat)} + V_\gamma.$$

As in the logic inverter circuit, this region is very narrow: the BJT is in the forward-active region for only a small range ($V_{BE(sat)} - V_{BE(on)} \approx 0.2\,\text{V}$) of input voltage values.

3. Once the input diodes turn off, the input is essentially disconnected from the circuit. Further increases in the value of V_{in} produce no change in V_o, which remains at:

$$V_o \approx V_{CE(sat)} = 0.2\,\text{V}.$$

To summarize the operation of this gate:

$$\begin{aligned} \text{One or more inputs LOW} \quad &\Rightarrow \quad \text{HIGH output } = V_{CC} \\ \text{All inputs HIGH} \quad &\Rightarrow \quad \text{LOW output } = V_{CE(sat)}. \end{aligned}$$

This behavior forms a logic NAND gate.

It has been shown that input voltages are not necessarily a particular value: they can vary over in a range of values without changing the output of the logic circuit:

$$\begin{aligned} \text{LOW input range } - 0 \le V_{iL} \le V_{BE(on)} + V_\gamma \\ \text{HIGH input range } - V_{BE(sat)} + V_\gamma \le V_{iH} \le V_{CC}. \end{aligned}$$

The susceptibility of this any gate to noise is dependent on these input level range and the nominal logic level that is expected. For this particular logic circuit the nominal logic levels are:

$$\begin{aligned} \text{nominal LOW} &= V_L = V_{CE(sat)} \\ \text{nominal HIGH} &= V_H = V_{CC}. \end{aligned}$$

The noise voltage at an input that will cause the circuit to function improperly is called the *noise margin*. Noise margin is typically different for each logic level and is defined as the difference between the edge of the level range and the nominal level. The DTL NAND gate under consideration has noise margins of:

$$\begin{aligned} \text{NM(LOW)} &= V_{BE(on)} + V_\gamma - V_{CE(sat)} \\ \text{NM(HIGH)} &= V_{BE(sat)} + V_\gamma - V_{CC}. \end{aligned}$$

It should be noted that the magnitude of NM(LOW) is significantly smaller than the magnitude of NM(HIGH): the LOW input signal is much more susceptible to noise than the HIGH input signal. Moving the transition voltage nearer to the center of the range of input values can equalize the noise rejection properties between HIGH and LOW inputs. This equalization of noise margin magnitudes (rarely done in DTL gates) can be accomplished by inserting additional diodes in series with D_{2a} and D_{2b}, or, as in a closely related logic family, the High-Threshold Logic (HTL) family, the two diodes can be replaced by a reverse-biased Zener diode with an appropriate Zener voltage greater than 2 V_γ. The Zener diode allows an increase in the transition voltage values without an increase in component quantity. The HTL family is particularly useful in high-noise environments.

Another important factor when considering using logic gates, is the quantity of gates of similar properties that can be connected in parallel to the output. The number of gates that can be driven by a single gate, without changing the value of the output voltages, is called the *fan-out* of the gate. The load that "slave" gates (the gates being driven) apply to the "master" gate (the gate driving the slaves) takes the form of a load current. When the input to a slave gate is HIGH, the input diodes to the slave are off: the slave gates draw no current and present no load to the master gate. When the input to a slave gate is LOW, current flows out of the slave gate into the output of the master gate: sufficient current added into the collector of the master gate BJT will force the BJT out of saturation and therefore change the output voltage level.

Example 3.5
For the circuit of Figure 3.11 assume the following circuit parameters:

$$V_{CC} = 5\,\text{V} \qquad R_a = 3.9\,\text{k}\Omega \qquad R_b = 5.6\,\text{k}\Omega \qquad R_c = 2.2\,\text{k}\Omega\,.$$

Silicon diodes and a Silicon BJT with $\beta_F = 50$.

Determine the fan-out of the NAND gate.

Solution:

It has already been determined which quantities are significant in fan-out computations for this DTL logic gate (only LOW slave inputs present a load) :

- the input current for a slave gate with LOW input
- the BJT base and collector currents for a master gate with LOW output

First calculate the input current for a slave gate with LOW input. The LOW output of the master gate will form the input voltage for the slave gate. Therefore,

$$V_{in} \approx V_{CE(sat)} = 0.2\,\text{V}.$$

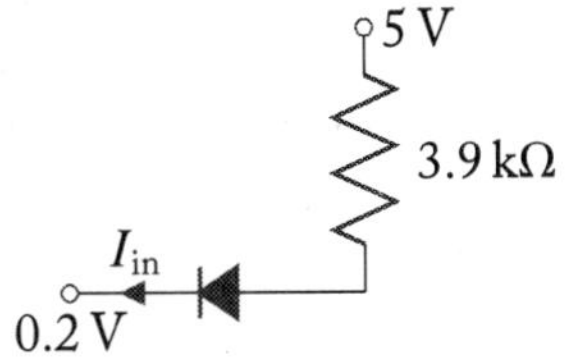

With this input voltage, the diodes, D_2, are both off, and the input current to the slave gate can be obtained as:

$$I_{in} = \frac{5 - 0.7 - 0.2}{3.9\,\text{k}} = 1.051\ \text{mA}.$$

The worst-scenario of all the current exiting only one of the slave input diodes has been considered.

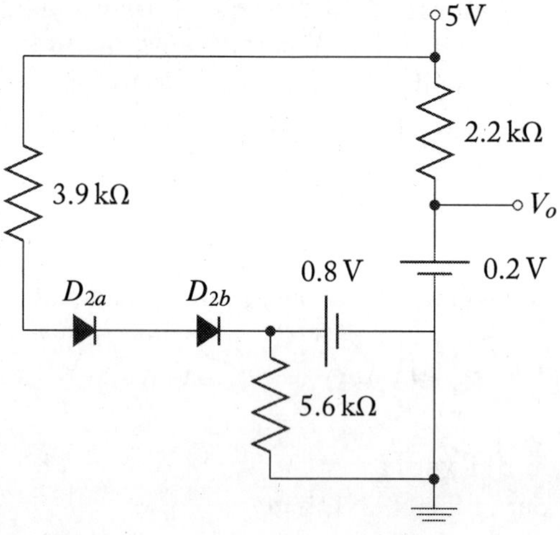

The BJT base and collector (no load) currents for the master gate with LOW output can be determined as:

$$I_B = \frac{5 - 0.8 - 2(0.7)}{3.9\text{k}} - \frac{0.8}{5.6\text{k}} = 575\,\mu\text{A}$$
$$I_{C(nl)} = \frac{5 - 0.2}{2.2\,\text{k}} = 2.182\,\text{mA}.$$

For the master gate BJT to remain in saturation it is necessary that the total collector current be less than β_F times the base current. This total collector current is the sum of the no load current and the input currents from N slave gates.

$$I_C = I_{C(nl)} + NI_{in} < \beta_F I_B,$$

or

$$2.182\,\text{mA} + N(1.051\,\text{mA}) < 50(575\,\text{mA})$$

therefore

$$N < 25.27.$$

Only integer numbers of gates can be driven. Thus,

$$\text{fan-out} = 25\text{ gates}.$$

The back-to-back arrangement in DTL logic gates of the input diodes and D_{2a} indicate that a possible replacement by an *npn* BJT might be possible. In order to accommodate multiple inputs, the transistor can be fabricated as a multiple-emitter structure as shown in Figure 3.12 (here shown as Q_1 with three emitters). This multiple-emitter transistor will not function exactly the same as the diodes in a DTL logic gate, but will switch between the various modes of BJT operation as described in earlier sections of this chapter. When diode D_{2b} of the DTL gate is also replaced by a BJT (Q_2 in Figure 3.12), the resultant circuit is a full transistor implementation of a NAND gate. The family of logic gates to which this all-transistor NAND belongs is called *transistor-transistor logic* or TTL.

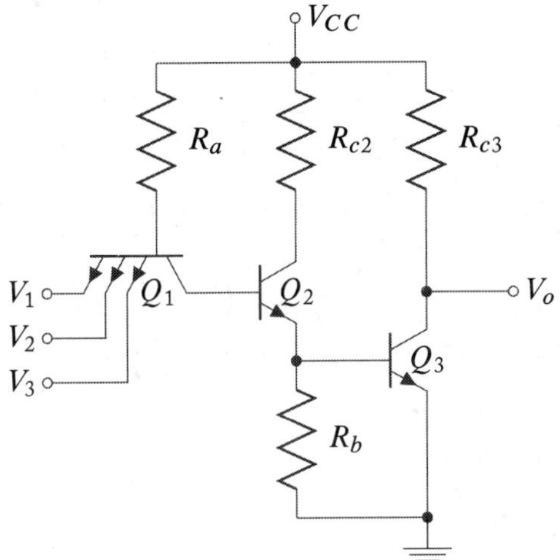

Figure 3.12: A TTL NAND gate.

3.5.3 TRANSISTOR-TRANSISTOR LOGIC GATE

The operation of this basic logic circuit can be described as:

1. For small values of *any one or more* input voltage, the input transistor, Q_1, base-emitter junction will be forward-biased. Since currents coming out of the base of Q_2 (this current is also the collector current of Q_1) are negligible, $I_{C1} < \beta_F I_{B1}$ and Q_1 is in saturation. The voltage at the base of Q_2 is given by:

$$V_{B2} = V_{in} + V_{CE(sat)}.$$

If this voltage is sufficiently small, that is if

$$V_{B2} < V_{BE(on)2} + V_{\gamma 3}$$

or equivalently,

$$V_{in} < V_{BE(on)2} + V_{\gamma 3} - V_{CE(sat)1}$$

the output transistor, Q_3, will be in the cut-off region and the output voltage,

$$V_o = V_{CC}.$$

2. Let the minimum value of *both* inputs be V_{in}. As V_{in} increases beyond the constraints of the above region of operation, i.e., when

$$V_{B2} > V_{BE(on)2} + V_{\gamma 3}$$

or equivalently when,

$$V_{in} > V_{BE(on)2} + V_{\gamma 3} - V_{BE(sat)1},$$

the transistor Q_1, with both its *p–n* junctions forward-biased, begins to have current flowing out its collector. This outward-flowing current brings Q_2 into the forward-active region, allowing current to flow through R_b. As V_{in} is increased farther, sufficient current flows through R_b to bring Q_3 into the active and finally the saturation region. Much of the action in this region is internal to the gate circuitry and not visible at the output. The output voltage transitions from HIGH to LOW over a small range on input voltages: typically of width 0.2 V or less.

3. Finally, when V_{in} increases sufficiently, that is when

$$V_{in} > V_{BE(sat)3} + V_{BE(sat)2} + V_{CE(sat)1} = 2V_{BE(sat)} + V_{CE(sat)},$$

first Q_3 and then Q_2 have their base-emitter junctions sufficiently forward-biased to enter the saturation region. As the bias on the base-emitter junction of Q_1 becomes less negative, Q_1 transitions through the saturation region (smaller inputs have the base-emitter junction more strongly forward-biased; larger inputs force the base-collector junction to be more strongly forward-biased) to the inverse-active region. It should be noted that Q_1 in the inverse-active region implies a current load on the input source. That load can certainly be present; however, Q_1 could also be in the saturation region (with the base-collector junction more strongly forward-biased) to achieve the same output voltage. Q_1 self-limits the amount of current that it draws to no more than is available from the source. The output voltage for HIGH inputs is:

$$V_o \approx V_{CE(sat)} = 0.2\,\text{V}.$$

To summarize the operation of this TTL gate:

$$\begin{array}{lcl} \text{One or both inputs LOW} & \Rightarrow & \text{HIGH output} = V_{cc} \\ \multicolumn{3}{c}{(Q_1\text{—saturation; } Q_2\,\&\,Q_3\text{—cut-off})} \\ \text{Both inputs HIGH} & \Rightarrow & \text{LOW output} = V_{CE(sat)} \\ \multicolumn{3}{c}{(Q_1\text{—inverse-active; } Q_2\,\&\,Q_3\text{—saturation})} \end{array}$$

This behavior forms a logic NAND gate.

Example 3.6
Determine the fan-out of the TTL gate of Figure 3.12 with the following circuit parameters:

$$V_{CC} = 5\,\text{V} \qquad R_a = 3.9\,\text{k}\Omega \qquad R_b = 1.0\,\text{k}\Omega$$
$$R_{c2} = 1.5\,\text{k}\Omega \quad R_{c3} = 3.9\,\text{k}\Omega$$

using Silicon BJTs with the properties

$$\beta_F = 50 \qquad \beta_R = 2.$$

Solution:

As in the DTL gate, it is necessary to find the following quantities:[9]

- the input current for a slave gate with LOW input
- the master gate output BJT base and collector currents when the master gate has a LOW output

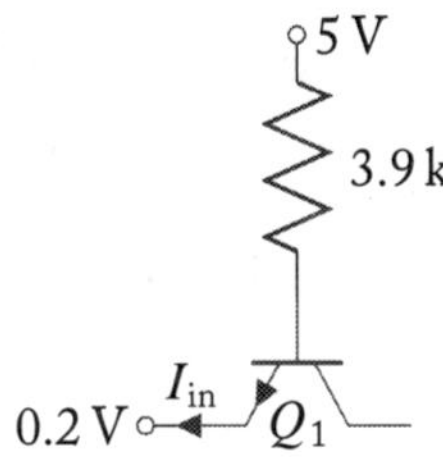

First calculate the input current for a slave gate with LOW input. The LOW output of the master gate will form the input voltage for the slave gate. Therefore,

$$V_{in} \approx V_{CE(sat)} = 0.2\,\text{V}.$$

Transistor Q_2 is off and Q_1 is in saturation, therefore:

$$I_{in} = \frac{5.0 - 0.2 - 0.8}{3.9\,\text{k}} = 1.026\text{ mA}.$$

The worst-scenario of all the current exiting only one of the emitters of the slave input transistor is considered. The master gate output BJT collector and base currents can be determined with the following process:

$$I_{B1} = \frac{5 - (0.8 + 0.8 + 0.7)}{3.9\text{k}} = 692\,\mu\text{A}.$$

[9]While a high input to a slave gate does draw current, it does not affect the proper operation of the slave gate. It does, however draw down the output voltage of the master gate. In extreme cases the input transistor of the slave gate will enter the inverse-saturation (base-collector junction more strongly forward-biased) region due to a limitation on the current available: the slave gate will self-limit the amount of current that it draws from the master gate.

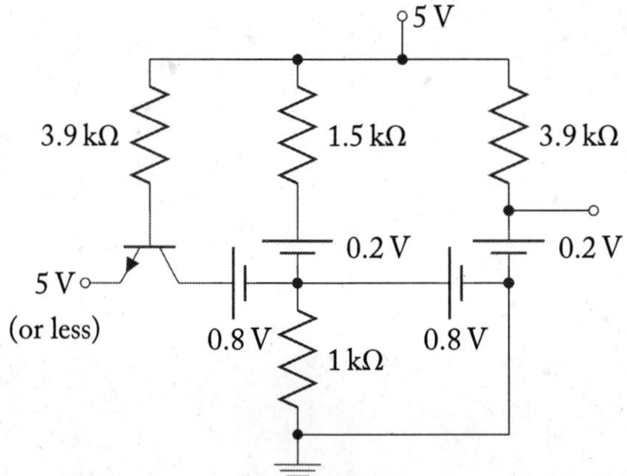

If Q_1 is in the inverse-active region, its emitter current is given by:

$$I_{E1} = I_{B1} = 2(692\,\text{mA}) = 1.384\,\text{mA}.$$

If the master gate is driven by other gates of the same type, it is unreasonable to assume that this large current is entering the emitter of Q_1 (it would draw the input voltage below zero). For fan-out calculations it is safer to assume the worst case scenario where the input current is approximately zero. Under that scenario

$$I_{B2} \approx I_{B1} = 692\,\text{mA}.$$

The collector current of Q_2 then becomes:

$$I_{C2} = \frac{5 - 0.8 - 0.2}{1.5\,\text{k}} = 2.667\,\text{mA}$$

and the base current of Q_3 is therefore:

$$I_{B3} = I_{B2} + I_{C2} - \frac{0.8}{1\,\text{k}} = 2.559\,\text{mA}.$$

The no-load collector current of Q_3 in the master gate is:

$$I_{C3(nl)} = \frac{5 - 0.2}{3.9\,\text{k}} = 1.231\,\text{mA}$$

The fan-out can now be calculated from:

$$I_{C3} < \beta_F I_{B3}$$

or

$$I_{C3(nl)} + N(I_{in}) < \beta_F I_{B3}$$

or

$$1.231\,\text{mA} + N(1.026\,\text{mA}) < 50(2.559\,\text{mA}) \Rightarrow N < 123.5.$$

The fan-out of this gate is 123 gates of similar construction.

3.5.4 EMITTER-COUPLED LOGIC GATE

Another common logic gate family is Emitter-Coupled Logic (ECL). A simple two input ECL gate is shown in Figure 3.13.

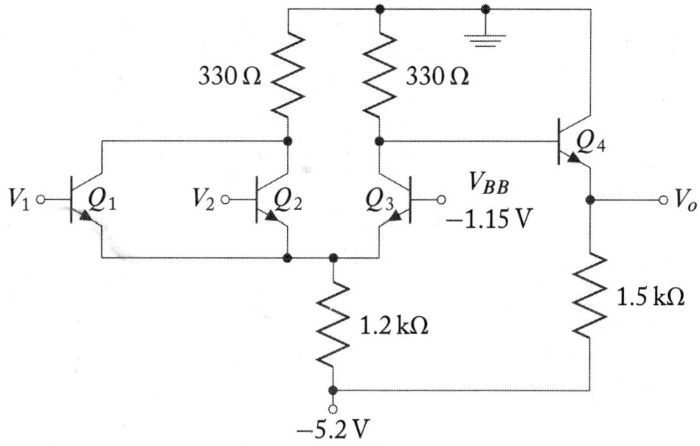

Figure 3.13: A simple ECL OR gate.

The operation of this gate can be described as (assume $\beta_F = 50$ for the calculations):

1. If V_1 and V_2 are both very negative (near −5.2 V), both Q_1 and Q_2 are in the cut-off region. Q_3 is in for forward-active region, which allows the following calculations:

$$-I_{E3} = \frac{-1.15 - 0.7 - (-5.2)}{1.2\,\text{k}} = 2.792\,\text{mA}$$
$$I_{C3} = -\alpha_F I_{E3} = \frac{50}{1+50}(2.792\,\text{mA}) = 2.737\,\text{mA}$$

Q_4 is being effectively driven by a Norton source described by the collector current of Q_3 and the 330 Ω resistor connected to its base. Replacing the Norton source by its Thévenin equivalent facilitates finding the output voltage:

$$V_{TH} = 0 - I_{C3}(2.737\,\text{mA}) = -0.903\,\text{V}$$
$$I_{B4} = \frac{V_{TH} - 0.7 - (-5.2)}{330 + (51)1.5\,\text{k}} = 46.8\,\mu\text{A}$$
$$V_o = -5.2 + 51 I_{B4} 1.5\,\text{k} = -1.62\,\text{V}.$$

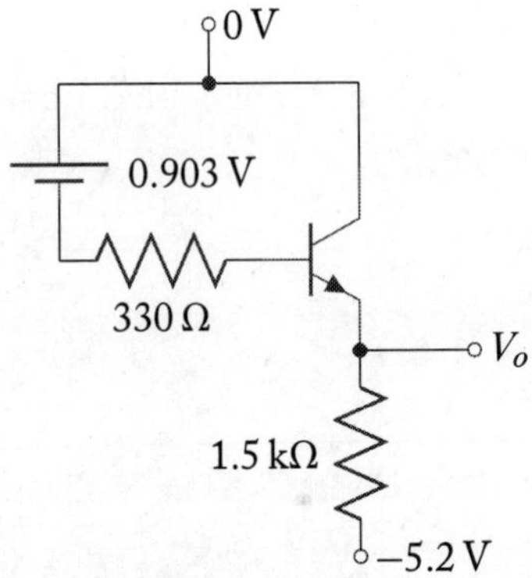

2. As either V_1 or V_2 increases to within $(V_\gamma - V_{BE(on)} \approx 0.1\,\text{V})$ of V_{BB}, the corresponding input transistor will begin to turn on. Since the current in R_b must remain relatively constant, Q_3 will supply R_b an ever decreasing portion of that current. I_{C3} will decrease and V_o will therefore increase. This process will continue until Q_3 enters the cut-off region. This linear region forms the basis for an amplifier type to be discussed in Chapter 6 (Book 2), of the series.

3. When either V_1 or V_2 becomes $(V_\gamma - V_{BE(on)} \approx 0.1\,\text{V})$ greater than V_{BB}, Q_3 will enter the cut-off region. V_o can be calculated, by considering Q_4 disconnected from Q_3, as follows:

$$330 I_B + 0.7 + 1.5\,\text{k}(51) I_B = 5.2$$
$$I_B = 58.6\,\text{mA}$$
$$V_o = 0 - 330 I_B - V_\gamma = -0.72\,\text{V}.$$

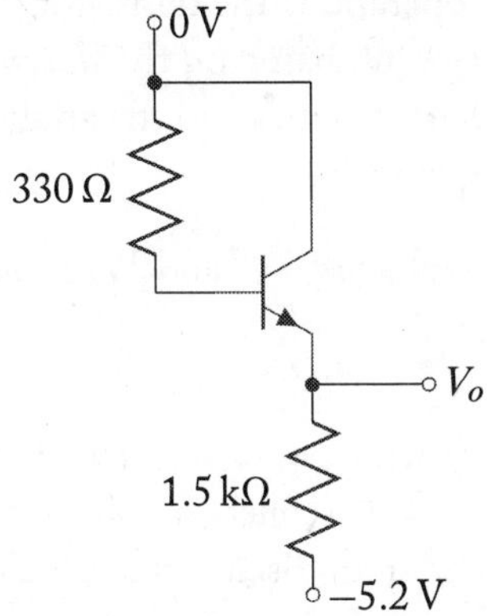

The gate operates as an OR gate with the following logic levels:

$$VH = -0.72\,\text{V}$$
$$VL = -1.62\,\text{V}.$$

The voltage transfer relationship for this gate is given in Figure 3.14.

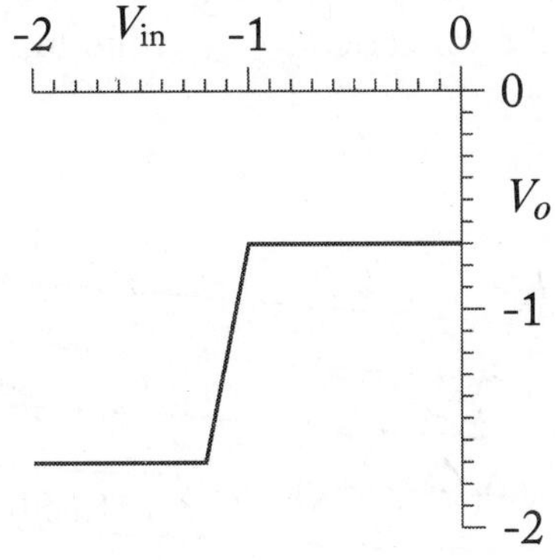

Figure 3.14: The voltage transfer relationship for an ECL OR gate.

3.6 BIASING THE BIPOLAR JUNCTION TRANSISTOR

In the previous section, transistors were used in digital (non-linear) circuits. In the digital applications considered, the transistors operated in one of two states that resulted in an output of either a logic 1 or a logic 0: the transistors transitioned between the saturation and cut-off regions. In linear applications (e.g., in the design of linear amplifiers) the transistors are *biased* to operate in only the forward-active region of operation. The transistor is biased at a quiescent operating point, which is commonly called the Q-point, based on the *dc conditions* of the transistor. The Q-point is determined by the transistor characteristics and the applied external currents and voltages. It is commonly described by four dc quantities:

- the two transistor terminal voltages, V_{BE} and V_{CE}, and
- the two transistor currents, I_B and I_C.

Once the Q-point is established, a time varying excursion of the input signal (for example, a base current) will cause an output signal (collector voltage or current) of the same waveform. The amplifier design and analysis techniques discussed in Book 2 of this series may be employed to determine the gain of the circuit.

If the output waveform is not a reproduction of the input signal (e.g., the waveform is clipped on one side), the Q-point is unsatisfactory and must be relocated. The selection of the Q-point in the forward-active region is also subject to the various transistor ratings that limit the range of useful operation. The manufacturers' specifications sheets for transistors list the maximum collector dissipation (sometimes listed as maximum power dissipation) $P_{C,\max}$, maximum collector current $I_{C,\max}$, maximum collector-emitter voltage V_{CEO}, maximum emitter-base voltage V_{EBO}, and maximum collector-base voltage V_{CB}.

A graphical representation of the operating limits of the transistor due to maximum power dissipation, $P_{C,\max}$, maximum collector current, $I_{C,\max}$, and maximum collector-emitter voltage V_{CEO} is shown in Figure 3.15.

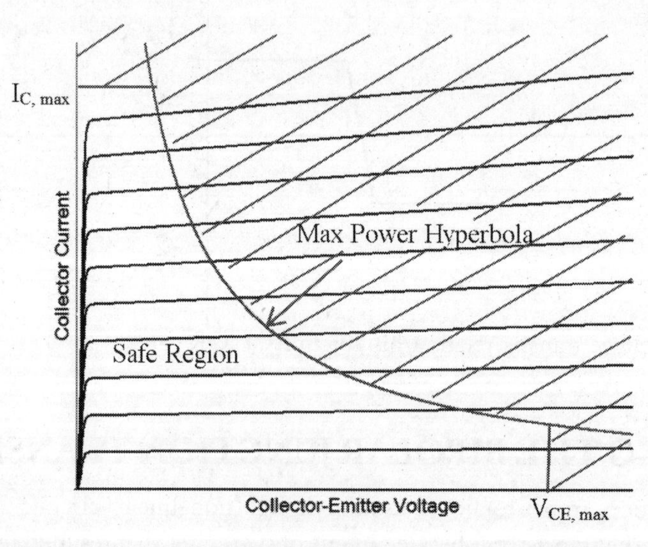

Figure 3.15: Safe operating region defined by maximum power dissipation hyperbola.

The safe operating region is not shaded and lies below the so-called maximum power dissipation hyperbola defined by $P_{C,\max}$, where

$$P_{C,\max} = I_{C,\max} V_{CEO}. \tag{3.18}$$

In an increased temperature environment, the maximum power dissipation hyperbola encroaches into the safe operating region. Therefore, the transistor should be load line and the operating point should lie well within the unshaded safe operating region to avoid the possibility of transistor thermal failure.

To help in the design and analysis of transistor biasing circuits, a few key terminal current relationships are re-iterated here. The direction of transistor terminal current flow is defined in Figure 3.1: all currents flow into the transistor. Three important transistor terminal current relationships from the previous sections of this chapter are:

$$I_C = \beta_F I_B, \qquad I_C = -\alpha_F I_E, \qquad I_B = -(I_C + I_E).$$

From these equations, the emitter current, I_E, can be related to the collector current, I_C, and the base current, I_B.

To find the relationship between the base and the emitter currents, Kirchhoff's Current Law (KCL) is applied to the BJT:

$$I_B = -(I_C + I_E)$$

rearranging the equation,

$$-I_E = I_B + I_C.$$

By substituting $I_C = \beta_F I_B$ for I_C in the previous equation,

$$-I_E = I_B(\beta_F + 1), \tag{3.19}$$

or

$$I_B = \frac{-I_E}{\beta_F + 1}. \tag{3.20}$$

Using Equation (3.4),

$$I_C = -\alpha_F I_E,$$

and

$$\beta_F = \frac{\alpha_F}{1 - \alpha_F} \Rightarrow \alpha_F = \frac{\beta_F}{\beta_F + 1},$$

the expression for the collector current is,

$$I_C = -\left(\frac{\beta_F}{\beta_F + 1}\right) I_E. \tag{3.21}$$

In real transistors, the output characteristic curves in the forward-active region slope slightly upward, increasing I_C with V_{CE} for a constant I_B. The slope of the curve is determined by the BJT Early voltage, V_A. The Early voltage is that voltage which is the point of intersection of the $I_C = 0$ line and the extended line from the characteristic curves in the forward-active region, with values typically in the 75 to 100 V range. Figure 3.16 provides a pictorial definition of the BJT Early voltage.

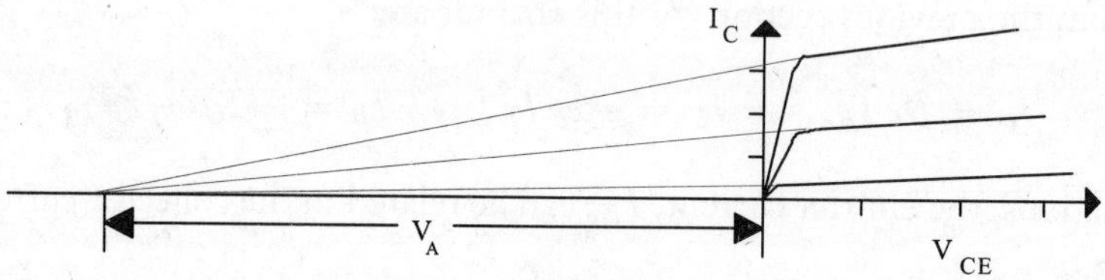

Figure 3.16: The Early voltage of the BJT.

The *.model* statement in the SPICE model of the BJT can be altered to include an Early voltage. For instance, the model statement for a BJT with $V_A = 75$ V,

.model NPXEX NPN(BF=200 VA=75) ,

is used to create the load-line analysis plot in Figure 3.17.

Using resistive networks, several common BJT biasing methods that achieve the desired Q-point will be discussed in this section. Current source biasing methods for achieving the desired Q-point will be discussed in Chapter 6 (Book 2). For linear applications, the transistor Q-point must be established in the forward-active region.

3.6.1 FIXED-BIAS CIRCUIT

One method of biasing a transistor to operate at a desired Q-point is illustrated in Figure 3.18a which shows the fixed-bias circuit (sometimes called the base-bias circuit). It is convenient to use the forward-active model of the npn BJT as shown in Figure 3.18b for analyzing bias circuits.

The collector-emitter voltage $V_{CE} = V_C - V_E$ is equal to the power supply voltage minus the voltage drop across the collector resistor R_C. That is:

$$V_{CE} = V_{CC} - I_C R_C \tag{3.22}$$

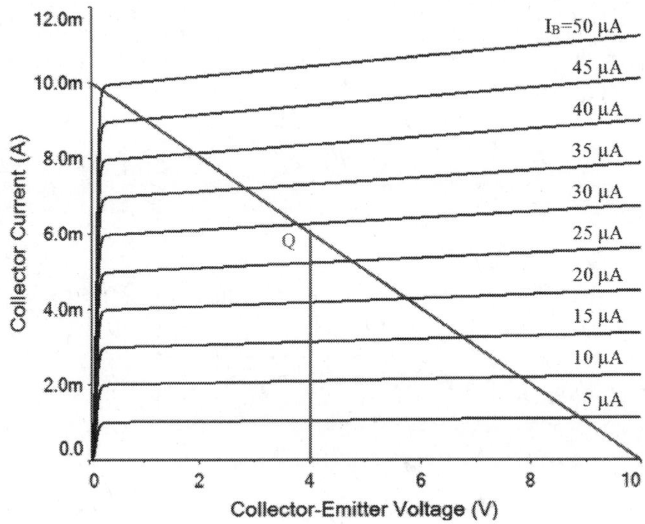

Figure 3.17: Load line analysis plot for a BJT with an Early voltage of 75 V.

where

$$\begin{aligned} V_{CE} &= \text{DC collector-emitter voltage} \\ V_{CC} &= \text{collector supply voltage} \\ I_C &= \text{DC collector current} \\ R_C &= \text{load resistance seen from the collector.} \end{aligned}$$

Kirchhoff's Voltage Law applied to the base-emitter loop, yields the following expression for the base current:

$$I_B = \frac{V_{CC} - V_{BE}}{R_B}. \tag{3.23}$$

Substituting $I_C = \beta_F I_B$ and Equation (3.23) into Equation (3.22), the collector-emitter voltage becomes

$$V_{CE} = V_{CC} - \beta_F \left(\frac{V_{CC} - V_{BE}}{R_B} \right) R_C. \tag{3.24}$$

The Q-point is defined by I_C and V_{CE} for a specified I_B.

The Q-point may also be found through graphical methods. The base-emitter voltage is determined by performing a load line analysis on the common-emitter input characteristic of the BJT. The slope of the input load line is $-1/R_B$ from Equation (3.23). The load line intersects the I_B axis at V_{CC}/R_B when $V_{BE} = 0$ and intersect the V_{BE} axis when at V_{CC}. Note that relative to the range of possible V_{BE} values of the load line, the intersection of the load line and the

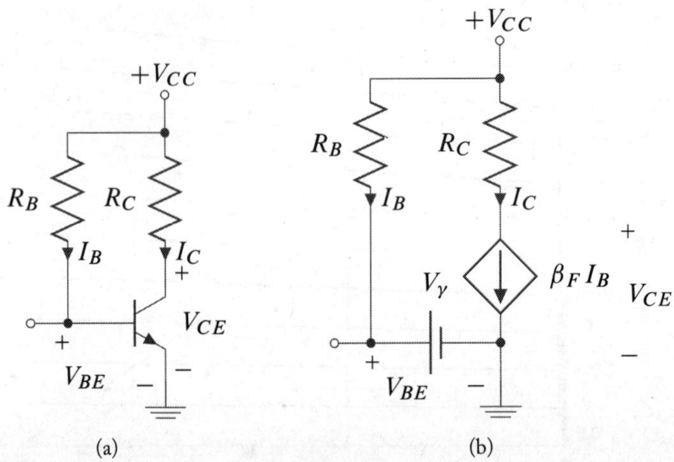

Figure 3.18: (a) The fixed-bias circuit, and (b) the equivalent circuit using the npn BJT model for the forward-active region.

characteristic curves for $V_{CE} = 0\,\text{V}$ and $V_{CE} = V_{CC}$ are approximately the same. Therefore, it is common practice to use the approximation: $V_{BE} = V_\gamma$ (therefore, $V_{BE} = 0.7\,\text{V}$ for silicon BJTs). Unless a more accurate determination of V_{BE} is required, this approximate value will be used for the remainder of this book.

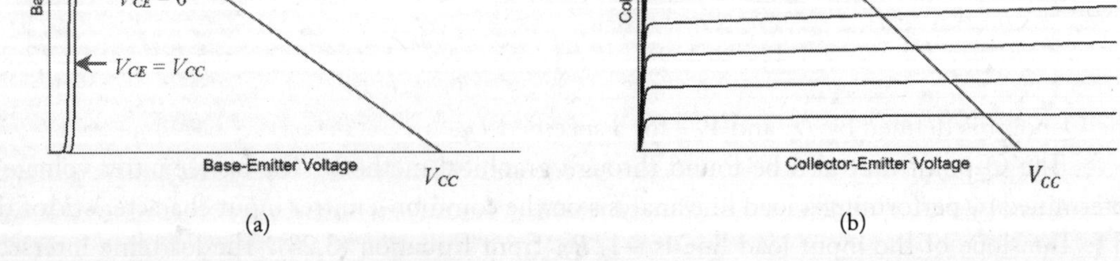

Figure 3.19: Fixed-bias transistor circuit: (a) input load line analysis, (b) output load analysis line.

A load line representing R_C is superimposed on the transistor output characteristics as shown in Figure 3.19b. As in the load line analysis performed for diode circuits in Chapter 2, the slope of the input load line is the negative of the inverse of the load resistance ($-1/R_C$) as is evident in Equation (3.22). The intersection of this load line with the common-emitter transistor output characteristic curve for the desired I_B determines the Q-point. This Q-point is identified by the resultant transistor collector current and the collector-emitter voltage.

Example 3.7

Complete the design of the fixed-bias transistor circuit shown by determining R_C and R_B for a Q-point of $I_C = 6\,\text{mA}$ and $V_{CE} = 4\,\text{V}$. The transistor forward current gain is $\beta_F = 200$ with a negligible β_R. Let $V_{BE} = 0.7\,\text{V}$.

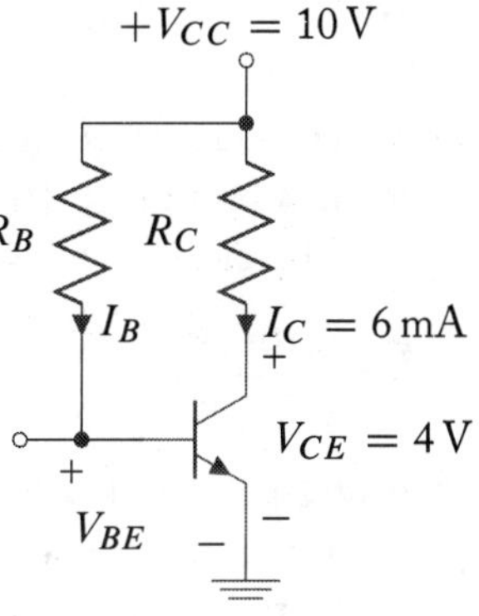

Solution #1 (Analytical):

The load resistor R_C is calculated by applying KVL to the collector-emitter loop,

$$R_C = \frac{V_{CC} - V_{CE}}{I_C} = \frac{10 - 4}{6 \times 10^{-3}} \quad \Rightarrow \quad R_C = 1\,\text{k}\Omega.$$

The base resistor R_B is calculated by applying KVL to the base-emitter loop,

$$R_B = \frac{V_{CC} - V_{BE}}{I_B} = \frac{V_{CC} - V_{BE}}{\frac{I_C}{\beta_F}} = \frac{10 - 0.7}{\frac{6\times10^{-3}}{200}} \quad \Rightarrow \quad R_B = 310\,\text{k}\Omega.$$

Solution #2 (Graphical):

In this example, the BJT was assumed to have a flat I_C vs. V_{CE} curve for each I_B: that is I_C was invariant with changes in V_{CE}. The variation in I_C with V_{CE} in the forward-active region is commonly described by a parameter called the Early voltage. No variation in I_C (a common first-order approximation) is modeled by infinite Early voltage. The BJT output characteristic curve for this example can be generated in SPICE with a net list similar to Example 3.2 Solution

#2. The BJT is designated in the SPICE net list as:

Qname collector_node base_node emitter_node model_name.

The *npn* transistor *.model* statement for this example is as follows:

.model NPXEX NPN(BF=200).

The default values of the transport saturation current (IS), Early voltage (VA), reverse beta (BR), and forward current emission coefficient (NF) were used. The default values for these parameters are:

$$\text{IS} = \text{1E-16} \qquad \text{VA} = \infty$$
$$\text{BR} = 1 \qquad \text{NF} = 1.$$

First, the load line representing R_C must be found. This is accomplished by first establishing the desired Q-point on the transistor output characteristics. The desired Q-point is located knowing V_{CEQ}, I_{CQ}, and I_{BQ}, where the additional subscript Q denote the voltages and currents at the Q-point. From the graph, $I_{BQ} = 30\,\mu$A. Therefore, the base resistance is,

$$R_B = \frac{V_{CC} - V_{BE}}{I_B} = \frac{10 - 0.7}{30 \times 10^{-6}} \quad \Rightarrow \quad R_B{=}310\,\text{k}\Omega.$$

From the Q-point, a straight line that intersects the V_{CE} axis at $V_{CE} = V_{CC}$ is drawn. The line is extended from the Q-point to intersect the I_C axis to complete the load line. These two points on the load line establishes the extremes of the transistor operation. That is, when $V_{CE} = V_{CC}$, all of the voltage from the power supply is dropped across the transistor (collector and emitter). Therefore, there is no voltage drop across the load resistor R_C corresponding to zero current flowing through the resistor ($I_C = 0$). When the load line intersects the I_C axis, $V_{CE} = 0$. Therefore, the transistor (collector-emitter) acts as if it were a short circuit and the voltage drop across the load resistor equals the power supply voltage, V_{CC}. From Equation (3.22), the current flowing when through the resistor $V_{CE} = 0$ is V_{CC}/R_C. The load line analysis of the fixed-bias circuit is shown in Figure 3.20. The slope of the line (actually, the negative inverse slope of the line) is the desired load resistor, R_C, which in this case is 1 kΩ.

In the fixed-bias circuit, the base current is essentially fixed by the value of V_{CC} and R_B. The collector current is determined by I_B and β_F. Because β_F varies widely from one transistor to another and with temperature change, the collector current also varies widely with these changes. In Section 3.7, the fixed-bias circuit will be shown to be one of the worst ways to bias a transistor from the standpoint of stability of the Q-point.

Several variations of the fixed-bias circuit are commonly used. These variations increase the stability of the Q-point, making the circuit less susceptible to variations performance due to changes in transistor parameters.

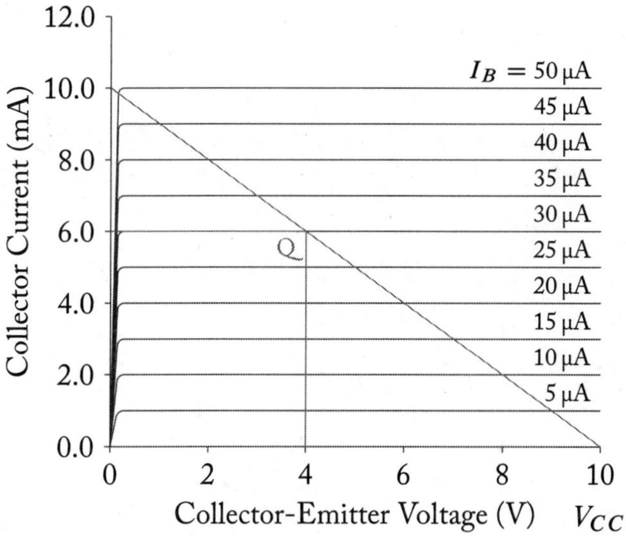

Figure 3.20: Load line analysis of Example 3.7.

In the fixed-bias circuit with emitter feedback, shown in Figure 3.21, an emitter resistor is added to the basic fixed-bias circuit.

The collector current is found by applying KVL to the base-emitter loop,

$$V_{CC} = I_B R_B + V_{BE} + I_{EE} R_E \tag{3.25}$$

where $I_{EE} = -I_E$, so that

$$I_{EE} = \frac{\beta_F + 1}{\beta_F} I_C.$$

Substituting $I_{EE} = \frac{\beta_F+1}{\beta_F} I_C$ and $I_B = \frac{I_C}{\beta_F}$ into Equation (3.25),

$$V_{CC} = V_{BE} + I_C \left[\frac{R_B}{\beta_F} + \left(\frac{\beta_F + 1}{\beta_F} \right) R_E \right]. \tag{3.26}$$

Equation (3.26) yields the collector current with respect to the power supply voltage, the transistor β_F, and the external resistors,

$$I_C = \frac{V_{CC} - V_{BE}}{\frac{R_B}{\beta_F} + \left(\frac{\beta_F+1}{\beta_F} \right) R_E}. \tag{3.27}$$

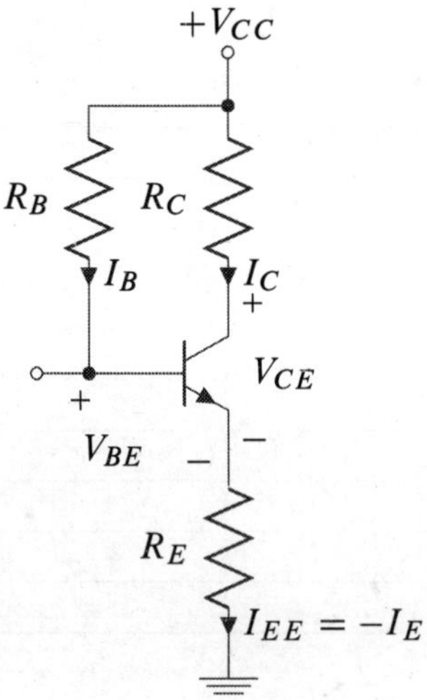

Figure 3.21: The fixed-bias circuit with emitter feedback.

The relationship between V_{CE} and I_C is found by applying KVL to the collector-emitter loop,

$$\begin{aligned} V_{CC} &= I_C R_C + V_{CE} + I_{EE} R_E \\ &= V_{CE} + I_C \left[R_C + \left(\frac{\beta_F + 1}{\beta_F} \right) R_E \right]. \end{aligned} \tag{3.28}$$

Solving Equation (3.28) to solve for V_{CE} yields,

$$\begin{aligned} V_{CE} &= V_{CC} - I_C R_C - I_{EE} R_E \\ &= V_{CC} - I_C \left(R_C + \left(\frac{\beta_F + 1}{\beta_F} \right) R_E \right). \end{aligned} \tag{3.29}$$

The slope of the load line for the fixed-bias circuit with emitter resistor on the BJT output characteristic curve is,

$$\text{Slope of the load line} = -\left[R_C + \left(\frac{\beta_F + 1}{\beta_F} \right) R_E \right]^{-1}. \tag{3.30}$$

If $\beta_F \gg 1$, then the slope of the load line is simplified as,

$$\text{Slope of the load line} \approx -\left(R_C + R_E \right)^{-1}. \tag{3.31}$$

The load line described by Equation (3.29) is superimposed on the transistor output characteristics as shown in Figure 3.22. As usual, the Q-point is determined by the collector current and the collector-emitter voltage for a given base current of the transistor.

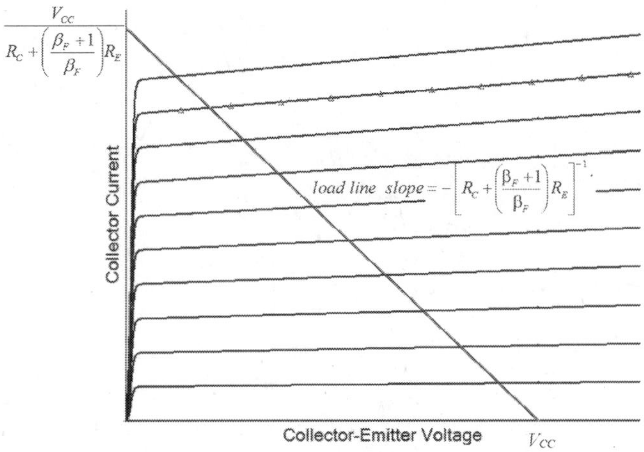

Figure 3.22: Load line analysis for the fixed-bias emitter feedback circuit.

To draw the load line on the transistor output characteristics, the two extremes of the load line are determined. From Equation (3.29), the point at which the load line intersects the I_C axis occurs when $V_{CE} = 0$ so that,

$$I_{C,V_{CE}=0} = \frac{V_{CC}}{R_C + \left(\frac{\beta_F+1}{\beta_F}\right) R_E}. \tag{3.32}$$

The other extreme of the load line occurs when the transistor collector-emitter voltage drops all of the voltage provided by the power supply. If all of the supply voltage is dropped across the transistor, the collector current must be zero ($I_C = 0$). Therefore, the intersection of the load line with the V_{CE} axis must occur at V_{CC}.

The fixed-bias circuit with collector feedback is shown in Figure 3.23. The circuit is similar to the fixed-bias circuit with the exception that the base resistor is connected directly to the collector of the BJT. The result is that the current provided by the power supply, I_{CC}, is not equal to the BJT collector current I_C. The current provided by the power supply is,

$$I_{CC} = I_C + I_B. \tag{3.33}$$

The collector current is found by applying KVL to the base-emitter loop,

$$\begin{aligned} V_{CC} &= I_{CC} R_C + I_B R_B + V_{BE} \\ &= (I_C + I_B) R_C + I_B R_B + V_{BE} \\ &= I_C \left[\left(\frac{\beta_F + 1}{\beta_F} \right) R_C + \frac{R_B}{\beta_F} \right] + V_{BE}. \end{aligned} \tag{3.34}$$

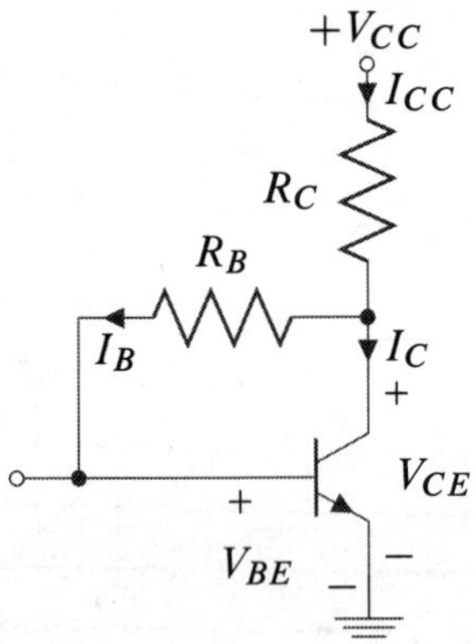

Figure 3.23: The fixed-bias circuit with collector feedback.

Solving for the collector current,

$$I_C = \left(\frac{\beta_F}{\beta_F + 1}\right) \frac{V_{CC} - V_{BE}}{R_C + \frac{R_B}{\beta_F+1}}. \tag{3.35}$$

To find the relationship between V_{CE} and I_C, the KVL equation for the collector-emitter loop which is identical to that of the fixed-bias circuit, is found,

$$\begin{aligned} V_{CC} &= I_{CC} R_C + V_{CE} \\ &= (I_C + I_B) R_C + V_{CE} \\ &= \left(\frac{\beta_F + 1}{\beta_F}\right) I_C R_C + V_{CE}, \end{aligned} \tag{3.36}$$

or,

$$V_{CE} = V_{CC} - I_C \left(\frac{\beta_F + 1}{\beta_F}\right) R_C. \tag{3.37}$$

From Equation (3.37), the slope of the load line for the fixed-bias circuit with collector feedback is,

$$\text{Slope of the load line} = -\frac{\beta_F}{(\beta_F + 1) R_C}. \tag{3.38}$$

The point at which the load line that intersects the I_C axis occurs when $V_{CE} = 0$ so from Equation (3.37),

$$I_{C,V_{CE}=0} = \frac{V_{CC}}{\left(\frac{\beta_F+1}{\beta_F}\right) R_C} = \frac{V_{CC}}{R_C}\left(\frac{\beta_F}{\beta_F + 1}\right). \tag{3.39}$$

The other extreme of the load line occurs when the transistor collector-emitter voltage drops all of the voltage provided by the power supply. If all of the supply voltage is dropped across the transistor, the collector current must be zero ($I_C = 0$). Therefore, the intersection of the load line with the V_{CE} axis must occur at V_{CC}. The BJT output characteristic with the load line for the fixed-bias circuit with collector feedback is shown in Figure 3.24.

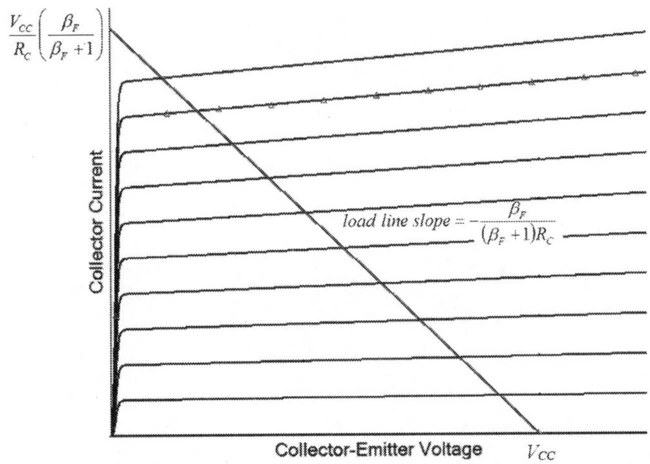

Figure 3.24: Load line analysis for the fixed-bias collector feedback circuit.

The last variation of base-biasing presented is the fixed-bias circuit with collector and emitter feedback shown in Figure 3.25.

Applying KVL to the base-emitter loop,

$$\begin{aligned} V_{CC} &= I_{CC}R_C + I_B R_B + V_{BE} + I_{EE}R_E \\ &= (I_C + I_B)\,R_C + I_B R_B + V_{BE} + I_{EE}R_E \\ &= V_{BE} + I_C\left[\left(\frac{\beta_F+1}{\beta_F}\right)(R_C + R_E) + \frac{R_B}{\beta_F}\right]. \end{aligned} \tag{3.40}$$

The collector current is found by re-arranging Equation (3.40),

$$I_C = \left(\frac{\beta_F}{\beta_F+1}\right)\frac{V_{CC} - V_{BE}}{\left(R_C + R_E + \frac{R_B}{\beta_F+1}\right)}. \tag{3.41}$$

The relationship between V_{CE} and I_C is found by applying KVL to the collector-emitter loop,

$$\begin{aligned} V_{CC} &= I_{CC}R_C + V_{CE} + I_{EE}R_E \\ &= I_C\left(\frac{\beta_F+1}{\beta_F}\right)(R_C + R_E) + V_{CE}. \end{aligned} \tag{3.42}$$

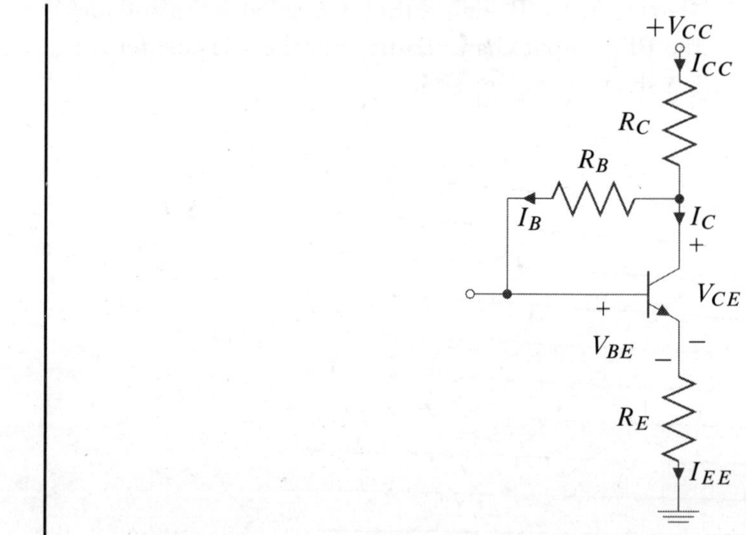

Figure 3.25: The fixed-bias circuit with collector and emitter feedback.

By rearranging Equation (3.42), the expression for V_{CE} for the fixed-bias circuit with collector and emitter feedback is found,

$$V_{CE} = V_{CC} - I_C \left(\frac{\beta_F + 1}{\beta_F} \right) (R_C + R_E). \tag{3.43}$$

The slope of the load line for the fixed-bias circuit with collector and emitter feedback is,

$$\text{Slope of the load line} = -\left(\frac{\beta_F}{\beta_F + 1} \right) \frac{1}{R_C + R_E}. \tag{3.44}$$

The point at which the load line that intersects the I_C axis occurs when $V_{CE} = 0$ so from Equation (3.43):

$$I_{C,V_{CE}=0} = \frac{V_{CC}}{\left(\frac{\beta_F+1}{\beta_F} \right) (R_C + R_E)} = \left(\frac{\beta_F}{\beta_F + 1} \right) \frac{V_{CC}}{R_C + R_E}. \tag{3.45}$$

The other extreme of the load line occurs when the transistor collector-emitter voltage drops all of the voltage provided by the power supply. If all of the supply voltage is dropped across the transistor, the collector current must be zero ($I_C = 0$). Therefore, the intersection of the load line with the V_{CE} axis must occur at V_{CC}. The BJT output characteristic with the load line for the fixed-bias circuit with collector feedback is shown in Figure 3.26.

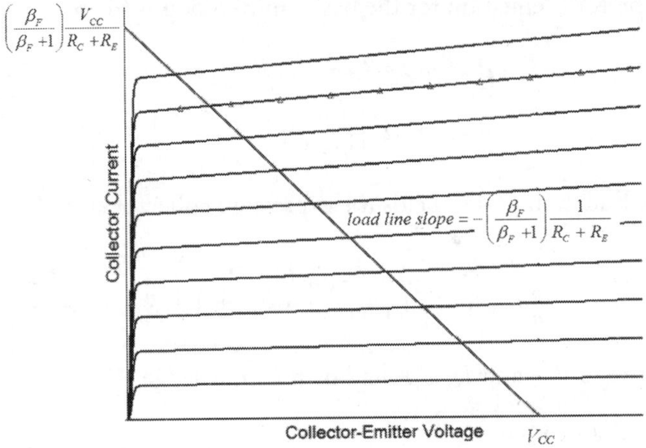

Figure 3.26: Load line analysis for the fixed-bias collector and emitter feedback transistor circuit.

3.6.2 EMITTER-BIAS CIRCUIT (WITH TWO POWER SUPPLIES)

The emitter-bias circuit is often used when two power supplies (positive and negative) are available. In this configuration, shown in Figure 3.27 the collector current can easily be made to be essentially independent of β_F, making the circuit less sensitive to variations in β_F due to temperature or transistor replacement.

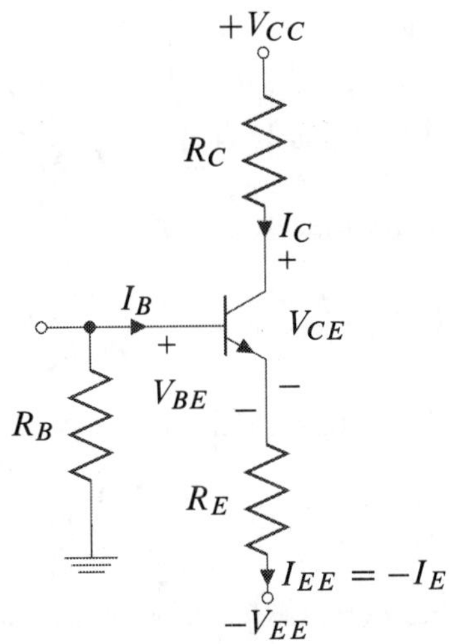

Figure 3.27: Emitter-bias of a transistor with two power supplies.

To find I_C, the KVL equation for the base-emitter loop is found,

$$\begin{aligned} V_{EE} &= V_{BE} + I_B R_B + I_{EE} R_E \\ &= V_{BE} + I_C \left[\frac{R_B}{\beta_F} + \left(\frac{\beta_F + 1}{\beta_F} \right) R_E \right]. \end{aligned} \tag{3.46}$$

Re-arranging Equation (3.45) yields the expression collector current,

$$I_C = \frac{V_{EE} - V_{BE}}{\frac{R_B}{\beta_F} + \left(\frac{\beta_F + 1}{\beta_F} \right) R_E} = \left(\frac{\beta_F}{\beta_F + 1} \right) \frac{V_{EE} - V_{BE}}{R_E + \frac{R_B}{\beta_F + 1}}. \tag{3.47}$$

The relationship between V_{CE} and I_C is found by the knowledge that,

$$V_{CE} = V_C - V_E. \tag{3.48}$$

The collector voltage with respect to ground is,

$$V_C = V_{CC} - I_C R_C. \tag{3.49}$$

The emitter voltage with respect to ground is found by applying KVL to the base-emitter loop,

$$V_E = -(V_{BE} + I_B R_B) = -\left(V_{BE} + I_C \frac{R_B}{\beta_F} \right). \tag{3.50}$$

By substituting Equations (3.49) and (3.50) into (3.48), the equation relating V_{CE} and I_C is,

$$\begin{aligned} V_{CE} &= (V_{CC} - I_C R_C) - \left[-\left(V_{BE} + \frac{R_B}{\beta_F} \right) \right] \\ &= V_{CC} + V_{BE} - I_C \left(R_C - \frac{R_B}{\beta_F} \right). \end{aligned} \tag{3.51}$$

The slope of the load line for the fixed-bias circuit with collector and emitter feedback is,

$$\text{Slope of the load line} = -\frac{\beta_F}{\beta_F R_C - R_B}. \tag{3.52}$$

The point at which the load line that intersects the I_C axis occurs when $V_{CE} = 0$ so from Equation (3.51),

$$I_{C,V_{CE}=0} = \beta_F \frac{V_{CC} + V_{BE}}{\beta_F R_C - R_B}. \tag{3.53}$$

The other extreme of the load line occurs when the transistor is cut-off ($I_C = 0$). By setting $I_C = 0$ in Equation (3.51), the intersection of the load line with the V_{CE} axis occurs at $V_{CC} + V_{BE}$. The BJT output characteristic with the load line for the fixed-bias circuit with collector feedback is shown in Figure 3.28.

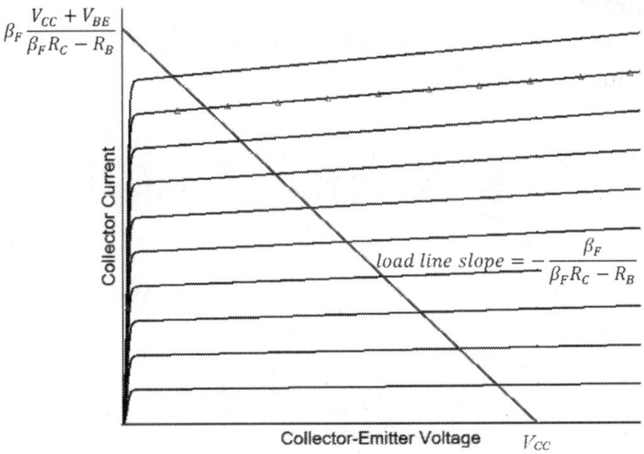

Figure 3.28: Load line analysis for the two supply emitter-bias transistor circuit.

Example 3.8

Find the operating point (V_{CE}, I_B, and I_C) for the emitter-biased circuit with two power supplies shown. Let $V_{BE} = 0.84$ V, $\beta_F = 200$, and $V_A = 75$ V.

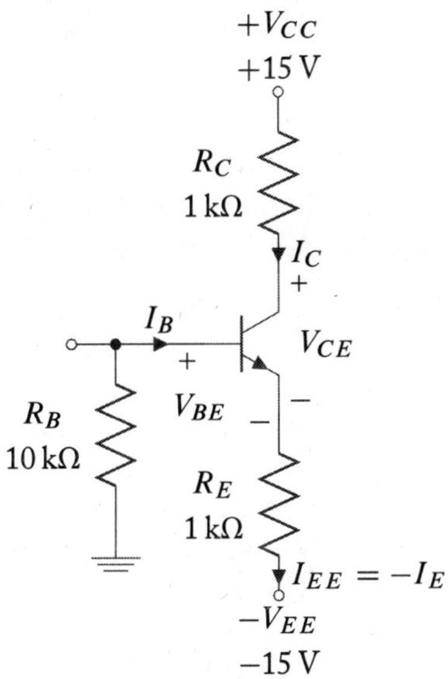

Solution #1(Analytical):

Hand analysis typically ignores the effect of the early voltage, V_A. Thus, from Equation (3.47) the collector current is

$$I_C = \frac{V_{EE} - V_{BE}}{\frac{R_B}{\beta_F} + \left(\frac{\beta_F+1}{\beta_F}\right) R_E} = \frac{15 - 0.84}{\frac{10\,\mathrm{k}}{200} + \left(\frac{201}{200}\right) 1\,\mathrm{k}}$$

$$\Rightarrow I_C = 13.4\,\mathrm{mA}.$$

The base current is then,

$$I_B = \frac{I_C}{\beta_F} = \frac{13.4 \times 10^{-3}}{200}$$

$$\Rightarrow I_B = 67\,\mu\,\mathrm{A}.$$

From Equation (3.51) the collector-emitter current is,

$$V_{CE} = V_{CC} + V_{BE} - I_C\left(R_C - \frac{R_B}{\beta_F}\right) = 15 + 0.84 - (13.4 \times 10^{-3})\left(1\,\mathrm{k} - \frac{10\,\mathrm{k}}{200}\right)$$

$$\Rightarrow V_{CE} = 3.1\,\mathrm{V}.$$

Solution # 2 (Simulation):

The circuit is typically drawn in the input schematic workspace (shown below using Multisim). A default npn transistor is used (identified as a *virtual BJT*) and the two parameters of interest: BF = 200 and VAF = 75, are altered in the default transistor parameter list. There are typically two approaches to determining the DC operating point:

1. Perform a DC operating point analysis

2. Use circuit probes in the schematic workspace

Each technique produces performs the same operation and produces the same results. Below is shown a Multisim output using circuit probes in the schematic workspace. Notice good agreement with hand calculation for the base (−2.54%) and collector (same) currents. Here V_{CE} can easily be determined to be

$$V_{CE} = V_C - V_E = 1.56 - (-1.49) = 3.03\,\mathrm{V}:$$

also in good agreement (−1.6%) with the hand results.

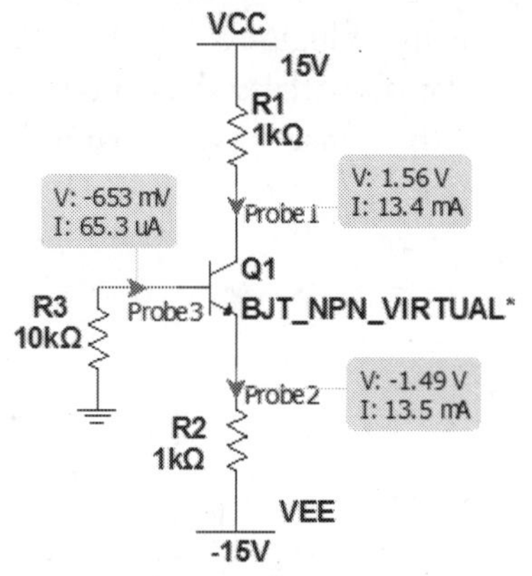

3.6.3 SELF-BIAS CIRCUIT (EMITTER-BIAS WITH ONE POWER SUPPLY)

In many instances, two power supply voltages are not available to the designer to implement the emitter-bias circuit in Section 3.6.2. In this case, a modified emitter-bias configuration called the self-bias circuit shown in Figure 3.29 is used.

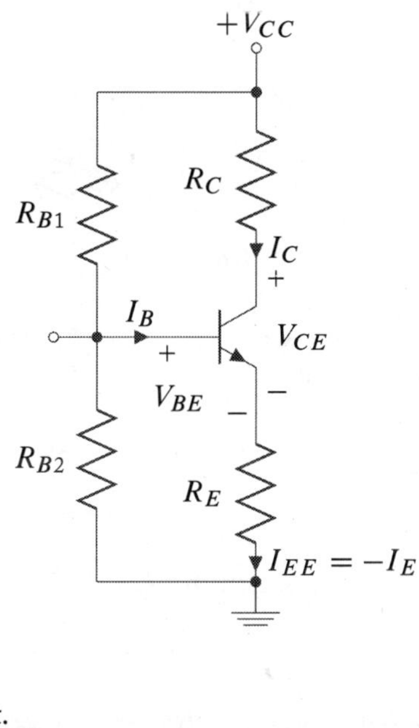

Figure 3.29: The self-bias circuit.

The analysis of the self-bias circuit of Figure 3.29 is facilitated by replacing the circuit to the left between the base and ground terminals with its Thévenin Equivalent. The Thévenin equivalent of the circuit attached to the base of the transistor is shown in Figure 3.30.

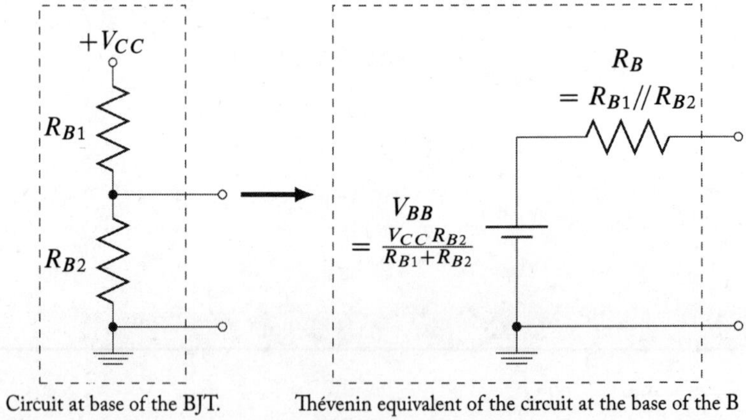

Figure 3.30: Thévenin equivalent circuit at the base of the BJT.

The self-bias circuit with the simplified base circuit using the Thévenin equivalent is shown in Figure 3.31.

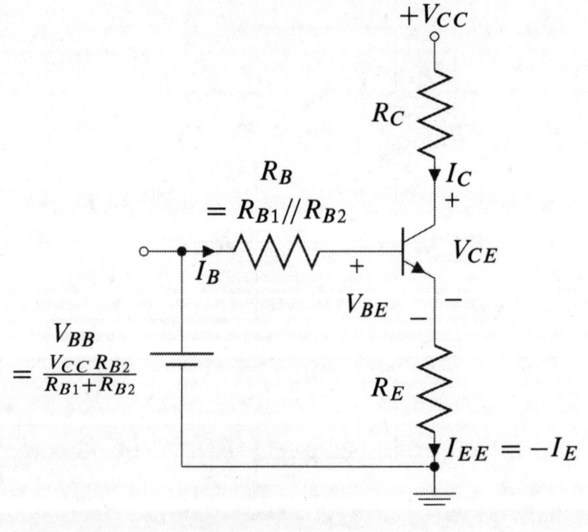

Figure 3.31: Simplification of the self-bias circuit through the use of Thévenin's theorem.

The analysis of the diagram of Figure 3.31 is similar to that of the fixed-bias circuit. Applying KVL on the base-emitter loop,

$$\begin{aligned} V_{BB} &= I_B R_B + V_{BE} + I_{EE} R_E \\ &= V_{BE} + I_C \left[\frac{R_B}{\beta_F} + \left(\frac{\beta_F + 1}{\beta_F} \right) R_E \right]. \end{aligned} \tag{3.54}$$

Solving for the collector current,

$$\begin{aligned} I_C &= \frac{V_{BB} - V_{BE}}{\frac{R_B}{\beta_F} + \left(\frac{\beta_F+1}{\beta_F}\right) R_E} = \left(\frac{\beta_F}{\beta_F + 1} \right) \frac{V_{BB} - V_{BE}}{R_E + \frac{R_B}{\beta_F+1}} \\ &= \left(\frac{\beta_F}{\beta_F + 1} \right) \frac{\left(\frac{V_{CC} R_{B2}}{R_{B1}+R_{B2}} - V_{BE} \right)}{R_E + \frac{R_{B1} R_{B2}}{(R_{B1}+R_{B2})(\beta_F+1)}}. \end{aligned} \tag{3.55}$$

The equation relating V_{CE} to I_C (Equation (3.28) repeated as Equation (3.56) below) and the load line analysis are identical to the fixed-bias circuit with emitter feedback,

$$\begin{aligned} V_{CE} &= V_{CC} - I_C R_C - I_{EE} R_E \\ &= V_{CC} - I_C \left(R_C + \left(\frac{\beta_F + 1}{\beta_F} \right) R_E \right). \end{aligned} \tag{3.56}$$

Example 3.9

For the self-bias circuit, find V_{CE} and I_B to achieve a Q-point of $I_C = 4\,\text{mA}$. Complete the design by finding R_{B1}. Let $\beta_F = 200$ and $V_\gamma = 0.7\,\text{V}$.

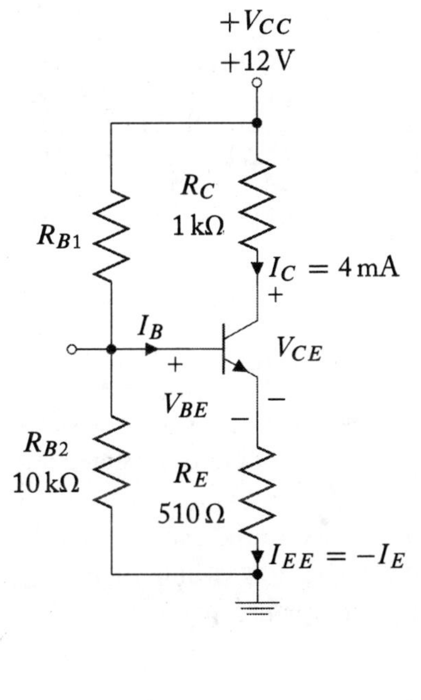

Solution

To find V_{CE} apply Equation (3.56),

$$\begin{aligned} V_{CE} &= V_{CC} - I_C R_C - I_{EE} R_E \\ &= V_{CC} - I_C \left(R_C + \left(\frac{\beta_F + 1}{\beta_F} \right) R_E \right) \\ &= 12 - (4 \times 10^{-3}) \left[1\,\text{k} + \left(\frac{201}{200} \right) 510 \right] \\ &\Rightarrow V_{CE} = 5.95\,\text{V}. \end{aligned}$$

The base current is

$$I_B = \frac{I_C}{\beta_F} = \frac{4 \times 10^{-3}}{200} = 20\,\mu\text{A}.$$

To find R_{B1}, the Thévenin equivalent circuit (Figure 3.31) at the base of the BJT is used. Applying KVL to the base-emitter loop (Equation (3.54)),

$$\begin{aligned} V_{BB} &= I_B R_B + V_{BE} + I_{EE} R_E \\ &= V_{BE} + I_C \left[\frac{R_B}{\beta_F} + \left(\frac{\beta_F + 1}{\beta_F} \right) R_E \right] \end{aligned}$$

or,

$$\frac{V_{CC} R_{B1}}{R_{B1} + R_{B2}} = V_{BE} + I_C \left[\frac{\frac{R_{B1} R_{B2}}{R_{B1} + R_{B2}}}{\beta_F} + \left(\frac{\beta_F + 1}{\beta_F} \right) R_E \right].$$

Solving for R_{B1},

$$\begin{aligned} R_{B1} &= \frac{V_{CC} R_{B2} - \left(V_{BE} + \left(\frac{\beta_F + 1}{\beta_F} \right) I_C R_E \right)}{\left(V_{BE} + \left(\frac{\beta_F + 1}{\beta_F} \right) I_C R_E \right) + I_B R_{B2}} \\ &\Rightarrow R_{B1} = 31.4\,\text{k}\Omega \approx 33\,\text{k}\Omega \quad \text{(common value)}. \end{aligned}$$

3.6.4 BIASING *PNP* TRANSISTORS

When pnp transistors are used, the polarity of all dc sources must be reversed. For operation in the forward-active region, the *pnp* transistor emitter voltage is greater than the collector voltage. The *pnp* characteristic curves are shown in Figure 3.32. Note that the polarity of the base and collector currents are negative in relation to the convention chosen in Figure 3.1. Because of the polarity reversal of the terminal voltages, it is customary to use positive values of V_{EC} and V_{EB} in the DC analysis and design of *pnp* transistor circuits.

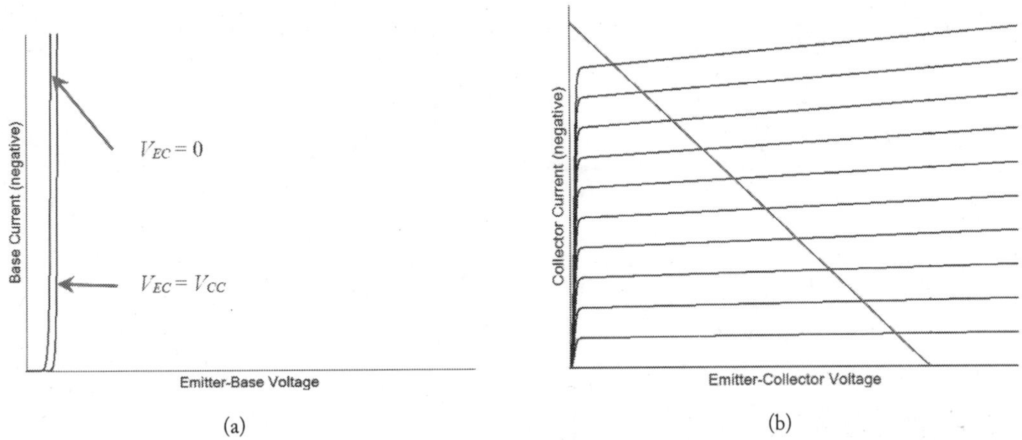

Figure 3.32: (a) input and (b) output characteristic curves for *pnp* BJT.

Consider the *pnp* BJT fixed-bias circuit with collector feedback shown in Figure 3.33. The collector power supply voltage is a negative value, $-V_{CC}$.

The KVL expression for the base-emitter loop is,

$$V_{CC} = V_{CB} - I_B R_B + I_{CC} R_C \tag{3.57}$$

where

$$I_{CC} = -I_B - I_C \quad \text{and} \quad I_C = \beta_F I_B.$$

By substituting the current relationships above into Equation (3.57),

$$V_{CC} = V_{EB} - \frac{I_C}{\beta_F} R_B - I_C \left(1 + \frac{1}{\beta_F}\right) R_C. \tag{3.58}$$

By rearranging Equation (3.58), the expression for the collector current is found,

$$I_C = \left(\frac{\beta_F}{\beta_F + 1}\right) \frac{V_{EB} - V_{CC}}{R_C + \frac{R_B}{\beta_F + 1}}. \tag{3.59}$$

To find V_{EC}, KVL is used in the collector-emitter loop,

$$V_{EC} = V_{CC} - I_{CC} R_C. \tag{3.60}$$

Substituting $I_{CC} = -I_B - I_C$ and $I_C = \beta_F I_B$ into Equation (3.60) yields the expression for V_{EC},

$$V_{EC} = V_{CC} + I_C \left(\frac{\beta_F + 1}{\beta_F}\right) R_C. \tag{3.61}$$

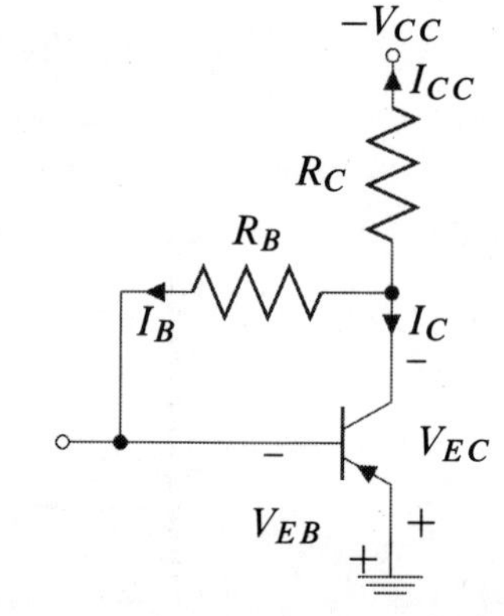

Figure 3.33: *pnp* BJT fixed-bias circuit with collector feedback.

The load line derived from Equation (3.61) is in the fourth quadrant of the BJT output characteristic curve. Since the characteristic curve is plotted with the first quadrant with the ordinate axis representing $-I_C$, the slope of the load line for the pnp BJT fixed-bias circuit with collector feedback is,

$$\text{Slope of the load line} = -\frac{\beta_F}{(\beta_F + 1)\, R_C}. \tag{3.62}$$

The point at which the load line that intersects the I_C axis occurs when $V_{EC} = 0$ so from Equation (3.61),

$$I_{C,V_{EC}=0} = -\left(\frac{\beta_F}{\beta_F + 1}\right)\frac{V_{CC}}{R_C}. \tag{3.63}$$

The other extreme of the load line occurs when the transistor is cut-off ($I_C = 0$). By setting $I_C = 0$ in Equation (3.61), the intersection of the load line with the V_{EC} axis occurs at V_{CC}. Like the npn transistor, the emitter-base voltage is assumed to be approximately equal to V_γ as is evident in the input load line analysis shown in Figure 3.34a. The BJT output characteristic with the load line for the fixed-bias circuit with collector feedback is shown in Figure 3.34b.

Similar analysis can be performed for the other biasing configurations discussed.

Example 3.10

Complete the design of the *pnp* self-bias circuit. Let $V_{EB} = 0.7\,\text{V}$, $\beta_F = 200$, and $V_{EC} = 7\,\text{V}$. What is the collector current, I_C?

Solution:

The Thévenin equivalent voltage V_{BB} and the Thévenin equivalent resistance at the base of the circuit is,

$$V_{BB} = \frac{-V_{CC} R_{B2}}{R_{B1} + R_{B2}} = -4.69\,\text{V},$$

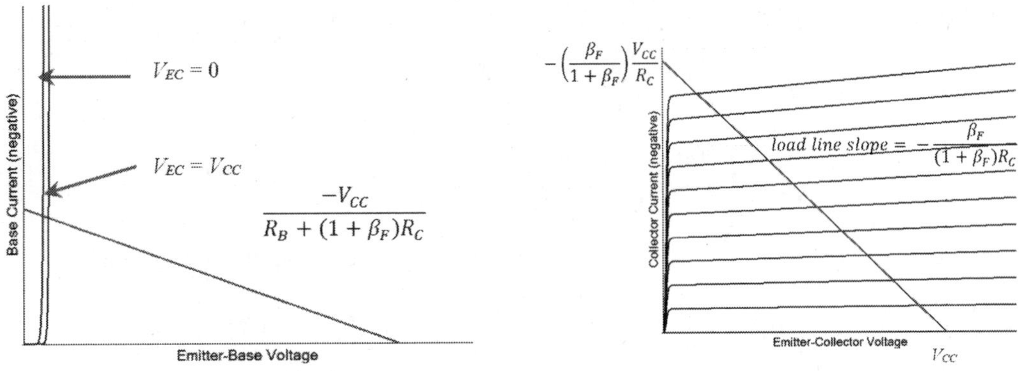

(a) (b)

Figure 3.34: Input (a) and output (b) load line analysis for the *pnp* BJT fixed-bias circuit with collector feedback.

and

$$R_B = \frac{R_{B1}R_{B2}}{R_{B1} + R_{B2}} = 6.875\,\mathrm{k\Omega}.$$

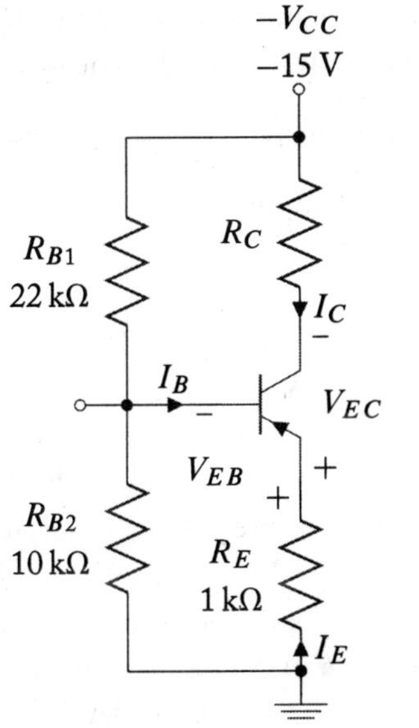

The collector current is found by applying KVL on the base-emitter loop,

$$I_C = \left(\frac{\beta_F}{\beta_F + 1}\right) \frac{V_{EB} - V_{BB}}{R_E + \frac{R_B}{\beta_F + 1}}$$
$$= \left(\frac{200}{210}\right) \frac{0.7 - 4.69}{1\,\mathrm{k} + \frac{6.875\,\mathrm{k}}{201}} \quad \Rightarrow \quad I_C = -3.84\ \mathrm{mA}.$$

The collector resistor R_C is calculated by applying KVL on the collector-emitter loop,

$$V_{CC} = I_E R_E + V_{EC} - I_C R_C = -\left(\frac{\beta_F + 1}{\beta_F}\right) I_C R_E + V_{EC} - I_C R_C.$$

Rearranging the above equation yields a solution for R_C,

$$R_C = \frac{V_{EC} - V_{CC} - \left(\frac{\beta_F+1}{\beta_F}\right) I_C R_E}{I_C} = \frac{7 - 15 - \left(\frac{201}{200}\right)\left(-3.84 \times 10^{-3}\right)(1\,\mathrm{k})}{-3.84 \times 10^{-3}}$$
$$\Rightarrow \quad R_C = 1.08\,\mathrm{k}\Omega \approx 1\,\mathrm{k}\Omega \qquad \text{(common value)}.$$

3.7 BIAS STABILITY

The quiescent operating point (Q-point) of a BJT in the forward-active region is dependent on the reverse saturation current, base-emitter voltage and the current gain of the transistor. This Q-point of a BJT circuit can change due to variations in operating temperature or parameter variations that occur when interchanging individual transistors with slightly different characteristics in the biasing circuit. A stable Q-point is desirable for the following reasons:

- Ensures that the transistor will operate over a specified range of DC voltages and currents
- The desired amplifier gain, and input and output resistances, which are all dependent on the bias condition, are achieved
- The maximum power hyperbola is not violated

As an example of the parameter variation, an input transistor characteristics in Figure 3.35 shows a decrease in V_γ (for $V_{CE} = 0$) for a rise in operating temperature from 27°C to 50°C. For this particular example, the change in V_γ is approximately –70 mV.

Figure 3.36 shows the variation in the *npn* BJT output characteristic curve with load line for a rise in operating temperature from 27°C to 50°C. At elevated temperatures, the change in the collector current, ΔI_C, increases for higher base currents. Therefore, for a constant I_B, the Q-point of the BJT is shifted by some increment of both I_C and V_{CE}. The change in the output characteristic is caused by increases in both the transistor β_F and the reverse saturation current.

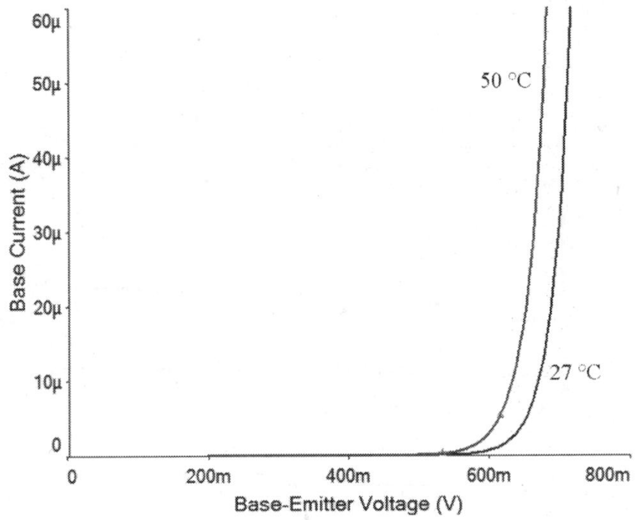

Figure 3.35: *npn* BJT input characteristic curves for $V_{CE} = 0$ for operating temperatures of 27°C and 50°C.

In order to quantitatively determine the variation in quiescent conditions, it is necessary to examine the Ebers-Moll Equations (3.3)a and (3.3)b in the forward-active region:

$$I_E = -I_{ES}\left[e^{\frac{V_{BE}}{\eta V_t}} - 1\right] + \alpha_R I_{CS}\left[e^{\frac{V_{BE}-V_{CE}}{\eta V_t}} - 1\right]$$

$$I_C = -I_{CS}\left[e^{\frac{V_{BE}-V_{CE}}{\eta V_t}} - 1\right] + \alpha_F I_{ES}\left[e^{\frac{V_{BE}}{\eta V_t}} - 1\right].$$

The collector current is found as a function of the emitter current by solving for $I_{ES}\left[e^{\frac{V_{BE}}{\eta V_t}} - 1\right]$ from Equation (3.3)a,

$$I_{ES}\left[e^{\frac{V_{BE}}{\eta V_t}} - 1\right] = -I_E + \alpha_R I_{CS}\left[e^{\frac{V_{BE}-V_{CE}}{\eta V_t}} - 1\right]. \tag{3.64}$$

Substitution of Equation (3.64) into Equation (3.3)b yields I_C as a function of I_E:

$$I_C = -I_{CS}\left[e^{\frac{V_{BE}-V_{CE}}{\eta V_t}} - 1\right] + \alpha_F\left\{-I_E + \alpha_R I_{CS}\left[e^{\frac{V_{BE}-V_{CE}}{\eta V_t}} - 1\right]\right\} \tag{3.65a}$$

$$I_C = -\alpha_F I_E + (\alpha_F\alpha_R - 1)\, I_{CS}\left[e^{\frac{V_{BE}-V_{CE}}{\eta V_t}} - 1\right]. \tag{3.65b}$$

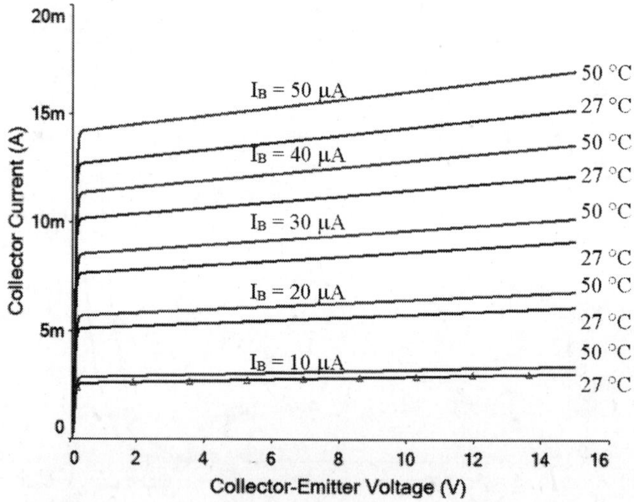

Figure 3.36: *npn* BJT output characteristic curves with a load line for operating temperatures of 27°C and 50°C.

When the BJT is in the forward-active region, $V_{BE} - V_{CE} \ll -\eta V_t$. Then,

$$I_{CS}\left[e^{\frac{V_{BE}-V_{CE}}{\eta V_t}} - 1\right] \approx -I_{CS}.$$

Therefore, the collector current for an *npn* BJT in the forward-active region is,

$$I_C = -\alpha_F I_E + (1 - \alpha_F \alpha_R) I_{CS} \tag{3.66a}$$

$$I_C = -\alpha_F I_E + I_{CO} \tag{3.66b}$$

where the reverse saturation current I_{CO} is defined as,

$$I_{CO} \approx (1 - \beta_F \beta_R) I_{CS}. \tag{3.67}$$

The collector current of Equation (3.66) can be written in terms of β_F as,

$$I_C = \beta_F I_B + (\beta_F + 1) I_{CO} \tag{3.68}$$

Equation (3.68) will be used to determine the dependence of the collector current of an *npn* BJT to changes in the reverse saturation current for different biasing arrangements.

A set of three stability factors is used to quantify the variation in the collector current with respect to the reverse saturation current, base-emitter voltage, and β_F. The stability factors are:

$$\begin{aligned} S_I &= \frac{\partial I_C}{\partial I_{CO}} \approx \frac{\Delta I_C}{\Delta I_{CO}} \\ S_V &= \frac{\partial I_C}{\partial V_{BE}} \approx \frac{\Delta I_C}{\Delta V_{BE}} \\ S_\beta &= \frac{\partial I_C}{\partial \beta_F} \approx \frac{\Delta I_C}{\Delta \beta_F}. \end{aligned} \tag{3.69}$$

The stability of the bias configuration is quantified with respect to the collector current for the following reasons:

- the collector current is dependent on the base current and the collector-emitter voltage, and determines the output signal of a BJT amplifier,
- small variations in β_F, I_{CO}, and V_{BE} can result in a large change in I_C.

The total incremental change in I_C for small changes in I_{CO}, V_{BE}, and β_F is,

$$\Delta I_{CT} = S_I \Delta I_{CO} + S_V \Delta V_{BE} + S_\beta \Delta \beta_F. \tag{3.70}$$

Equation (3.70) clearly shows that in order to keep the change in I_C small, the magnitude of the stability factors must also be kept small.

An accurate method of determining β_F at the operating point would be to include the effects of the Early voltage. Increase in accuracy is attained by using the slope of the output characteristics caused by the Early voltage, V_A, shown in Figure 3.37.

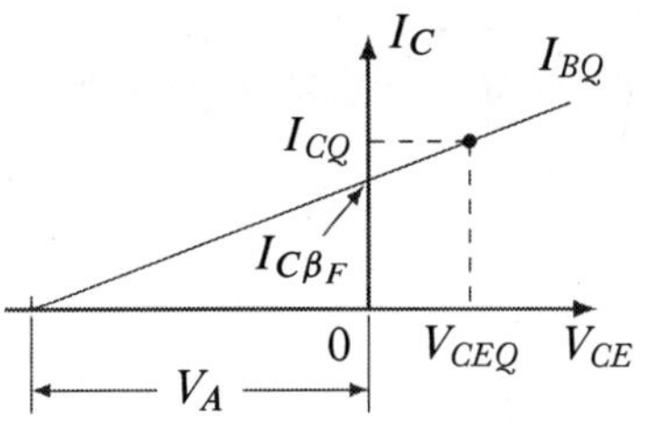

Figure 3.37: The actual Q-point of the BJT taking into account the Early voltage V_A.

The Q-point given the base current I_{BQ} is I_{CQ}. The SPICE output file provides β_F for $V_{CE} = 0$ V. The collector current using I_{BQ} and β_F provided by SPICE yields the y-axis intercept, $I_{C\beta F} = \beta_F I_{BQ}$. The collector-emitter voltage for a fixed-bias circuit is defined in Equation (3.18),

$$V_{CEQ} = V_{CC} - I_{CQ} R_C. \tag{3.71}$$

Using the expression for the equation of a line, the collector current can be written as,

$$I_{CQ} = \left(\frac{I_{C\beta_F}}{V_A}\right) V_{CEQ} + I_{C\beta_F}. \tag{3.72}$$

Solving for I_{CQ},

$$I_{CQ} = \frac{\frac{I_{C\beta_F}}{V_A} V_{CC} + I_{C\beta_F}}{1 + \frac{I_{C\beta_F}}{V_A} R_C}. \tag{3.73}$$

In order to demonstrate the effect of circuit element values on quiescent point stability, the fixed-bias and self-bias circuits are examined in the remainder of this section.

3.7.1 FIXED-BIAS CIRCUIT STABILITY

For the fixed-bias circuit shown in Figure 3.15, the reverse saturation current stability factor, S_I, is found by applying Equation (3.68) to the expression for the base current in Equation (3.19):

$$I_B = \frac{V_{CC} - V_{BE}}{R_B}. \tag{3.74}$$

Substituting Equation (3.74) into Equation (3.68) yields,

$$I_C = \beta_F \left(\frac{V_{CC} - V_{BE}}{R_B}\right) + (\beta_F + 1)\, I_{CO}. \tag{3.75}$$

The reverse saturation stability factor is,

$$\begin{aligned} S_I &= \frac{\partial I_C}{\partial I_{CO}} = \frac{\partial}{\partial I_{CO}} \left[\beta_F \left(\frac{V_{CC} - V_{BE}}{R_B}\right) + (\beta_F + 1)\, I_{CO}\right] \\ &= (\beta_F + 1)\,. \end{aligned} \tag{3.76}$$

It is apparent from Equation (3.76) that S_I is very large. Therefore, a small change in I_{CO} leads to a large change in I_C.

The base-emitter voltage stability factor, S_V, is found by using Equation (3.74),

$$\begin{aligned} S_V &= \frac{\partial I_C}{\partial V_{BE}} = \frac{\partial}{\partial V_{BE}} \left(\beta_F \frac{V_{CC} - V_{BE}}{R_B}\right) \\ &= -\frac{\beta_F}{R_B}. \end{aligned} \tag{3.77}$$

The β_F stability factor, S_β, for the fixed-bias circuit is,

$$\begin{aligned} S_\beta &= \frac{\partial I_C}{\partial \beta_F} = \frac{\partial}{\partial \beta_F} \left(\beta_F \frac{V_{CC} - V_{BE}}{R_B} + (\beta_F + 1)\, I_{CO}\right) \\ &= \frac{V_{CC} - V_{BE}}{R_B} + I_{CO} \approx \frac{I_C}{\beta_F}. \end{aligned} \tag{3.78}$$

The change in collector current due to a change in β_F is found by using S_β,

$$\begin{aligned}\Delta I_C &= S_\beta \Delta\beta_F \\ &= \frac{I_{CQ1}}{\beta_{FQ1}}\Delta\beta_F,\end{aligned} \tag{3.79a}$$

or

$$\frac{\Delta I_C}{I_{CQ1}} = \frac{\Delta\beta_F}{\beta_{FQ1}}, \tag{3.79b}$$

where I_{CQ1} and β_{FQ1} are the collector current and β_F at the known Q-point.

Equation (3.79b) states that there is a one-for one correspondence between a percentage change in β_F to I_C. Therefore, the fixed-bias circuit is not a stable biasing arrangement. The desire is to reduce the percentage change in I_C for a given change in β_F.

Example 3.11

For the fixed-bias transistor circuit shown, find the collector current at 50°C. Assume $V_{BE} \approx$ 0.7 V.

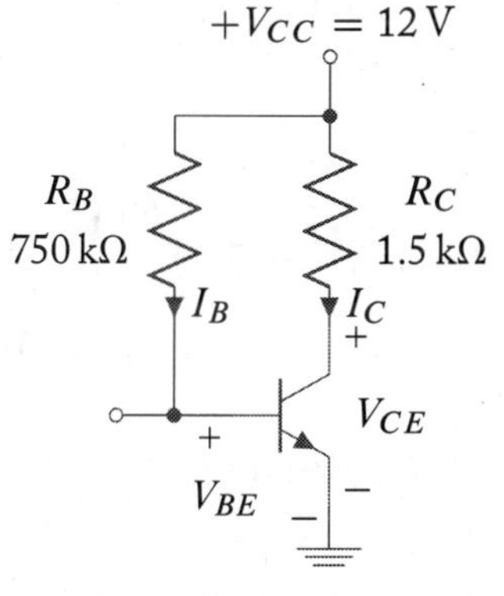

A 2N2222 BJT is used with the following parameters and governing equations:

BF ≡ $\beta_F = 255$
BR ≡ $\beta_R = 6$
VA ≡ Early Voltage = 75
IS ≡ Transport Saturation Current = $I_S = 14.34\text{E} - 15$

$$I_{CO} \approx I_S \frac{(1-\alpha_R\alpha_F)}{\alpha_R} = I_S\left(\frac{\beta_R+1}{\beta_R} - \frac{\beta_F}{\beta_F+1}\right)$$

XTB ≡ Forward and reverse β temperature coefficient = 1.5 which is typically used for small signal BJT transistors where,

$$\beta_F(T_2) = \beta_F(T_1)\left(\frac{T_2}{T_1}\right)^{XTB} \quad \text{and} \quad \beta_R(T_2) = \beta_R(T_1)\left(\frac{T_2}{T_1}\right)^{XTB}.$$

The I_S temperature coefficient, *XTI* (default value of = 3) is used in the determination of the variation of I_S with temperature:

$$I_S(T_2) = I_S(T_1)\, e^{\left(\frac{T_2}{T_1}-1\right)\left(\frac{12855}{T_2}\right)} \left(\frac{T_2}{T_1}\right)^{XTI}.$$

Solution:

The nominal value of I_B is determined in the usual fashion:

$$I_B = \frac{V_{CC} - V_{BE}}{R_B} = \frac{12 - 0.7}{750 \times 10^3} = 15\,\mu\text{A} \qquad I_C = \beta_F I_B = 255\, I_B = 3.825\,\text{mA}.$$

The parameter changes with respect to temperature must be found. The variation with I_{CO} and β_F, can be found using the governing equations given above, and the variation with V_{BE} can be found using a SPICE simulation.

Variation in β_F with temperature can be determined from the SPICE parameter *XTB*:

$$\beta_F(T_2) = \beta_F(T_1)\left(\frac{T_2}{T_1}\right)^{XTB} = 255\left(\frac{273.2+50}{273.2+27}\right)^{1.5} = 285 \quad \Rightarrow \quad \Delta\beta_F = 30.$$

Similarly, the variation in β_R is:

$$\beta_R(T_2) = \beta_R(T_1)\left(\frac{T_2}{T_1}\right)^{XTB} = 6\left(\frac{273.2+50}{273.2+27}\right)^{1.5} = 6.7 \quad \Rightarrow \quad \Delta\beta_R = 0.7.$$

Variation in I_{CO} can be found using the variation in β_F and the value of I_S at 50°C:

$$I_S(50) = 14.43 \times 10^{-15} e^{\left(\frac{323.2}{300.2}-1\right)\left(\frac{12860}{323.2}\right)} \left(\frac{323.2}{300.2}\right)^{1.5} = 379.7 \times 10^{-15}.$$

Substitution of these values into the given equation for ΔI_{CO} yields:

$$\Delta I_{CO} = 58.0 \times 10^{-15} - 2.45 \times 10^{-15} = 55.55 \times 10^{-15}.$$

To determine the variation in V_{BE}, the input characteristic curves must be generated. The curves of Figure 3.38 were generated by creating two circuits, one with $V_{CE} = 0\,\text{V}$ and the other with $V_{CE} = 12\,\text{V}$ in the same Multisim worksheet. A temperature sweep analysis was then performed with 27°C and 50°C as the only two temperatures in the sweep. From the two graphs shown in Figure 3.38a and b, ΔV_{BE} is nearly identical for $V_{CE} = 0$ and $V_{CE} = 12\,\text{V}$, where

$$\Delta V_{BE} \approx 30\,\text{mV}.$$

From Equation (3.70), the total change in collector current is,

$$\Delta I_{CT} = S_I \Delta I_{CO} + S_V \Delta V_{BE} + S_b \Delta\beta_F.$$

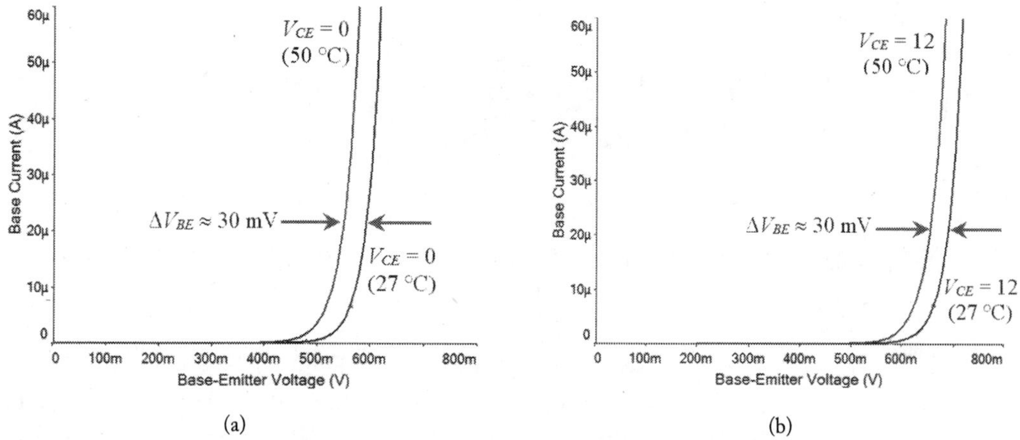

Figure 3.38: Input characteristic curves for the 2N2222 BJT for (a) $V_{CE} = 0$ and (b) $V_{CE} = 12\,\text{V}$ where $\Delta V_{BE} \approx 30\,\text{mV}$.

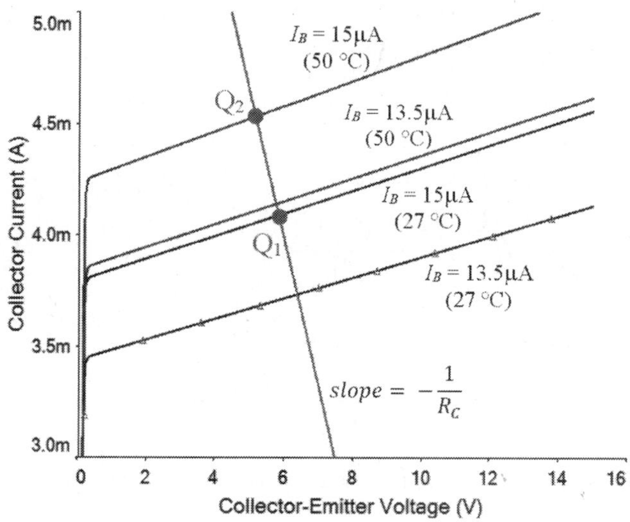

Figure 3.39: Output characteristic curves for the 2N2222 BJT.

For a fixed-bias BJT circuit,

$$S_I = \beta_F + 1 = 255 + 1 = 256$$
$$S_V = -\frac{\beta_F}{R_B} = -\frac{255}{750\,\text{k}} = -0.36 \times 10^{-3} S$$
$$S_\beta = \frac{V_{CC} - V_{BE}}{R_B} = \frac{12 - 0.7}{750\,\text{k}} \approx 15\,\mu\text{A}.$$

Therefore,

$$\Delta I_C = 256\left(55.54 \times 10^{-15}\right) + \left(-0.36 \times 10^{-3}\right)(-0.030) + \left(15 \times 10^{-6}\right)(30)$$
$$= 0.46\,\text{mA}.$$

The analytical result is in agreement with the load line result in Figure 3.39 where ΔI_C is slightly larger than 0.5 mA. The discrepancy is caused by the Early Voltage effect. Although β_F changes due to temperature, SPICE does not alter the Early voltage (VA). Therefore, β_F increases as V_{CE} is increased for a given I_B.

For $I_B = 15\,\mu$A at 27°C and $I_{C\beta f} = \beta_F I_{BQ} = 255\ (15 \times 10^{-6}) = 3.625$ mA, $I_{CQ} =$ 4.12 mA. For the most accurate determination of I_{CQ}, SPICE should be used.

3.7.2 SELF-BIAS CIRCUIT STABILITY

The bias stability factors for the self-bias circuit, shown in Figure 3.15, can be derived using the analysis for the fixed-bias circuit. S_I for the self-bias circuit is found by applying Equation (3.68) to find the base and emitter currents:

$$I_B = \frac{I_C - (\beta_F + 1)\, I_{CO}}{\beta_F} \tag{3.80a}$$

and

$$I_{EE} = -I_E = \frac{(\beta_F + 1)}{\beta_F}\,(I_C - I_{CO})\,. \tag{3.80b}$$

The collector current is found by using the base-emitter loop equation of Equation (3.18),

$$V_{BB} = V_{BE} + I_B R_B + I_{EE} R_E\,. \tag{3.81}$$

Substituting Equations (3.80a) and (3.80b) into Equation (3.81) yield the expression for I_C,

$$I_C = \frac{\beta_F\,(V_{BB} - V_{BE}) + I_{CO}\,(\beta_F + 1)\,(R_B + R_E)}{R_B + (\beta_F + 1)\,R_E}. \tag{3.82}$$

S_I is found by taking the derivative of Equation (3.82) with respect to I_{CO},

$$\begin{aligned} S_I &= \frac{\partial I_C}{\partial I_{CO}} = \frac{\partial}{\partial I_{CO}} \left[\frac{\beta_F (V_{BB} - V_{BE}) + I_{CO} (\beta_F + 1)(R_B + R_E)}{R_B + (\beta_F + 1) R_E} \right] \\ &= \frac{(\beta_F + 1)(R_B + R_E)}{R_B + (\beta_F + 1) R_E}. \end{aligned} \tag{3.83}$$

S_I can be reduced by choosing R_B as small as possible and R_E as large as possible. The range of values of R_B and R_E are limited by the required input resistance and the limitation of the power supply current output. That is, if a large R_E is chosen, the power supply must be able to provide a large collector current to the transistor to yield the required emitter voltage.

S_V for the self-biased circuit is also found by using the expression for the base-emitter loop of Equation (3.81). By re-arranging Equation (3.81), the expression for I_C (for $I_{CO} \approx 0$) is found:

$$I_C = \frac{\beta_F (V_{BB} - V_{BE})}{R_B + (\beta_F + 1) R_E}. \tag{3.84}$$

From Equation (3.84) the base-emitter voltage stability factor is,

$$\begin{aligned} S_V &= \frac{\partial I_C}{\partial V_{BE}} = \frac{\partial}{\partial V_{BE}} \left[\frac{\beta_F (V_{BB} - V_{BE})}{R_B + (\beta_F + 1) R_E} \right] \\ &= -\frac{\beta_F}{R_B + (\beta_F + 1) R_E}. \end{aligned} \tag{3.85}$$

For maximum base-emitter stability, a large R_E is desirable.

The β_F stability factor is found by taking the derivative of Equation (3.84) with respect to β_F,

$$\begin{aligned} S_\beta &= \frac{\partial I_C}{\partial \beta_F} = \frac{\partial}{\partial \beta_F} \left[\frac{\beta_F (V_{BB} - V_{BE})}{R_B + (\beta_F + 1) R_E} \right] \\ &= (V_{BB} - V_{BE}) \frac{R_B + R_E}{[R_B + (\beta_F + 1) R_E]^2}. \end{aligned} \tag{3.86}$$

Since S_β is a function of varying β_F, the expression does not indicate whether β_{FQ1} or β_{FQ2} should be used in Equation (3.86). This uncertainty is solved through the use of an alternate derivation of S_β. By taking finite differences rather than taking the derivative,

$$S_\beta \approx \frac{I_{CQ2} - I_{CQ1}}{\beta_{FQ2} - \beta_{FQ1}} = \frac{\Delta I_C}{\Delta \beta_F}. \tag{3.87}$$

From the equation describing the collector current (Equation (3.19)),

$$\frac{I_{CQ2}}{I_{CQ1}} = \frac{\beta_{FQ2}}{\beta_{FQ1}} \frac{R_B + (\beta_{FQ1} + 1) R_E}{R_B + (\beta_{FQ2} + 1) R_E}. \tag{3.88}$$

Subtracting unity from both sides of Equation (3.88),

$$\frac{I_{CQ2}}{I_{CQ1}} - 1 = \left(\frac{\beta_{FQ2}}{\beta_{FQ1}} - 1\right) \frac{R_B + (\beta_{FQ1} + 1) R_E}{R_B + (\beta_{FQ2} + 1) R_E} \tag{3.89}$$

yields an expression for ΔI_C as a function of $\Delta \beta_F$:

$$\begin{aligned} \Delta I_C &= I_{CQ2} - I_{CQ1} = I_{CQ1} \left(\frac{\beta_{FQ2} - \beta_{FQ1}}{\beta_{FQ1}}\right) \frac{R_B + R_E}{R_B + (\beta_{FQ2} + 1) R_E} \\ &= \frac{I_{CQ1}}{\beta_{FQ1}} \frac{R_B + R_E}{R_B + (\beta_{FQ2} + 1) R_E} \Delta \beta_F, \end{aligned} \tag{3.90}$$

or

$$\begin{aligned} S_\beta &= \frac{\Delta I_C}{\Delta \beta_F} = \frac{I_{CQ1}}{\beta_{FQ1}} \frac{R_B + R_E}{R_B + (\beta_{FQ2} + 1) R_E} \\ &= \frac{(V_{BB} - V_{BE}) R_B + R_E}{\left[R_B + (\beta_{FQ1} + 1) R_E\right]\left[R_B + (\beta_{FQ2} + 1) R_E\right]}. \end{aligned} \tag{3.91}$$

If a 1 percent change in I_C is desired for a 10 percent change in β_F, the ratio R_B/R_E can be determined for the self-bias circuit using Equation (3.90):

$$\begin{aligned} \frac{\Delta I_C}{I_{CQ1}} &= \frac{\Delta \beta_F}{\beta_{FQ1}} \frac{R_B + R_E}{R_B + (\beta_{FQ2} + 1) R_E} \\ 0.01 &= 0.1 \frac{R_B + R_E}{R_B + (\beta_{FQ2} + 1) R_E}. \end{aligned} \tag{3.92}$$

The ratio R_B/R_E to achieve a bias stability of 1 percent change in I_C for a 10 percent change in β_F is,

$$\frac{R_B}{R_E} \le \frac{0.1\, \beta_{F\,\text{min}} - 0.9}{0.9} = \frac{\beta_F}{9} - 1. \tag{3.93}$$

The ratio R_B/R_E in Equation (3.93) can be used as a rule of thumb when designing self-bias BJT circuits. It is evident from Equation (3.93) that increasing R_E results in increased stability. The value of R_E is determined by many factors including the amplifier gain and maximum allowable swing which will both be discussed in Chapter 5 (Book 2).

The bias stability factors for various npn BJT bias arrangements are shown in Table 3.5. In general, the addition of R_E to the bias network decreases all three stability factors with increases the overall stability of the Q-point of the BJT .

Since the β_F variation of a BJT is, in general, the dominant factor that determines Q-point stability, the an approximate relationship for the change in collector current with respect to a change in β_F can be established. The change in collector current, ΔI_C, for a change in β_F for the different bias configurations are shown in Table 3.6.

Table 3.5: Stability factors for BJT bias circuit configurations

Bias Configuration	S_I, I_C—Reverse Saturation Current Bias Stability Factor	S_V, I_C—Base-Emitter Voltage Bias Stability Factor	S_B, I_C—β Bias Stability Factor
Fixed-Bias	$\beta_F + 1$	$-\frac{\beta_F}{R_B}$	$\frac{V_{CC} - V_{BE}}{R_S} + I_{CO} \approx \frac{I_C}{\beta_F}$
Fixed Bias with Emitter Resistor	$\frac{(\beta_F+1)(R_B+R_E)}{R_B+(\beta_F+1)R_E}$	$\frac{-\beta_F}{R_B+(\beta_F+1)R_E}$	$\frac{(V_{CC}-V_{BE})(R_B+R_E)}{[R_B+(\beta_{FQ1}+1)R_E][R_B+(\beta_{FQ2}+1)R_E]}$
Fixed Bias with Collector Feedback	$\frac{(\beta_F+1)(R_B+R_E)}{R_B+(\beta_F+1)R_C}$	$\frac{-\beta_F}{R_B+(\beta_F+1)R_C}$	$\frac{(V_{CC}-V_{BE})(R_B+R_C)}{[R_B+(\beta_{FQ1}+1)R_C][R_B+(\beta_{FQ2}+1)R_C]}$
Fixed Bias with Collector and Emitter Feedback	$\frac{(\beta_F+1)(R_B+R_C+R_E)}{R_B+(\beta_F+1)(R_C+R_E)}$	$\frac{-\beta_F}{R_B+(\beta_F+1)(R_C+R_E)}$	$\frac{(V_{CC}-V_{BE})(R_B+R_C+R_E)}{[R_B+(\beta_{FQ1}+1)(R_C+R_E)][R_B+(\beta_{FQ2}+1)(R_C+R_E)]}$
Emitter Bias with Two Power Supplies	$\frac{(\beta_F+1)(R_B+R_E)}{R_B+(\beta_F+1)R_E}$	$\frac{-\beta_F}{R_B+(\beta_F+1)R_E}$	$\frac{(V_{EE}-V_{BE})(R_B+R_E)}{[R_B+(\beta_{FQ1}+1)R_E][R_B+(\beta_{FQ2}+1)R_E]}$
Self-Bias	$\frac{(\beta_F+1)(R_B+R_E)}{R_B+(\beta_F+1)R_E}$	$\frac{-\beta_F}{R_B+(\beta_F+1)R_E}$	$\frac{(V_{BB}-V_{BE})(R_B+R_E)}{[R_B+(\beta_{FQ1}+1)R_E][R_B+(\beta_{FQ2}+1)R_E]}$

Table 3.6: Change in I_C for a change in β

Bias Configuration	ΔI_C *as function of* $\Delta\beta$
Fixed-Bias	$\Delta I_C = \frac{I_{CQ1}\Delta\beta_F}{\beta_{FQ1}}$
Fixed Bias with Emitter Resistor	$\Delta I_C = \frac{I_{CQ1}\Delta\beta_F}{\beta_{FQ1}} \frac{R_B+R_E}{R_B+(\beta_{FQ2}+1)R_E}$
Fixed Bias with Collector Feedback	$\Delta I_C = \frac{I_{CQ1}\Delta\beta_F}{\beta_{FQ1}} \frac{R_B+R_E}{R_B+(\beta_{FQ2}+1)R_E}$
Fixed Bias with Collector and Emitter Feedback	$\Delta I_C = \frac{I_{CQ1}\Delta\beta_F}{\beta_{FQ1}} \frac{R_B+R_C+R_E}{R_B+(\beta_{FQ2}+1)(R_C+R_E)}$
Emitter Bias with Two Power Supplies	$\Delta I_C = \frac{I_{CQ1}\Delta\beta_F}{\beta_{FQ1}} \frac{R_B+R_E}{R_B+(\beta_{FQ2}+1)R_E}$
Self-Bias	$\Delta I_C = \frac{I_{CQ1}\Delta\beta_F}{\beta_{FQ1}} \frac{R_B+R_E}{R_B+(\beta_{FQ2}+1)R_E}$

Example 3.12

Find the change in the collector current from 27° to 50°C for the self-bias BJT circuit of Example 3.7 shown. Assume a nominal $V_{BE} = 0.7$ V. The circuit is designed for $I_B = 15\,\mu$A. Compare the result to the fixed-bias circuit in Example 3.11.

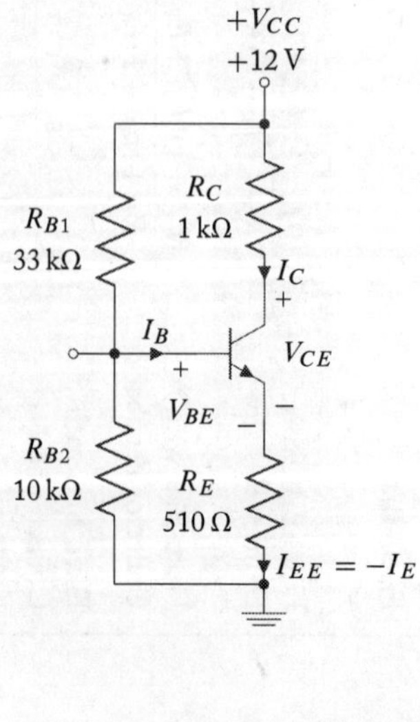

A 2N2222 BJT is used with the following parameters (identical to Example 3.11):

$$\begin{aligned} \text{BF} &\equiv \beta_F = 255 \\ \text{BR} &\equiv \beta_R = 6 \\ \text{VA} &\equiv \text{Early Voltage} = 75 \\ \text{IS} &\equiv \text{Transport Saturation Current} = I_S = 14.34\text{E} - 15 \end{aligned}$$

Solution:

The variation in I_{CO} and β_F are determined in the same manner as in Example 3.11, however the stability factors are different for the two circuit topologies.

Unlike the fixed-bias circuit where the base current was invariant with parameter changes, in the self-bias circuit, the base current decreases with increasing β_F (the base current increases with decreasing β_F). In this example, at 27°C, the base current is,

$$I_B = \frac{V_{BB} - V_{BE}}{R_B + (\beta_F + 1)\, R_C},$$

where

$$V_{BB} = \frac{V_{CC} R_{B2}}{R_{B1} + R_{B2}} = 2.79\,\text{V} \quad \text{and} \quad R_B = R_{B1}//R_{B2} = 7.67\,\text{k}\Omega.$$

The base current for 27°C ($\beta_F = 255$) and 50°C ($\beta_F = 285$) are $I_{BQ1} = 15\text{mA}$ and $I_{BQ2} = 13.6\,\mu\text{A}$, respectively.

From Equation (3.70), the total change in collector current is,

$$\Delta I_{CT} = S_I \Delta I_{CO} + S_V \Delta V_{BE} + S_b \Delta \beta_F.$$

For a self-bias BJT circuit,

$$\begin{aligned} S_I &= \frac{(\beta_F + 1)(R_B + R_E)}{R_B + (\beta_F + 1)\, R_E} = \frac{256\,(7.67\,\text{k} + 510)}{7.67\,\text{k} + 256\,(510)} = 15.2 \\ S_V &= -\frac{\beta_F}{R_B + (\beta_F + 1)\, R_E} = -\frac{255}{7.67\,\text{k} + 256\,(510)} = -1.85\text{ mS} \\ S_\beta &= \frac{(V_{BB} - V_{BE})(R_B + R_E)}{\left[R_B + (\beta_{FQ1} + 1)\, R_E\right]\left[R_B + (\beta_{FQ2} + 1)\, R_E\right]} \\ &= \frac{(2.79 - 0.7)\,(7.67\,\text{k} + 510)}{[7.67\,\text{k} + 256\,(510)]\,[7.67\,\text{k} + 286\,(510)]} = 0.81\ \mu\text{A}. \end{aligned}$$

Therefore,

$$\begin{aligned} \Delta I_C &= 15.2\left(55.54 \times 10^{-15}\right) + \left(-1.85 \times 10^{-3}\right)(-0.030) + \left(0.81 \times 10^{-6}\right)(30) \\ &= 79.8\ \mu\text{A}. \end{aligned}$$

The analytical result is in general agreement with the load line result in Figure 3.40 where ΔI_C is slightly larger than 0.06 mA. Note that the collector current change due ΔI_{CO} is negligible, and in most cases, can be ignored.

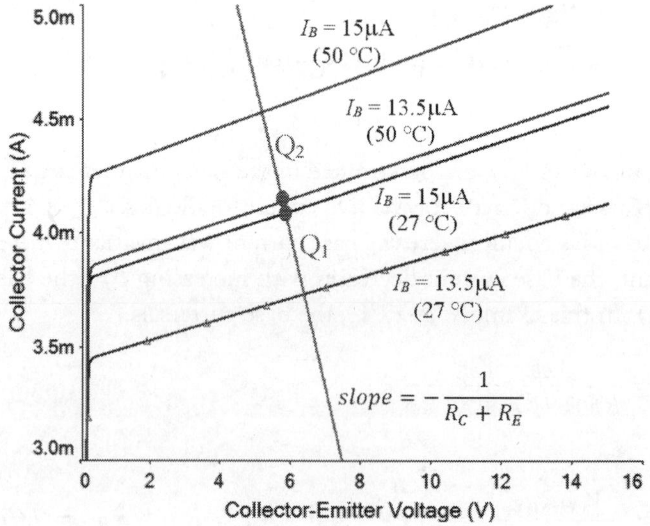

Figure 3.40: Output characteristic curves for the 2N2222 BJT with load line.

The calculated variation in collector current is 79.8 μA for the self-bias circuit as opposed to 460 μA for the fixed-bias circuit. Therefore, the self-bias circuit has better bias stability than the fixed-bias arrangement.

Some SPICE software packages allow for worst-case or Monte-Carlo analysis of the circuit with variations in operating parameters. These packages are useful when analyzing bias stability when individual BJTs are replaced in a bias circuit without varying the operating temperature. For instance, in PSpice, the .WCASE command is used for worst-case analysis: Multisim has a worst case analysis available under the simulation tab.

For instance, if β_F varies for an *npn* BJT from 200 to 220, the simple .model statement for PSpice is:

.model *device_name* NPN(BF=200 DEV 20)

where DEV is the deviation from the nominal (BF=200).

To perform a worst-case analysis to determine the impact of the change in β_F on the collector current on the transistor Q1, the netlist includes the following statement:

.WCASE DC IC(Q1) YMAX HI VARY DEV DEVICES Q

This statement performs a worst-case analysis to find the greatest difference from the nominal (YMAX), with BF = Nominal BF + 20 (HI), varying the parameter denoted by DEV (VARY and DEV), on the transistor only (Q). For BF = Nominal BF – 20, replace HI with LO.

A sample netlist in PSpice for finding the BJT output characteristic curve of an npn BJT is shown below:

```
Worst-case Analysis with PSpice
VCE   1  0   15V
IB    0  2   60u
Q1    1  2   0   NPNBJT
.model NPNBJT NPN(BF=200 DEV 20 VA=75)
.DC DEC VCE 0.001 15 10 IB 0 50u 10u
.WCASE DC IC(Q1) YMAX HI VARY DEV DEVICES Q
.PROBE
.END
```

The resulting BJT output characteristic curve is shown in Figure 3.41.

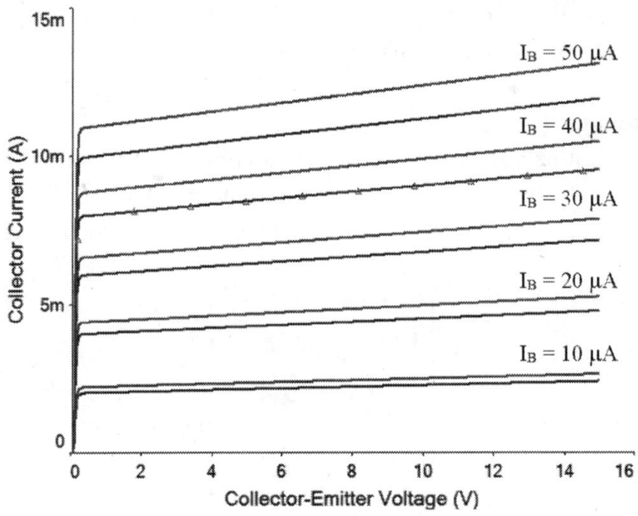

Figure 3.41: Output characteristics for *npn* BJT using netlist for varying BF.

3.8 CONCLUDING REMARKS

The Bipolar Junction Transistor has been described in this chapter as a commonly used semiconductor device with four basic regions of operation: the saturation, forward-active, inverse-active, and cut-off regions. The BJT operation into these regions is controlled by the bias conditions on the transistor base-emitter and base-collector junctions. Typical applications lead to the use of the transistor base current and the collector-emitter voltage as more accessible quantities for region verification.

The Ebers-Moll transistor model was presented to quantify the current-voltage relationships of the BJT in all four regions of operation. As with the semiconductor diode, adequate representation of BJT performance can be obtained with piece-wise linear approximations of the transistor characteristics. A set of simple linear models, one for each region of operation, was developed using transistor characteristics as expressed by the Ebers-Moll model and its corresponding set of equations.

BJT logic gate applications have provided a good example of transistor circuitry using a variety of regions of operation. The two-state output necessary for binary gates is often achieved by the output BJT transitioning between the saturation or cut-off region. In addition, Transistor-Transistor Logic Gates, provide an example of inverse-active region operation of the input BJT. Linear BJT applications utilize the forward-active region. Here, the operation of the transistor is nearly linear about a quiescent operating point (Q-point) achieved with external bias circuitry. Several biasing circuits were developed, and the bias stability due to variations in transistor parameters was quantified. Based on the foundations developed in this chapter, additional linear and non-linear applications will be examined in later chapters.

Summary Design Example

Bipolar Junction Transistors can often be used as a controlled current shunt to reduce power consumption in sensitive or expensive electronic components. A Zener diode voltage regulator is one device that can benefit from a shunting BJT. The basic topology of such a voltage regulator is shown below.

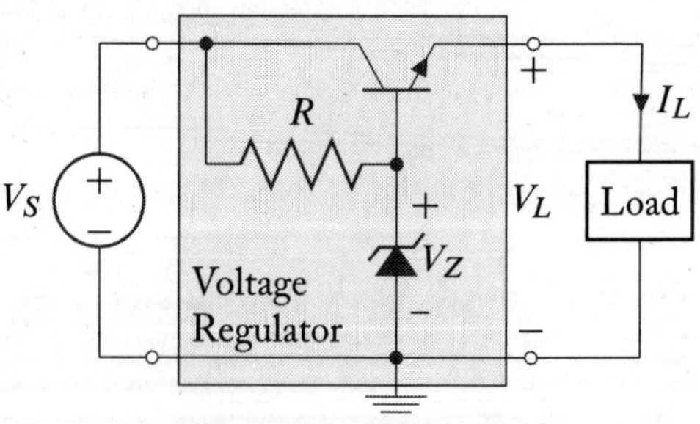

Without the use of a shunting BJT, the regulator resistor must be capable of carrying currents in excess of the of the load current. Similarly, the Zener diode also carries very large currents. The shunting BJT in this design topology reduces the current through both these devices, thereby reducing power consumption and component cost.

Design a Zener voltage regulator with BJT shunt to meet the following design requirements:

- Regulated load voltage, $V_L = 10\,\text{V}$
- Load current, $0\,\text{A} \leq I_L \leq 5\,\text{A}$
- Source voltage, $12\,\text{V} \leq V_S \leq 15\,\text{V}$

Determine the appropriate ratings for all components. Assume the minimum Zener current for proper regulation is 2 mA and a power BJT with typical forward current gain, $\beta_F = 50$.

Solution:

In order to maintain 10 V at the load, the diode Zener voltage must be:

$$V_Z = V_L + V_\gamma = 10.7\,\text{V}.$$

The remainder of the design process for this circuit topology is similar to that of a simple Zener diode regulator except the current through the resistor, R, is the Zener current plus the base current of the BJT:

$$I_R = I_z + \frac{I_L}{\beta_F + 1}.$$

The minimum design value for the resistor current, I_R, that will ensure regulation under all load and source variation is given by:

$$I_{R(\text{min})} = I_{z(\text{min})} + \frac{I_{L(\text{max})}}{\beta_F + 1} = 2\,\text{mA} + \frac{5\,\text{A}}{51} = 100\,\text{mA}.$$

This minimum resistor current must occur with minimum source voltage, V_s. The resulting maximum value for the resistor is therefore given by:

$$R = \frac{12\,\text{V} - 10.7\,\text{V}}{100\,\text{mA}} = 13\,\Omega.$$

With this choice of resistor value, the maximum power dissipated in the resistor is given by:

$$P_{R(\text{max})} = \frac{(15 - 10.7)^2}{13} = 1.423\,\text{W}.$$

The maximum current through the Zener diode is given by the maximum resistor current less the minimum BJT base current:

$$I_{Z(\text{max})} = \frac{15 - 10.7}{13} - \frac{I_{L(\text{min})}}{\beta_F + 1} = 0.331\,\text{A}.$$

The maximum power dissipated by the Zener diode is 3.542 W. The BJT must be capable of an emitter current equal to the maximum load current (5 A) at a maximum V_{CE} (5 V). Thus, the BJTs must be able to dissipate at least 25 W.

Total power consumption is given by:

$$P_T = V_S \left\{ \frac{V_S - 10.7}{13} + \frac{\beta_F}{\beta_F + 1} I_L \right\}.$$

Thus, total power consumption ranges between 1.2 W and 78.5 W depending on source and load conditions.

Summary of Required Components:

Zener diode:	$V_Z = 10.7\,\text{V}$,	$I_{Z(\max)} > .331\,\text{A}$,	$P_{(\max)} > 3.542\,\text{W}$
Resistor:	$R = 13\,\Omega$,		$P_{(\max)} > 1.432\,\text{W}$
BJT:	$\beta_F = 50$,	$I_{C(\max)} = 4.902\,\text{A}$,	$P_{(\max)} > 25\,\text{W}$

Comparison to Simple Zener Regulator:

A simple Zener diode regulator (without a shunting BJT) requires the following components:

Zener diode:	$V_Z = 10\,\text{V}$,	$I_{Z(\max)} > 12.6\,\text{A}$,	$P_{(\max)} > 126\,\text{W}$
Resistor:	$R = 0.397\,\Omega$,		$P_{(\max)} > 63\,\text{W}$

This simple design results in a power consumption of between 60.5 W ($V_S = 12\,\text{V}$) and 190 W ($V_S = 15\,\text{V}$) independent of the load current (maximum load power = 50 W). The difference in component specifications is striking: the BJT shunt provides for a more efficient and cost effective solution to voltage regulation in this case.

3.9 PROBLEMS

3.1. A Silicon *npn* BJT is described by the following parameters (remember $\eta = 1$ for Silicon BJTs):

$$I_S = 1\,\text{fA}$$
$$\alpha_F = 0.992$$
$$\alpha_R = 0.94$$

The BJT is operating at room temperature and its junctions are biased so that:

$$V_{BC} = -1.2\,\text{V}$$
$$V_{BE} = 0.6\,\text{V}$$

(a) Determine the base, collector, and emitter currents.

(b) The junction biasing is changed so that:

$$V_{BC} = 0.6\,\text{V}$$
$$V_{BE} = -1.2\,\text{V}.$$

Repeat part a).

3.2. A Silicon *pnp* BJT is described by the following parameters (remember $\eta = 1$ for Silicon BJTs):

$$I_S = 1\,\text{fA}$$
$$\alpha_F = 0.995$$
$$\alpha_R = 0.91$$

The BJT is operating at room temperature and its junctions are biased so that:

$$V_{BC} = 0.4\,\text{V}$$
$$V_{BE} = 0.6\,\text{V}$$

Determine the base, collector, and emitter currents.

3.3. A typical 2N4401 npn BJT is described by the following parameters:

$$I_S = 26\,\text{fA}$$
$$\alpha_F = 0.994$$
$$\alpha_R = 0.75$$

(a) Generate the output characteristic curve using PSpice. Increment the base current from 0 to 100 μA in 10 μA increments.

(b) Generate the input characteristic curve using PSpice.

3.4. For the given circuit, determine the following transistor currents and voltages using load line analysis:

- the collector current
- the base current
- the base-emitter voltage
- the collector-emitter voltage

Assume the transistor is a 2N2222A npn BJT as is described in Figure 3.2.

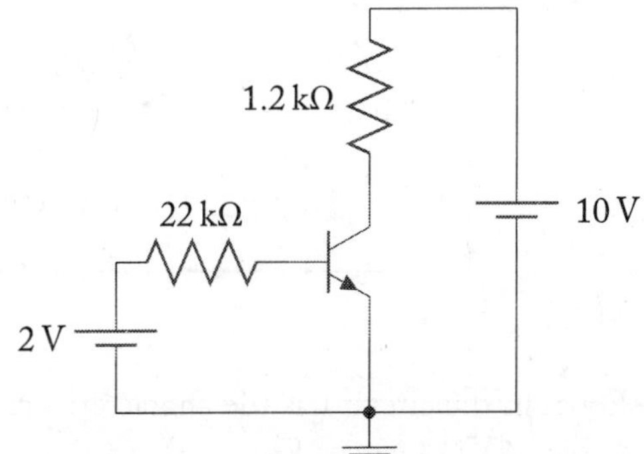

3.5. For the circuit shown, find the transistor currents and voltages using load line analysis:

- Collector current
- Base current
- Base-emitter voltage
- Collector-emitter voltage

Assume that the transistor is a 2N2222A npn BJT as shown in Figure 3.2.

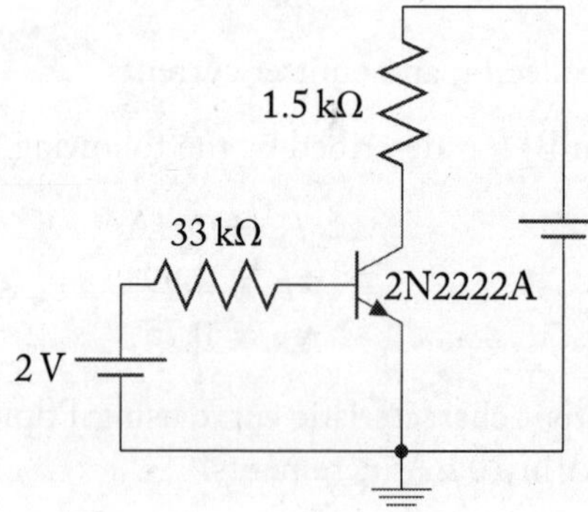

3.6. For the given circuit, determine the following transistor currents and voltages using load line analysis:

- the collector current
- the emitter current
- the base-emitter voltage
- the base-collector voltage

Assume the transistor is a 2N2222A npn BJT as is described in Figure 3.2.

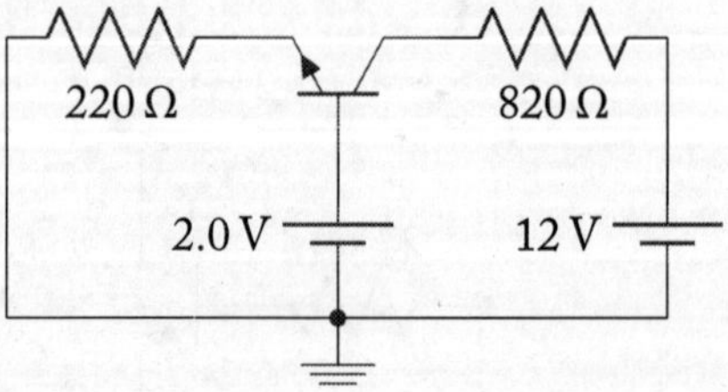

3.7. The transistor shown in the circuit has the characteristics given in Figure 3.2. Plot over the range, $0\,V \le V_i \le 4\,V$, V_o versus V_i.

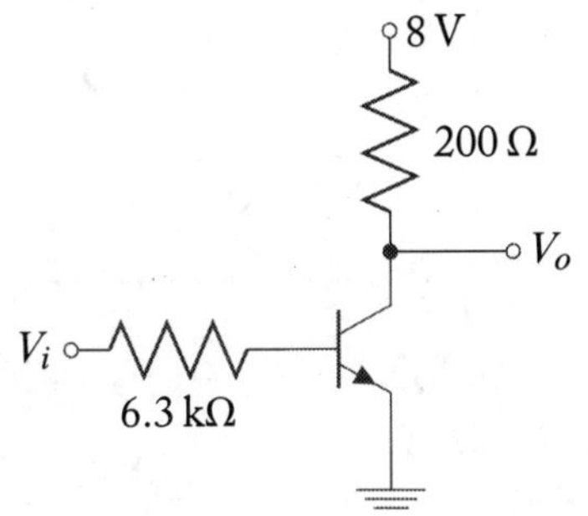

3.8. For the circuit shown, draw the transfer curve, V_O vs. V_I, for:

$$-15\,\text{V} \leq V_I \leq 15\,\text{V} \quad \text{and} \quad \beta_F = 180.$$

Confirm the transfer curve using SPICE.

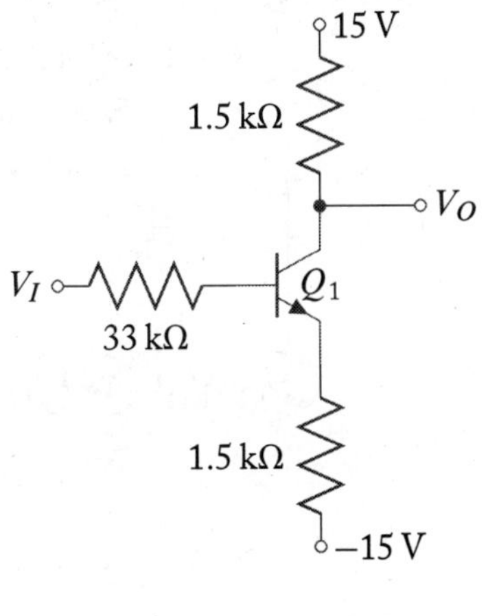

3.9. For the circuit shown,

(a) Draw the transfer curve, V_O vs. V_I, for

$$-9\,\text{V} \leq V_I \leq 9\,\text{V}.$$

(b) Determine the value(s) of V_I that will saturate the transistor.

(c) Confirm the transfer curve using SPICE.

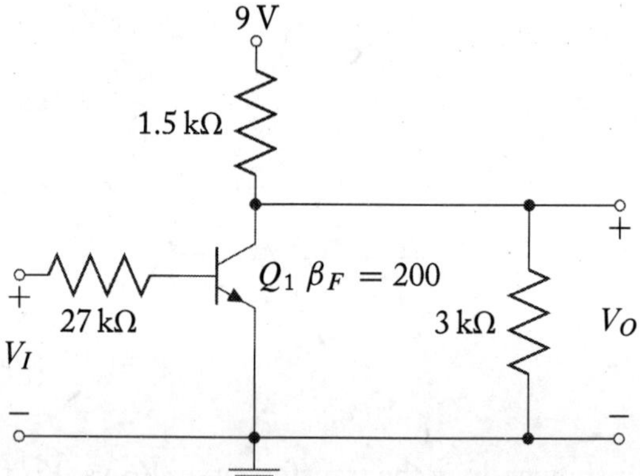

3.10. Calculate the collector and base currents in the Silicon transistor shown. Assume $\beta_F = 75$. Hint: make a Thévenin equivalent of the circuit connected to the base of the BJT.

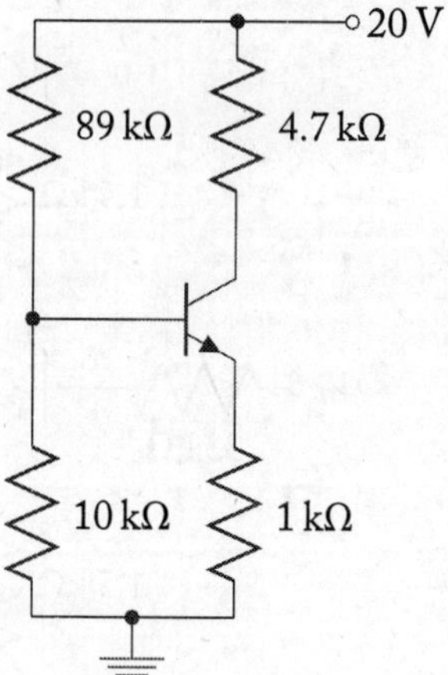

3.11. For the circuit shown, find the transistor currents and voltages using load line analysis:

- Collector current
- Base current
- Base-emitter voltage
- Collector-emitter voltage

Assume that the transistor is a 2N3906 pnp BJT. Use the V-I characteristics found in the Appendix.

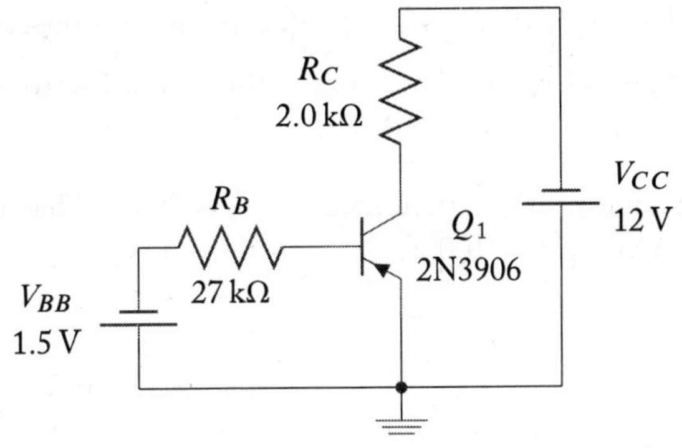

3.12. Determine the maximum value of the resistor R_b for which the transistor remains in saturation. Assume a Silicon BJT with $\beta_F = 150$.

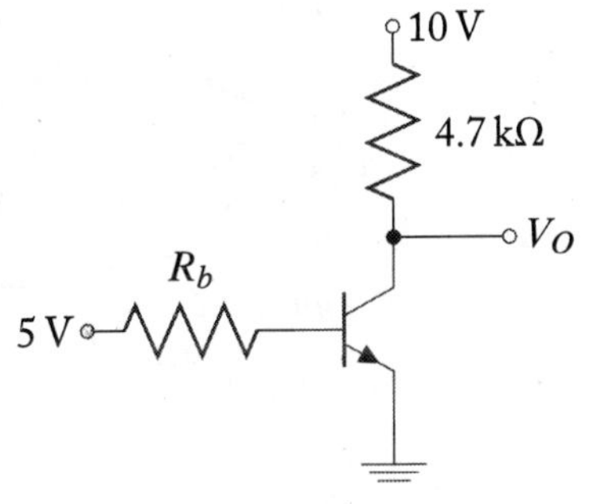

3.13. For the circuit shown, find the minimum V_{BB} for transistor saturation. Assume $\beta_F =$ 210.

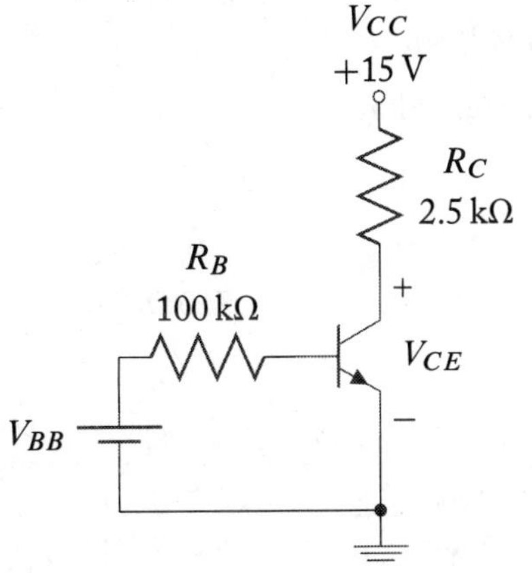

3.14. For the circuit shown:

(a) Find V_{BB} for saturation. Use SPICE to confirm the result.

(b) Find V_{BB} and V_{CE} for operation in the forward active region with $I_C = 0.5(I_C)_{sat}$. Use SPICE to confirm the result.

Assume the Silicon BJT is described by $\beta_F = 200$ and has additional SPICE parameters $IS = 6.7\,\text{fA}$ and $VA = 100\,\text{V}$.

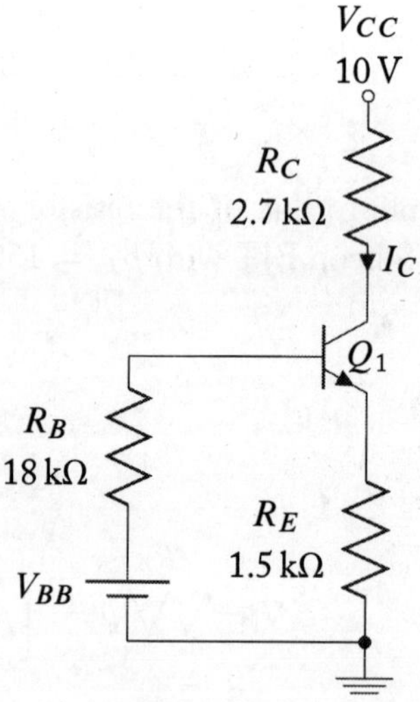

3.15. Complete the design of the circuit below by finding R_B for $I_C = 11\,\text{mA}$. Determine the following voltages and currents:

- Base current
- Base-emitter voltage
- Collector-emitter voltage

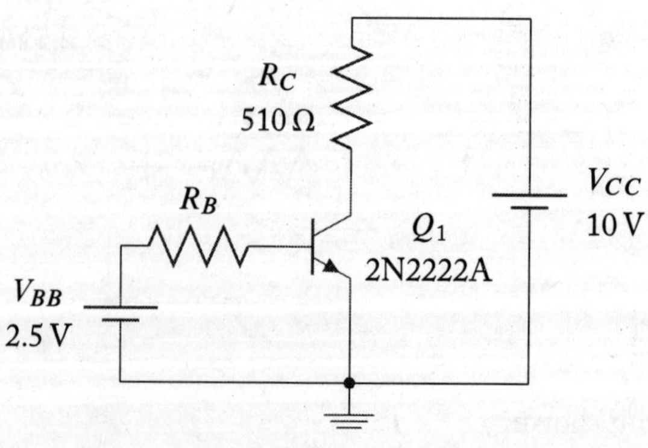

3.16. Complete the design of the fixed-bias BJT circuit shown, given $I_B = 60\,\mu$A.

Determine the resistance value for R_E. What is the transistor Q-point? That is, find:

- Collector-emitter voltage
- Collector current
- Base-emitter voltage

Assume that the transistor is a 2N2222A npn BJT as shown in Figure 3.2.

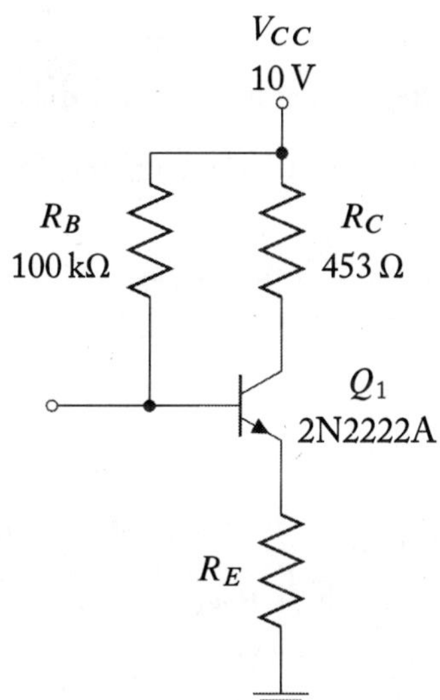

3.17. Complete the design of the self-biased BJT circuit shown given the following:

$$V_{EBQ} = 0.7\,\text{V}, \quad \beta_F = 150, \quad I_C = -2\,\text{mA}.$$

Determine the resistance value for R_{B2}. What is the transistor Q-point? That is, find:

- Base current
- Collector-emitter voltage

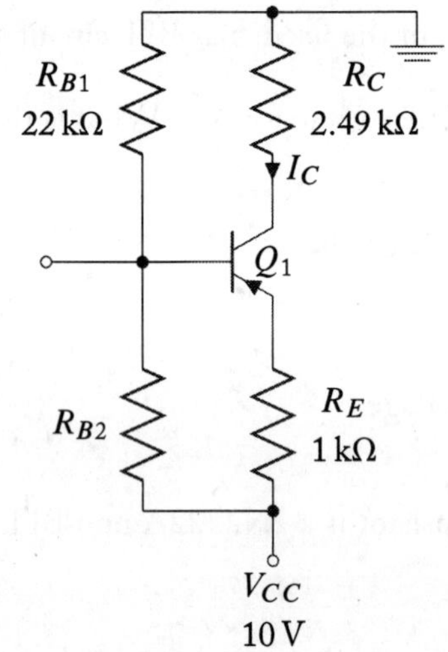

3.18. The simple logic inverter shown is constructed using a Silicon BJT with $\beta_F = 75$.

(a) What is the maximum input voltage, v_i, for which the output will be HIGH ($\approx$ 5 V).

(b) What is the minimum input voltage, v_i, for which the output will be LOW ($<$ 0.2 V).

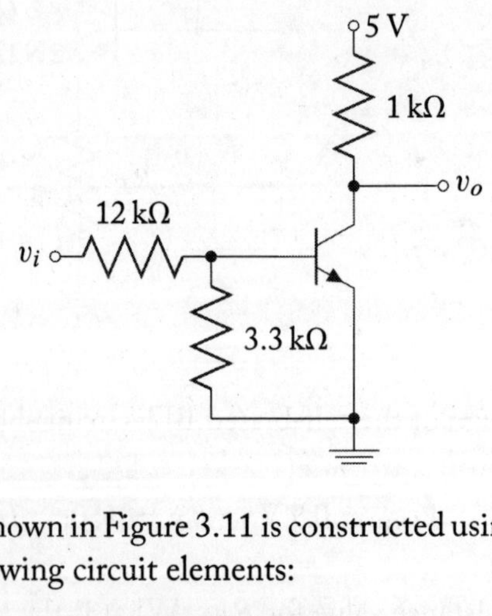

3.19. The DTL NAND Gate shown in Figure 3.11 is constructed using a Silicon BJT described by $\beta_F = 75$ and the following circuit elements:

$$R_a = 3.6\,\text{k}\Omega$$
$$R_b = 6.2\,\text{k}\Omega$$
$$R_c = 1.8\,\text{k}\Omega$$

Determine the fan-out of this gate.

3.20. Use SPICE to verify the operation of the TTL gate of Example 3.6.

3.21. Determine the fan-out of the simple ECL OR gate shown in Figure 3.13.

3.22. Given the input circuit for a TTL gate (with only one input) shown. The transistor parameters of interest are:

$$\beta_F = 200, \qquad \beta_R = 6, \qquad V_{BE(ON)} = 0.6\,\text{V},$$
$$V_{BE(sat)} = 0.8\,\text{V}, \quad V_{CE(sat)} = 0.2\,\text{V}.$$

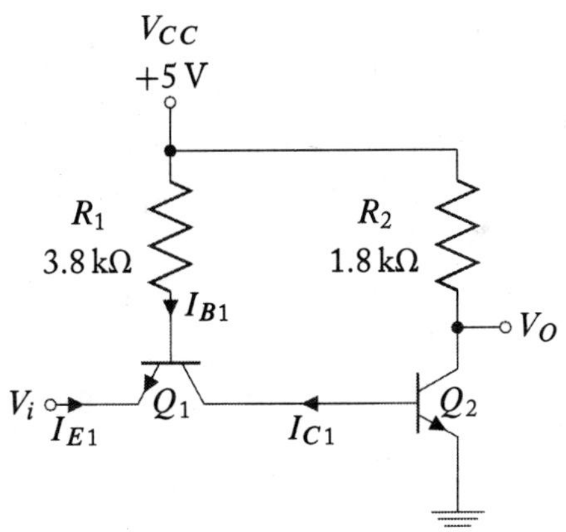

Assume that Q_1 and Q_2 are identical.

(a) find $V_{B1}, I_{B1}, I_{C1}, I_{E1}, V_{B2}$ and V_O for input LOGIC 1 (5 V) and LOGIC 0 (0 V).

(b) Find the *fan-out* of the circuit.

(c) Compare the results of parts (a) and (b) and comment on the potential design advantages of using the circuit analyzed in Example 3.6.

3.23. In an effort to reduce the power consumption for the TTL gate of Example 3.6, V_{CC} is reduced to 3.3 V.

(a) Verify that the gate operates properly and calculate the noise margins and fan-out.

(b) Compare average power consumption of the gate with $V_{CC} = 5\,\text{V}$ and 3.3 V. Hint: find the power supply currents for a ZERO and a ONE output and then average them.

3.24. The circuit shown is a form of high-threshold logic (HTL) gate. Analytically determine the following gate properties:

(a) The logic function performed

(b) The logic levels

(c) The fan out

(d) The noise margins

(e) The average power consumption.

Assume:

- Silicon diodes and BJTs
- $\beta_F = 100$
- $V_Z = 5.6\,\text{V}$

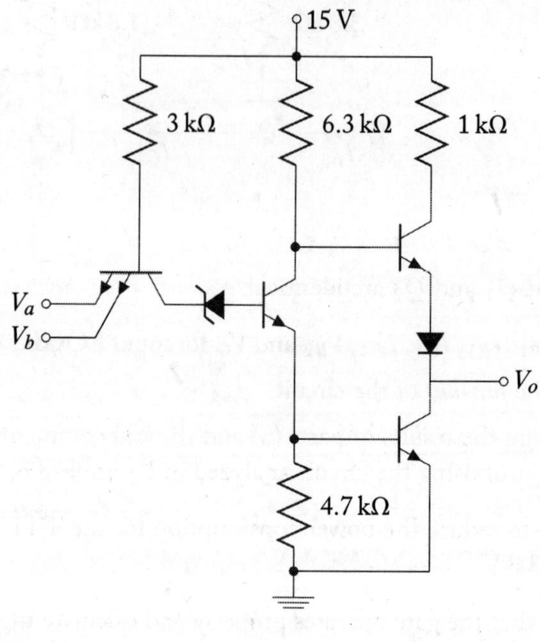

3.25. Complete the design of the self-biased BJT circuit shown given the input and output characteristic curves shown in Figure 3.2 and $I_C = 7\,\text{mA}$ and $V_{CE} = 3\,\text{V}$.

(a) What is I_B and V_{BE} at the Q-point?

(b) What is β_F?

(c) Find the resistance values R_C and R_{B1}.

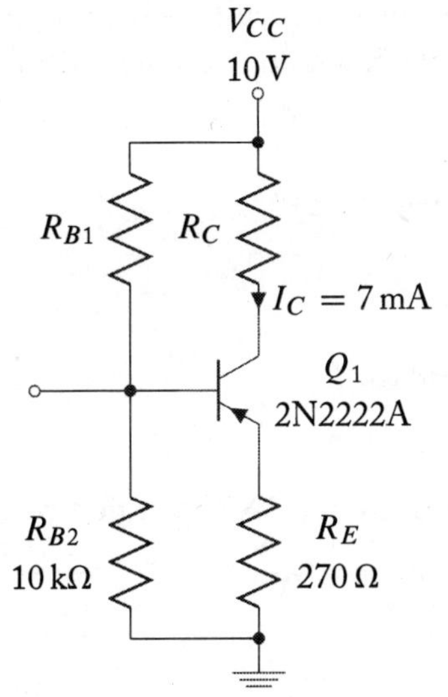

3.26. Complete the design of the BJT biasing circuit shown given the input and output characteristic curves and $I_C = 3.5\,\text{mA}$ and $V_{CE} = 4\,\text{V}$.

(a) What is I_B and V_{BE} at the Q-point?

(b) What is β_F ?

(c) Find the resistance values R_C and R_B.

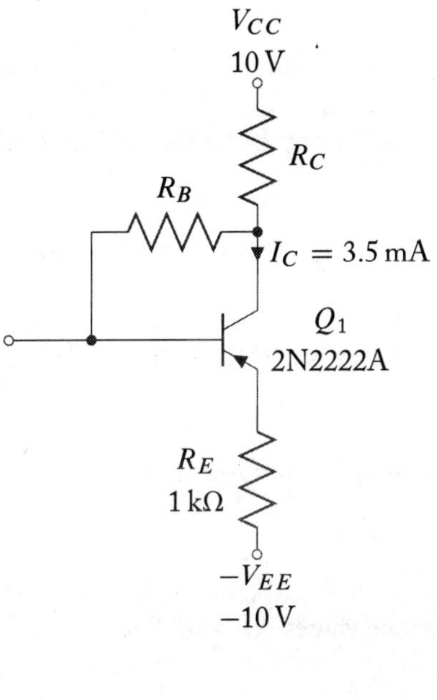

3.27. Complete the design of the circuit shown for $I_C = 5.3\,\text{mA}$: determine the resistance values for R_B. Find the transistor currents and voltages using load line analysis:

- Collector-emitter voltage
- Base current
- Base-emitter voltage

Assume that the transistor is a 2N2222A npn BJT as shown in Figure 3.2.

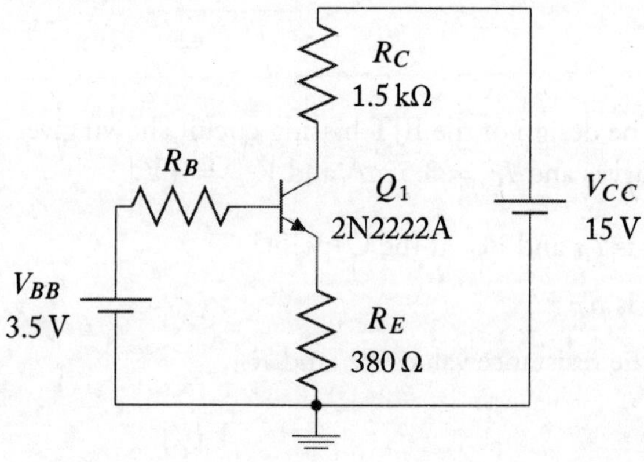

3.28. Complete the design of the pnp BJT biasing circuit below to achieve the desired quiescent conditions:

$$I_C = -4\,\text{mA} \quad \text{and} \quad V_{CE} = -4\,\text{V}.$$

Assume that the transistor is a 2N2222A npn BJT as shown in Figure 3.2.

(a) What is I_B and V_{BE} at the Q-point?

(b) What is β_F?

(c) Find the resistance values R_C and R_B.

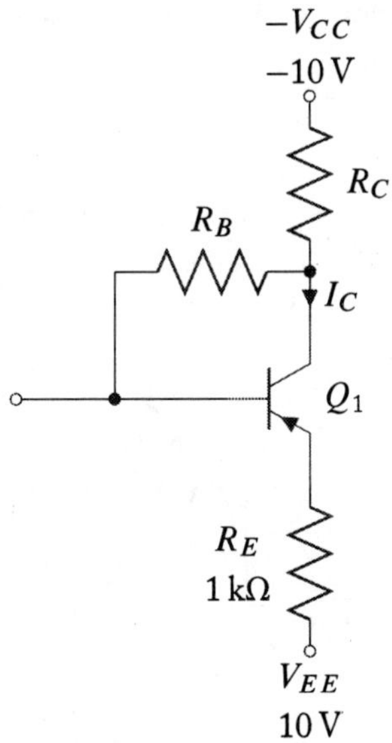

3.29. Complete the design of the circuit below: determine the resistance values for R_B. Find the transistor currents and voltages using load line analysis if $V_{CE} = -5$ V:

- Collector current
- Base current
- Base-emitter voltage

Assume that the transistor is a 2N3906 *pnp* BJT with V-I characteristics found in the Appendix.

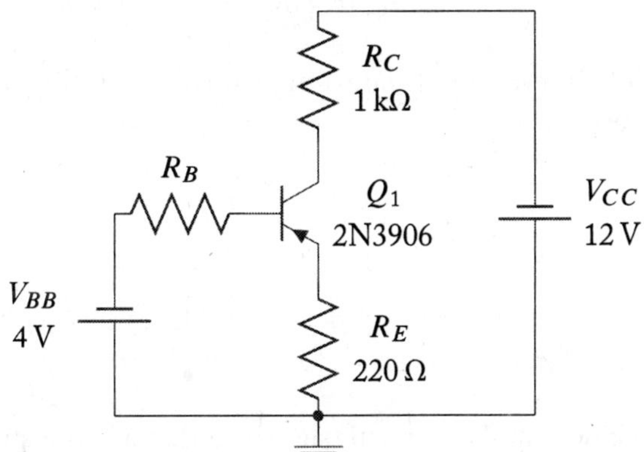

3.30. Complete the design of the circuit shown by determining the base bias voltage V_{BB} for $V_{CE} = 3$ V. Find all transistor terminal currents and voltages using load line analysis where applicable (the input and output characteristics are attached). Assume the BJT is a 2N2222A.

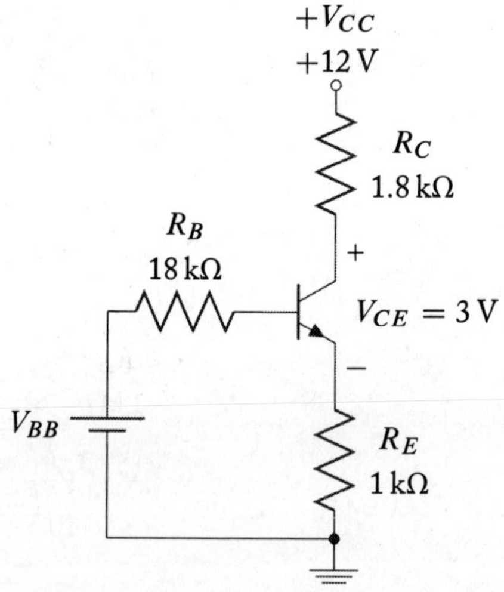

3.31. Design a fixed-bias circuit to achieve the following BJT Q-point:

$$I_C = 4\,\text{mA}; \qquad V_{CE} = 8\,\text{V}.$$

Use a 20 V DC power supply and a Silicon BJT with $\beta_F = 100$.

3.32. Design a self-bias circuit to achieve the following BJT Q-point:

$$I_C = 4\,\text{mA}; \qquad V_{CE} = 8\,\text{V}.$$

Use a 20 V DC power supply and a Silicon BJT with $\beta_F = 100$. Additional constraints on the bias resistors are:

$$\frac{R_c}{R_e} = 5 \quad \text{and} \quad R_{B1}//R_{B2} = 15\,\text{k}\Omega.$$

3.33. For the self-bias circuit topology, determine the ratio R_B/R_E to achieve a bias stability that results in a 0.2% change in I_C for a 10% change in β_F.

3.34. Complete the design of the circuit shown so that a bias stability to achieve a 1% change in I_C for a 10% change in β_F operating over the temperature range of 15°C to 50°C. The BJT SPICE parameters are: $BF = 100$, $IS = 1.1$ fA, and $VA = 120$ V. Confirm the circuit stability using SPICE.

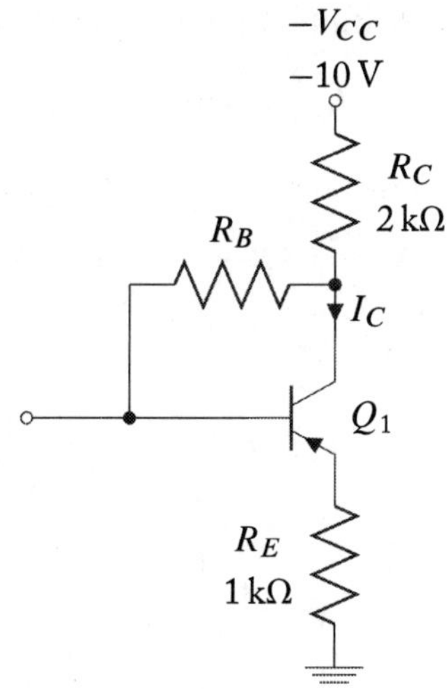

3.35. An *npn* BJT emitter-bias configuration with two power supplies must be designed so that it achieves a bias stability of 1% change in I_C for the variation in β_F found in the transistor used to manufacture the circuit. The value of β_F due to component tolerances ranges from 170 to 230 at 25°C.

(a) Design the circuit.

(b) Simulate the circuit using SPICE to confirm that the design meets the stability requirement.

3.36. Design a self-bias circuit to accomplish the following design goal: as β_F of a silicon BJT varies between 80 and 200 the collector current lies within the range 1.35 mA to 1.65 mA. Assume $V_{CC} = 15$ V, $R_C = 2.2$ kΩ, and no transistor variation in V_{BE} or I_{CO}.

3.37. A transistor with $\beta_F = 60$ and $V_{BE} = 0.8$ V is used in the self-bias circuit with $V_{CC} =$ 25 V. The quiescent point is $I_C = 2.5$ mA and $V_{CE} = 15$ V. The transistor is replaced by another with $\beta_F = 200$ and $V_{BE} = 0.65$ V. It is desired that the effect of the change in β_F does not increase I_C by more than 0.1 mA and that the same should be true for the change in V_{BE}. Determine the resistor values to accomplish these design goals.

3.38. A Silicon BJT with $\beta_F = 70$ and $V_{BE} = 0.75$ V produces a quiescent point of $I_C = 2$ mA and $V_{CE} = 10$ V when inserted into a self-bias circuit with a 20 V power supply. When the transistor is replaced by another Silicon BJT with $\beta_F = 180$ and $V_{BE} = 0.67$ (no change in I_{CO}) the effect of each change increases I_C by 0.8 mA (Final $I_C = 2.16$ mA). Determine the values of the four resistors in the self-bias configuration.

3.39. It is common to attempt to improve amplifier performance by connecting two transistors as shown. Assume identical Silicon BJTs with $\beta_F = 100$. Determine resistor values to accomplish a quiescent condition on transistor Q_2 of

$$I_{C2} = 4\,\text{mA}; \qquad V_{CE2} = 8\,\text{V}.$$

With the design restrictions that:

$$R_{CC} = 4R_{EE} \qquad R_{B1}//R_{B2} = 22\,\text{k}\Omega.$$

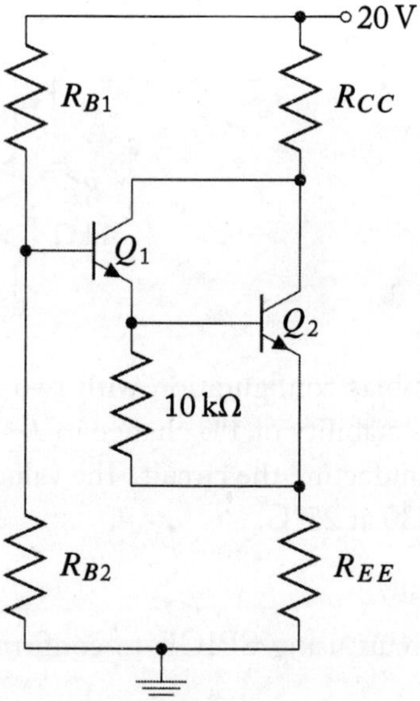

3.40. Design a Zener voltage regulator with BJT shunt to meet the following design requirements:

- Regulated load voltage, $V_L = 5\,\text{V}$
- Load current, $0\,\text{A} \leq I_L \leq 4\,\text{A}$
- Source voltage, $12\,\text{V} \leq V_S \leq 16\,\text{V}$

Determine the appropriate ratings for all components. Assume the minimum Zener current for proper regulation is 1.5 mA and a power BJT with typical forward current gain, $\beta_F = 75$.

3.10 REFERENCES

[1] Antognetti, P. and Massobrio, G., *Semiconductor Device Modeling with SPICE*, McGraw-Hill Book Company, New York, 1988.

[2] Colclaser, R. A. and Diehl-Nagle, S., *Materials and Devices for Electrical Engineers and Physicists*, McGraw-Hill Book Company, New York, 1985.

[3] Ghausi, M. S., *Electronic Devices and Circuits: Discrete and Integrated,* Holt, Rinehart and Winston, New York, 1985.

[4] Gray, P. R., and Meyer, R. G., *Analysis and Design of Analog Integrated Circuits*, 3rd. Ed., John Wiley & Sons, Inc., New York, 1993.

[5] Malvino, A. P., *Transistor Circuit Approximations*, 2nd. Ed., McGraw-Hill Book Company, New York, 1973.

[6] Millman, J., *Microelectronics, Digital and Analog Circuits and Systems*, McGraw-Hill Book Company, New York, 1979.

[7] Millman, J. and Halkias, C. C., *Integrated Electronics: Analog and Digital Circuits and Systems*, McGraw-Hill Book Company, New York, 1972.

[8] Sedra, A. S. and Smith, K. C., *Microelectronic Circuits*, 3rd. Ed., Holt, Rinehart, and Winston. Philadelphia, 1991.

[9] Tuinenga, P., *SPICE: A Guide to Circuit Simulation and Analysis Using PSpice*, 2nd. Ed., Prentice Hall, Englewood Cliffs, 1992.

CHAPTER 4

Field Effect Transistor Characteristics

In Chapter 3 Bipolar Junction Transistors were shown to be semiconductor devices that operate on carrier flow from the emitter to the base and then through to the collector. For example, npn BJTs are devices where the current flow from the collector to the emitter is regulated by the current injected into the base. Therefore, the BJT is a *current controlled* three-terminal semiconductor device.

Field Effect Transistors (FETs) are semiconductor devices that employ a channel between the *drain* and the *source* to transport carriers. An adjacent controlling surface, called the *gate*, regulates the current flow through the drain-source channel. This channel is controlled by a voltage applied to the gate of the FET. Therefore, the FET can be described as a *voltage controlled* three-terminal semiconductor device (see Figure 4.1). The physical properties of FETs make them suitable for amplification, switching, and other electronic applications.

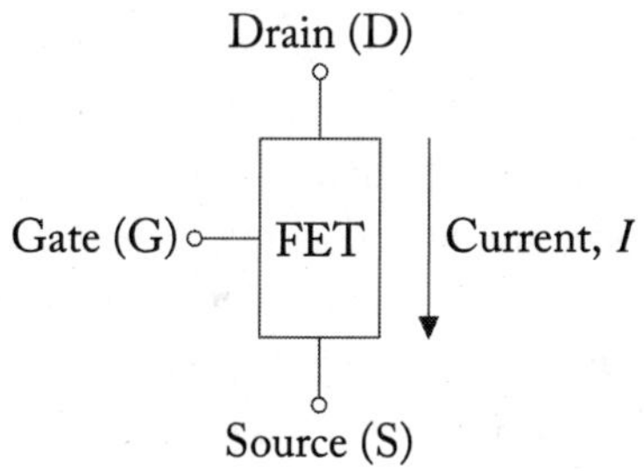

Figure 4.1: FET represented as a three terminal device.

The terminal characteristics of Junction Field Effect Transistors (JFETs) and Metal-Oxide-Semiconductor FETs (MOSFETs) are described in this chapter. Other types of FETs exist, but JFETs and MOSFET are the predominate FET types.[1] MOSFETs are used extensively in integrated circuits for digital applications: JFETs are most commonly found in analog applications.

[1]Other FET types include the Metal-Semiconductor FET (MESFET), Modulation-doped FET (MODFET), and Vertical MOSFET (VMOSFET). Analysis of JFETs and MOSFETs will allow for a general understanding of FET behavior that can be used with other FET devices.

Terminal characterization of FETs is sufficient for electronic analysis and design: for discussions on the device physics of FETs, the reader is referred to the references.

Within each FET type, further categorization is based upon two properties of the channel: channel doping and gate action on the channel. The FET channel may be fabricated from *n*- or *p-type* material. Therefore, the FETs are designated as *n*-channel JFETs, *p*-channel JFETs, *n*-channel MOSFETs (NMOSFETs), and *p*-channel MOSFETs (PMOSFETs). The equations governing the operation of the two channel types of FETs are identical with the exception that the current and voltages in the two types are of opposite polarities. That is, the *n*-channel positive current and voltages are replaced by negative current and voltages for *p*-channel devices. In addition to the two FET channel types, MOSFETs can either be depletion or enhancement mode MOSFETs. JFETs are depletion mode devices. The terms *depletion* and *enhancement* refer to the action of the gate control voltage on the carriers in the channel and the channel itself.

Depletion mode devices can be thought of has having a normally open channel for charge carriers between the drain and the source. With the application of a potential of the proper polarity across the gate and source, the carriers in the channels are essentially "depleted" which "pinches off," or squeezes, the channel between the drain and source disallowing additional charge carriers (current) to flow through the channel. The point at which the current between the drain and the source is "pinched-off" is regulated by decreasing gate voltage relative to the source for *n*-channel devices and increasing the voltage relative to the source for *p*-channel devices. For *n*channel devices, decreasing the gate potential relative to the source squeezes the drain-source channel closed.

The drain-source channel in an enhancement mode device is described as normally closed. In the enhancement mode case, the application of a potential of the proper polarity between the gate and source terminals causes the drain-source channel to become "enhanced," allowing additional carriers (current) to flow. For *n*-channel devices, increasing the gate potential relative to the source enhances the drain-source channel.

JFETs, whether *n*- or *p*-type, are depletion mode devices.[2] Depletion MOSFETs can operate either in the depletion mode or the enhancement mode depending on the gate potential relative to the source, allowing the drain-source channel to either open up to allow additional current flow (enhancement) or "squeezing" closed the channel and restricting further flow of current (depletion).

Like BJTs, FETs have different regions of operation. The regions are identified and characterized through the FET terminal characteristics. Simple circuits are developed that use the basic terminal characteristics of the various types of FETs in each region. Among the significant circuits analyzed include a FET constant current source, active resistive loads constructed with FETs, a CMOS inverter, FET switches, and voltage variable resistors. SPICE modeling parameters for the various FETs and the effect of those parameters on circuit design and analysis are discussed.

[2]Gallium Arsenide (GaAs) enhancement mode JFETs exist. The GaAs enhancement JFETs operate in a similar manner to other enhancement mode devices. GaAs JFETs are currently not widely used. Unless specifically stated, JFETs are always thought of as being depletion mode devices in this text.

4.1 JUCTION FIELD-EFFECT TRANSISTORS

Junction Field-Effect Transistors (JFETs) are either n-channel or p-channel devices. The terminal voltage and current relationships of n-channel JFETs are developed in this section. The p-channel JFET terminal voltage and current relationships are identical to that of the n-channel JFET with the exception that the polarities of the voltages and currents are reversed. A brief description of the p-channel JFET characteristics will be presented at the end of this section.

The circuit symbol for the n-channel JFET is shown in Figure 4.2a. The terminal connection to the gate has an arrow whose direction indicates the channel type of device depicted. For n-channel JFETs the arrow points to the source on the gate terminal. For p-channel JFETs, the arrow points away from the source as shown in Figure 4.2b.

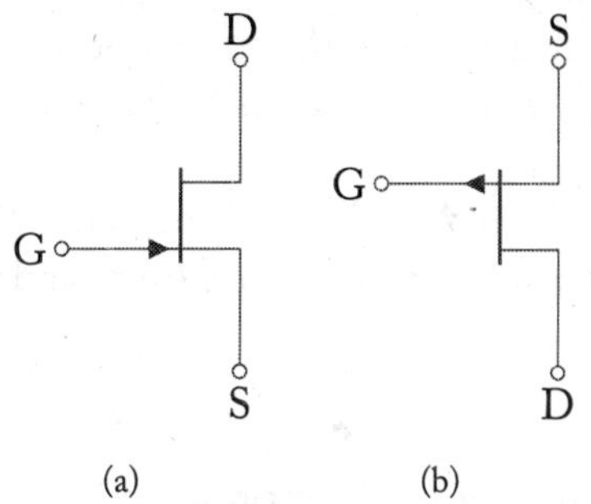

Figure 4.2: Circuit symbols for (a) n-channel JFET and (b) p-channel JFET.

All FETs are three terminal devices with the gate acting as the current regulating terminal between the drain and the source. The voltage and current sign conventions for n-channel JFETs are shown in Figure 4.3.

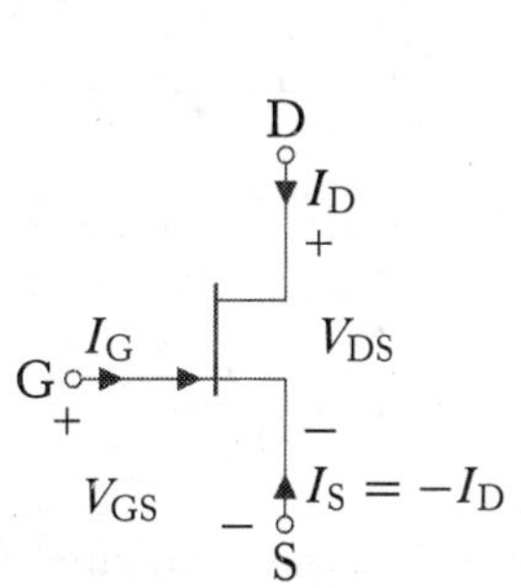

Figure 4.3: Voltage and current directions for the n-channel JFET.

When the gate-source junction is reverse-biased in n-channel JFETs, the conductivity of the drain-source channel is reduced with decreasing gate to source voltage, V_{GS}. The current through the drain-source channel is I_D. The source current is equal to $-I_D$,

$$I_S = -I_D. \tag{4.1}$$

4.1.1 *n*-CHANNEL JFET

Since JFETs normally operate with the gate junction reverse-biased, the gate current is essentially zero,

$$I_G = 0. \tag{4.2}$$

The gate-to-source voltage that "pinches off" the drain-source channel is called the pinch-off voltage, V_{PO},

$$V_{PO} = V_{GS}|_{I_D=0,\, V_{DS}small}. \tag{4.3}$$

For n-channel devices, V_{PO} is a negative voltage and is specific to the particular FET. If V_{GS} is positive (for n-channel JFETs), the gate junction is forward biased and the equations developed in this section do not apply.

The drain (common source) n-channel JFET characteristics are shown in Figure 4.4. The transfer characteristic is presented in Figure 4.5. The current and voltage reference directions were shown in Figure 4.3.

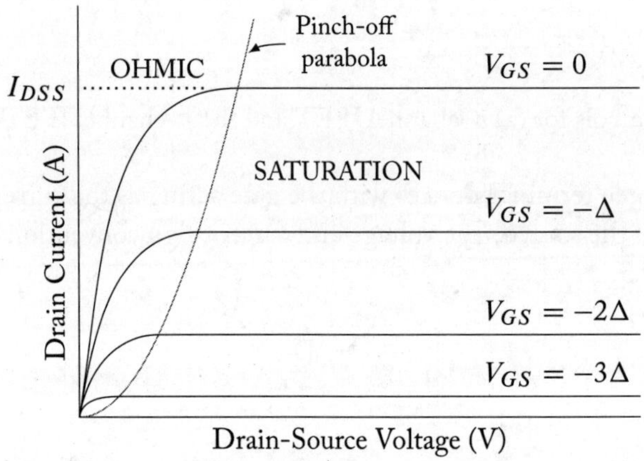

Figure 4.4: Output characteristic n-channel JFET.

The different regions of operation of the n-channel JFET can be illustrated by selecting one curve from the output characteristics (for example, the curve corresponding to $V_{GS} = 0$). For small values of V_{DS}, the channel allows current to readily flow with an initial resistance, R_{DS}. The output curve in this region can be approximated by a straight line of slope R_{DS}^{-1}. This approximate linear relationship of I_D vs. V_{DS} leads to the descriptive name for this region of operation, the *Ohmic Region*. In the ohmic region, the JFET is acts as a voltage variable resistor: the gate-source

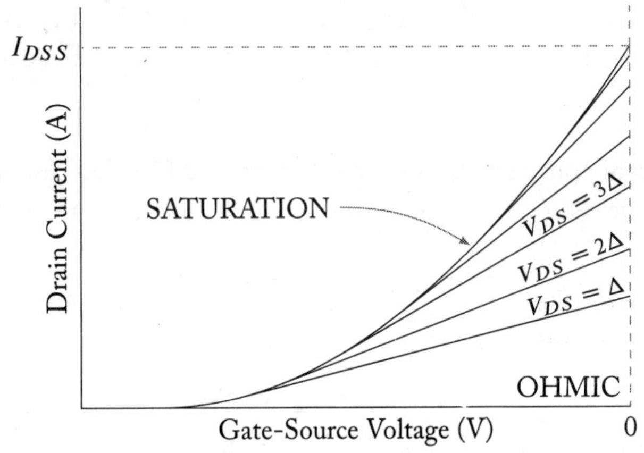

Figure 4.5: Transfer characteristic n-channel JFET.

voltage, V_{GS}, controls the value of the equivalent resistance. As V_{GS} increases, the channel narrows, causing an increase in resistance. The result is a decrease in the slope of the characteristic curve.

With increasing V_{DS}, when V_{GS} is held constant, the channel resistance increases as evidenced by the leveling of I_D. At pinch-off, the drain current, I_D, remains almost constant due to low conductivity through the channel. The resistance of the device in this region is very high. The drain-to-source voltage at pinch-off is,

$$V_{DS}\,(\text{at pinch-off}) = V_{GS} - V_{PO}. \tag{4.4}$$

The boundary between the *Ohmic* and *Saturation* region is the so called pinch-off parabola (refer to Figure 4.4) defined by the relationship in Equation (4.4). The JFET is said to be in Saturation when,

$$V_{DS} > V_{GS} - V_{PO}. \tag{4.5}$$

Saturation in FETs differs from saturation in BJTs. FETs operating in the saturation region are analogous to BJTs in the forward-active region. As can be seen in the FET output characteristic of Figure 4.4, operation in saturation allows the drain current to be adjusted by a varying a control voltage at the gate, in this case V_{GS}, independent of V_{DS}. Amplifiers designed using JFETs takes advantage of the small voltage variations in V_{GS} controlling the current flow through the device. JFETs can also be used as switches since large and abrupt changes in V_{GS} causes the current I_D to change from zero to a relatively large value.

The JFET does not require significant input gate current. Therefore, the input characteristic of the device provides limited and nearly useless information for circuit design. The transfer characteristic shown in Figure 4.5 is far more useful. Since the gate channel diode in JFETs must be reverse biased, only negative values of V_{GS} allow for operation in the ohmic and saturation regions.

From the transfer function, the current at $V_{GS} = 0$ is defined as I_{DSS}. I_{DSS} is the drain current at $V_{GS} = 0$ at pinch-off ($V_{DS} = -V_{PO}$). I_{DSS} is temperature dependent and decreases with increasing temperature.

The voltage and current relationships of n-channel JFETs can be described for the three most common regions of operation: Ohmic, Saturation, and Cut-Off. For n-channel JFETs operating in the ohmic or saturation regions, the following voltage and current conditions must hold:

$$I_D > 0 \qquad V_{PO} < V_{GS} \leq 0 \qquad V_{DS} > 0.$$

The n-channel JFETs have the following characteristic parameters:

$$V_{PO} < 0 \qquad I_{DSS} > 0.$$

In the negative quadrant, where $I_D < 0$ and $V_{DS} < 0$, the drain-gate junction becomes forward biased causing the drain current to increase rapidly. The characteristics in this region are similar to diode characteristics (except that $I_D < 0$ and $V_{DS} < 0$) with the turn-on voltage determined by V_{GS}.

Ohmic Region

The ohmic region is that portion of the curve between $V_{DS} = 0$ and pinch-off on the output characteristic curve in Figure 4.4. The mathematical expression defining this region is,

$$0 < V_{DS} \leq V_{GS} - V. \tag{4.6}$$

The V-I relationship in this region is,

$$I_D = I_{DSS}\left[2\left(\frac{V_{GS}}{V_{PO}} - 1\right)\frac{V_{DS}}{V_{PO}} - \left(\frac{V_{DS}}{V_{PO}}\right)^2\right]. \tag{4.7}$$

For small values of V_{DS}, the drain current in Equation (4.7) is approximately,

$$I_D \approx 2I_{DSS}\left(\frac{V_{GS}}{V_{PO}} - 1\right)\frac{V_{DS}}{V_{PO}}. \tag{4.8}$$

If V_{GS} is held constant, Equation (4.8) is a linearly varying function of I_D and V_{DS}.
Therefore, the output resistance in this region is found by taking the derivative of Equation (4.8) with respect to V_{DS}:

$$R_{DS}^{-1} = \frac{\partial I_D}{\partial V_{DS}} = \frac{2I_{DSS}}{V_{PO}}\left(\frac{V_{GS}}{V_{PO}} - 1\right). \tag{4.9}$$

Saturation Region

The saturation region occupies the portion of the output characteristic curve of Figure 4.4 where $I_D > 0$ and to the right of the pinch-off parabola. That is,

$$V_{DS} \geq V_{GS} - V_{PO}. \tag{4.10}$$

The drain current is virtually independent of V_{DS} in this region,

$$I_D = I_{DSS}\left(1 - \frac{V_{GS}}{V_{PO}}\right)^2. \tag{4.11}$$

Equation (4.11) is called the transfer characteristic and is shown in Figure 4.4. The values of V_{PO} and I_{DSS} are specified by the manufacturers.

The expression for the pinch-off parabola can be derived from Equation (4.11) by substituting $V_{GS} = V_{DS} + V_{PO}$,

$$I_D = I_{DSS}\left(\frac{V_{DS}}{V_{PO}}\right)^2. \tag{4.12}$$

Cut-off Region

The JFET is said to be in the cut-off region when,

$$V_{GS} < V_{PO}. \tag{4.13}$$

The drain current is zero when the JFET is cut-off,

$$I_D = 0. \tag{4.14}$$

4.1.2 THE *p*-CHANNEL JFET

The voltage and current sign conventions for *p*-channel JFETs is illustrated in Figure 4.6.

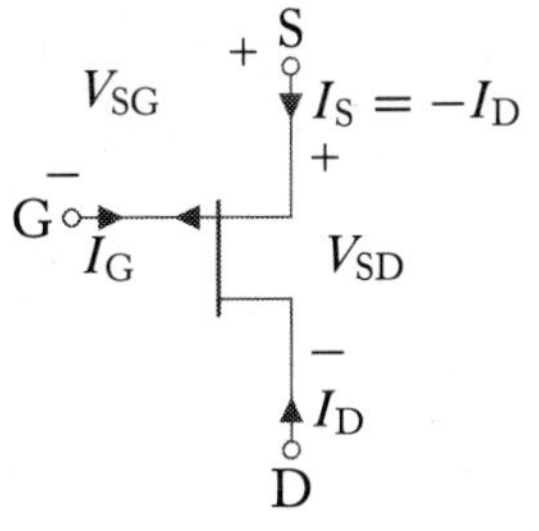

Figure 4.6: Voltage and current directions of *p*-channel JFETs.

The arrow on the gate is leaving the device. Also note that the direction of the drain current is opposite that for the *n*-channel JFET. In essence, all terminal voltages and currents have been reversed. The *p*-channel operating voltages and currents are given below:

$V_{PO} > 0 \qquad V_{SD} > 0 \qquad -V_{PO} < V_{SG} \leq 0$

$I_{DSS} < 0 \qquad I_D < 0.$

The conditions that identifies the regions of operation are,

$$\begin{array}{ll} \text{Ohmic Region:} & 0 < V_{SD} < V_{PO} + V_{SG} \\ \text{Saturation Region:} & V_{SD} \geq V_{PO} + V_{SG} \\ \text{Pinch-off Parabola:} & V_{SD} = V_{PO} + V_{SG}. \end{array} \tag{4.15}$$

The output characteristic curve and the transfer characteristic for a p-channel JFET are shown in Figures 4.7 and 4.8.

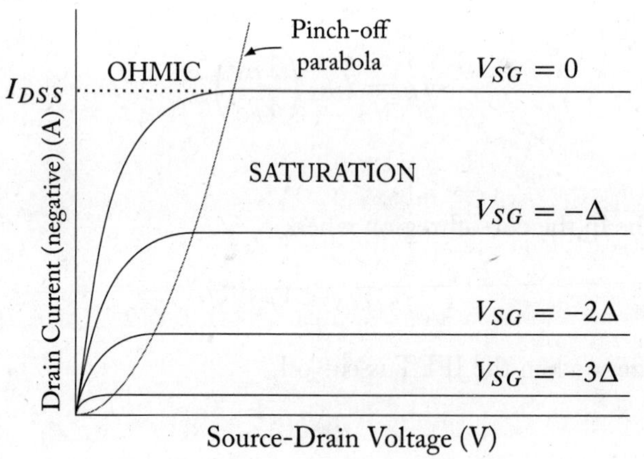

Figure 4.7: p-channel JFET output characteristic curve.

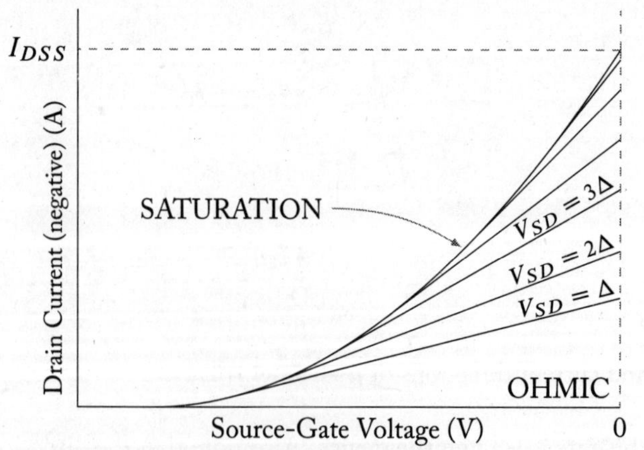

Figure 4.8: Transfer characteristic for p-channel JFET.

The V-I relationship in the three regions can be found by using the mathematical expressions for n-channel JFETs in Equations (4.6) to (4.14). In the p-channel JFET in the Ohmic and

Saturation regions, the gate-to-source and drain-to-source voltages are of the opposite polarity to the n-channel JFET voltage. That is, in the p-channel JFET, the gate-to-source voltage is positive ($V_{GS} > 0$ or $V_{SG} < 0$), and the drain-to-source voltage is negative ($V_{DS} < 0$ or $V_{SG} > 0$).

In both p- and n-channel JFETs, V_{PO} and I_D have opposite polarities.

Example 4.1
An n-channel JFET has the following characteristics:

$$V_{PO} = -3.5\,\mathrm{V}$$

and

$$I_{DSS} = 10\,\mathrm{mA}.$$

Find the minimum drain-to-source voltage, V_{DS}, for the JFET to operate at in saturation for a gate-to-source voltage, $V_{GS} = -2\,\mathrm{V}$. What is its resistance in the ohmic region?

Solution:

The condition for saturation is,

$$V_{DS} \geq V_{GS} - V_{PO}.$$

Substituting yields,

$$V_{DS} \geq -2 - (-3.5) = 1.5\,\mathrm{V}.$$

In the ohmic region the resistance is,

$$R_{DS}^{-1} = \frac{\partial I_D}{\partial V_{DS}} = \frac{2I_{DSS}}{V_{PO}}\left(\frac{V_{GS}}{V_{PO}} - 1\right).$$

Substituting $V_{GS} = -2\,\mathrm{V}$, $V_{PO} = -3.5\,\mathrm{V}$, and $I_{DSS} = 10\,\mathrm{mA}$ yields an output resistance of

$$R_{DS} = 408\,\Omega.$$

Example 4.2
Given a p-channel JFET with the following parameters:

$$V_{PO} = 4\,\mathrm{V},$$

and

$$I_{DSS} = -8\,\mathrm{mA}.$$

What is the drain current I_D for $V_{SG} = -3\,\mathrm{V}$ and $V_{SD} = 2\,\mathrm{V}$?

Solution:

Determine the region the operating region of the FET.

$$V_{PO} + V_{SG} = 4 - 3 = 1\,\text{V}.$$

But $V_{SD} = 2\,\text{V}$ so,

$$V_{SD} \geq V_{PO} + V_{SG},$$

which indicates that the transistor is in saturation. Therefore, the drain current is,

$$I_D = I_{DSS}\left(1 - \frac{V_{GS}}{V_{PO}}\right)^2.$$

Substituting and solving for I_D yields,

$$I_D = -8 \times 10^{-3}\left(1 - \frac{-(-3)}{4}\right)^2 = -0.5\,\text{mA}.$$

4.2 METAL-OXIDE-SEMICONDUCTOR FIELD-EFFECT TRANSISTORS

Metal-Oxide Semiconductor Field Effect Transistors (MOSFETs) are widely used in integrated circuits. Because MOS devices can be fabricated in very small geometries and are relatively simple to manufacture, most Very Large Scale Integrated (VLSI) circuits are fabricated from MOS devices.

MOSFETs come in either of two types:

- Depletion type.
- Enhancement type.

Both types of MOSFETs are either n- or p-channel devices, commonly abbreviated NMOSFET and PMOSFET, respectively. The depletion MOSFET can operate in either the depletion or enhancement modes. The enhancement MOSFET operates in the enhancement mode only.

This section considers the depletion- and enhancement-type MOSFETs separately.

4.2.1 DEPLETION-TYPE MOSFET

In this section, the terminal voltage and current relationships of depletion-type n-channel MOSFETs (NMOSFETs) will be developed. The depletion-type p-channel MOSFET (PMOSFET) terminal voltage and current relationships are identical to the NMOSFET with the exception that the polarities of the voltages and currents are reversed. The depletion MOSFET regions of operation are identical to those of the JFET. In fact, the V-I relationships of the depletion

MOSFET are identical in form to those of the JFET. Both types of transistors are depletion mode devices; that is, with the application of the appropriate polarity potential between the gate and the source (a negative potential for NMOSFETs), the carriers in the channel are depleted which squeezes off the channel through which charge carriers flow. Therefore, the same types of transistor parameters are used to describe the V-I characteristics of the depletion MOSFET as for the JFET. The operational characteristics of the depletion MOSFET does differ from the JFET by one important attribute, the depletion MOSFET can additionally operate in the enhancement mode by applying a positive gate-source potential, thus "enhancing" the channel through which the charge carriers flow.

The circuit symbol for the depletion NMOSFET is shown in Figure 4.9a. In addition to the three FET terminals of gate, drain, and source, the symbol depicts a terminal representing the substrate of the semiconductor ("B" for body) with an arrow pointing into the junction. It is common for the substrate to be electrically connected to the source: this connection does not affect the characteristics of the MOSFET. However, in integrated circuits using NMOSFETs, the substrate is commonly connected to the most negative supply voltage. With NMOSFETs connected in this fashion it is guaranteed that the substrate is at signal ground. Unfortunately, there is some possibility that circuit performance may be compromised.

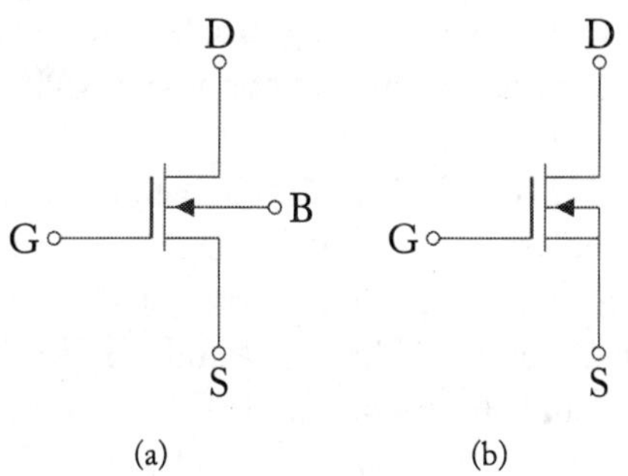

Figure 4.9: (a) Circuit symbol for the depletion type NMOSFET. (b) Depletion type NMOSFET with the substrate connected to the source.

A simplified circuit symbol for the depletion type NMOSFET is shown in Figure 4.10 with the standard voltage and current directions indicated. For NMOSFETs, the arrow points toward the source from the junction. The direction of the arrow corresponds to the direction of standard current flow toward the source terminal.

As in the n-channel JFET the source current is equal to the negative of the drain current when the gate-source junction is reverse biased,

$$I_S = -I_D. \tag{4.16}$$

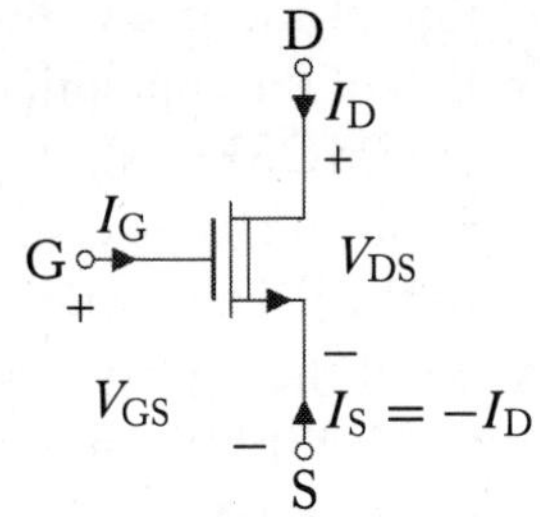

Figure 4.10: Simplified circuit symbol for the depletion NMOSFET.

The depletion type MOSFET operates similarly to the JFET. Like the JFET, the MOSFET gate current is essentially zero,

$$I_G = 0. \tag{4.17}$$

As in the JFET, the pinch-off voltage is the gate-to-source voltage that depletes or "squeezes off" the drain-source channel. For NMOSFETs, V_{PO} is negative and is transistor specific.

Unlike the n-channel JFET, the depletion type NMOSFET allows for enhancement mode operation with the application of a positive V_{GS}. Application of a positive gate-to-source potential "opens up" the drain-source channel by increasing the channel conductivity, allowing more current to flow. Enhancement mode operation of the depletion type NMOSFET allows for drain currents in excess of I_{DSS}.

Typical depletion type NMOSFET drain characteristics are shown in Figure 4.11. The transfer characteristic of the depletion NMOSFET is shown in Figure 4.12.

The depletion NMOSFET characteristic curves of Figures 4.11 and 4.12 are very similar to those of the n-channel JFET curves of Figure 4.4 and 4.5. The equations describing the curves for the depletion NMOSFET and the n-channel JFET are identical. As in the JFET, there are three common regions of operation: the Ohmic, Saturation, and Cut-Off regions.

For the depletion type NMOSFET operating in the depletion-mode ohmic or saturation regions, the following voltage and current conditions must hold:

$$I_D > 0 \qquad V_{PO} < V_{GS} \leq 0 \qquad V_{DS} > 0.$$

For the depletion type NMOSFET operating in the enhancement mode, V_{GS} changes sign. The voltage and current conditions become:

$$I_D > 0 \qquad V_{GS} > 0 \qquad V_{DS} > 0.$$

Depletion type NMOSFETs have the following characteristic parameters:

$$V_{PO} < 0 \qquad I_{DSS} > 0.$$

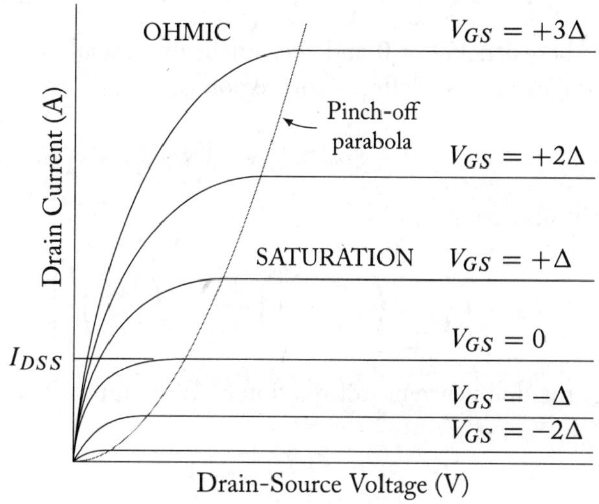

Figure 4.11: Depletion-type NMOSFET drain characteristics.

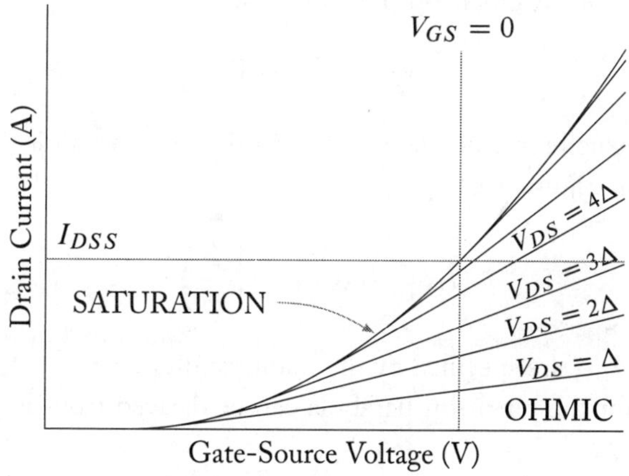

Figure 4.12: Depletion-type NMOSFET transfer characteristics.

As in the JFET operating in the negative quadrant, where $I_D < 0$ and $V_{DS} < 0$, the drain-gate junction becomes forward biased causing the drain current to increase rapidly. The characteristics in this region are similar to a diode (except that $I_D < 0$ and $V_{DS} < 0$) with the turn-on voltage determined by V_{GS}.

Ohmic region

The ohmic region lies between $V_{DS} = 0$ and the pinch-off parabola in the drain characteristic curve. The mathematical expression defining this region is,

$$0 < V_{DS} \le V_{GS} - V_{PO}. \tag{4.18}$$

The V-I relationship in this region is,

$$I_D = I_{DSS}\left[2\left(\frac{V_{GS}}{V_{PO}} - 1\right)\frac{V_{DS}}{V_{PO}} - \left(\frac{V_{DS}}{V_{PO}}\right)^2\right]. \tag{4.19}$$

For small values of I_D the drain current in Equation (4.19) is approximately,

$$I_D \approx 2I_{DSS}\left(\frac{V_{GS}}{V_{PO}} - 1\right)\frac{V_{DS}}{V_{PO}}. \tag{4.20}$$

Saturation Region

The saturation region occupies the portion of the output characteristic curve of Figure 4.11 where $I_D > 0$ and to the right of the pinch-off parabola. That is,

$$V_{DS} \ge V_{GS} - V_{PO}. \tag{4.21}$$

The drain current is virtually independent of V_{DS} in this region and is described by the transfer characteristic (shown in Figure 4.12),

$$I_D = I_{DSS}\left(1 - \frac{V_{GS}}{V_{PO}}\right)^2. \tag{4.22}$$

The values of V_{PO} and I_{DSS} are specified by the manufacturers.

The expression for the pinch-off parabola can be derived from Equation (4.22) by substituting $V_{GS} = V_{DS} + V_{PO}$,

$$I_D = I_{DSS}\left(\frac{V_{DS}}{V_{PO}}\right)^2. \tag{4.23}$$

Cut-off Region

The depletion type NMOSFET is said to be in the cut-off region when,

$$V_{GS} < V_{PO}. \tag{4.24}$$

When the depletion type NMOSFET is cut-off,

$$I_D = 0. \tag{4.25}$$

Example 4.3

For a depletion type NMOSFET with $V_{PO} = -3\,\text{V}$ and $I_{DSS} = 8\,\text{mA}$ what is the drain to source voltage required to saturate the transistor at $V_{GS} = +2\,\text{V}$? What is the drain current?

Solution:

The pinch-off condition is,

$$V_{DS} = V_{GS} - V_{PO}.$$

Substituting the values of the pinch-off voltage and the gate-source voltage,

$$V_{DS} = 2 - (-3) = 5\,\text{V}.$$

To find the drain current at pinch-off, the expression for the current in the saturation region is used,

$$I_D = I_{DSS}\left(1 - \frac{V_{GS}}{V_{PO}}\right)^2.$$

Substituting in the values for I_{DSS}, V_{GS}, and V_{PO} yields the drain current,

$$I_D = 8 \times 10^{-3}\left(1 - \frac{2}{-(-3)}\right)^2 = 22.2\,\text{mA}.$$

Since $I_D > I_{DSS}$, the MOSFET is in the enhancement mode of operation.

4.2.2 DEPLETION-TYPE PMOSFET

The circuit symbol of the depletion-type PMOSFET is shown in Figure 4.13. All of the voltages and currents of the depletion-type PMOSFET are the opposite polarity to those of the NMOSFET. The pinch-off voltage of the depletion-type PMOSFET is greater than zero,

$$V_{PO} > 0. \tag{4.26}$$

The simplified circuit symbol of the depletion-type PMOSFET is shown in Figure 4.14. The current and voltage sign conventions are also shown.

The V-I relationship in the three regions of the depletion PMOSFET can be found by using the mathematical expressions for depletion NMOSFET in Equations (4.18) to (4.25). In both p- and n-channel depletion MOSFETs, the polarity of V_{PO} is always opposite from I_D.

For the depletion type PMOSFET operating in the depletion mode ohmic or saturation regions, the following voltage and current relationships must hold:

$$I_D < 0 \qquad -V_{PO} < V_{SG} \leq 0 \qquad V_{SD} > 0.$$

For the depletion type PMOSFET operating in the enhancement mode, V_{SG} change sign and the conditions become:

$$I_D < 0 \qquad V_{SG} > 0 \qquad V_{SD} > 0.$$

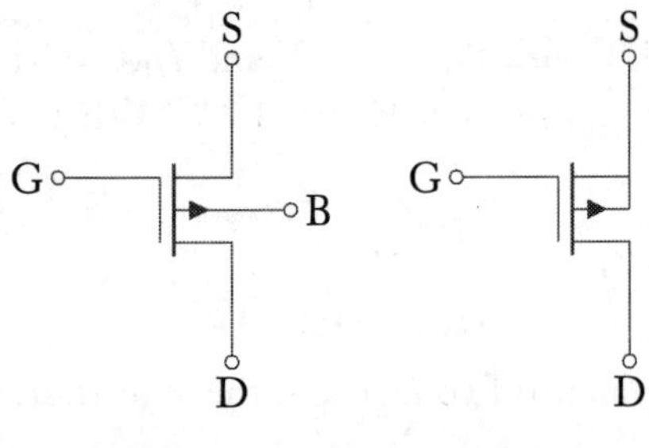

Figure 4.13: (a) Circuit symbol of the depletion type PMOSFET. (b) Symbol for depletion type PMOSFET with the substrate electrically connected to the source.

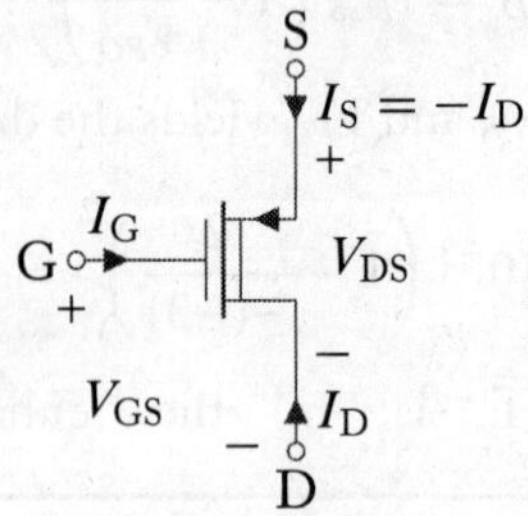

Figure 4.14: Simplified circuit symbol of the depletion type PMOSFET. Current directions and voltage polarities are also indicated.

Note that

$$V_{SG} = -V_{GS},$$

and

$$V_{SD} = -V_{DS}.$$

4.2.3 ENHANCEMENT TYPE MOSFETS

The enhancement type MOSFET is commonly used in integrated circuit design because of its ease of fabrication, small geometry, and low power dissipation. Without an applied voltage between the gate source terminals, the drain-source channel is closed. With the application of a gate-to-source potential, the channel becomes "enhanced" to conduct carriers. The enhancement type carrier conduction mechanism can be thought of as the antithesis of that of the depletion MOSFET.

An *n*-type channel layer (for the enhancement NMOSFET) is formed to conduct carriers from the drain to the source with the application of a positive V_{GS}. A *p*-type channel layer conducts carriers from the drain to the source with the application of a positive V_{SG} (negative V_{GS})

in enhancement PMOSFETs. The gate-to-source voltage that starts to form the drain-source channel is called the *threshold voltage*, V_T. When V_{GS} is less than V_T (in NMOSFETs), I_D is zero since the drain-source channel does not exist. The value of V_T is dependent on the specific MOSFET device and commonly ranges in value from 1 to 5 volts for enhancement NMOSFETs. Since $I_D = 0$ for $V_{GS} = 0$, the quantity I_{DSS} found in the depletion MOSFET and JFET is not pertinent to the enhancement type MOSFET.

Unlike the depletion MOSFETs that operate in both enhancement and depletion modes, enhancement MOSFETs can only operate in the enhancement mode.

4.2.4 ENHANCEMENT TYPE NMOSFET

The circuit symbol for the enhancement type NMOSFET is shown in Figure 4.15. The symbol is similar to that of the depletion NMOSFET with the exception of the three short lines representing the junction area. As with the depletion type NMOSFET, the symbol depicts a terminal representing the substrate with an arrow pointing into the junction.

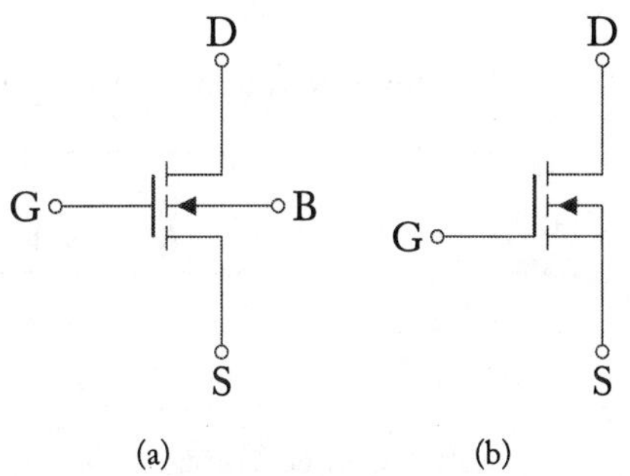

Figure 4.15: (a) Circuit symbol for the enhancement type NMOSFET. (b) Enhancement type NMOSFET with the substrate connected to the source.

A simplified circuit symbol for the enhancement type NMOSFET is shown in Figure 4.16 with the applicable voltage and current sign conventions. The arrow on the circuit symbol points away from the junction to the source. The direction of the arrow corresponds to the direction of the current flow relative to the source terminal.

In the enhancement NMOSFET, the FET channel allows charge flow only when the gate-source potential, V_{GS}, is greater than some threshold voltage V_T. When $V_{GS} > V_T$,

$$I_S = -I_D,$$

and

$$I_G = 0.$$

The enhancement NMOSFET characteristics are shown in Figure 4.17. The transfer characteristic of the enhancement NMOSFET is shown in Figure 4.18.

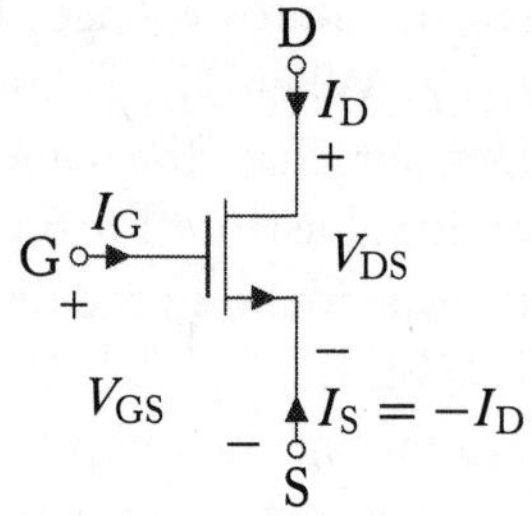

Figure 4.16: Simplified circuit symbol for the enhancement NMOSFET.

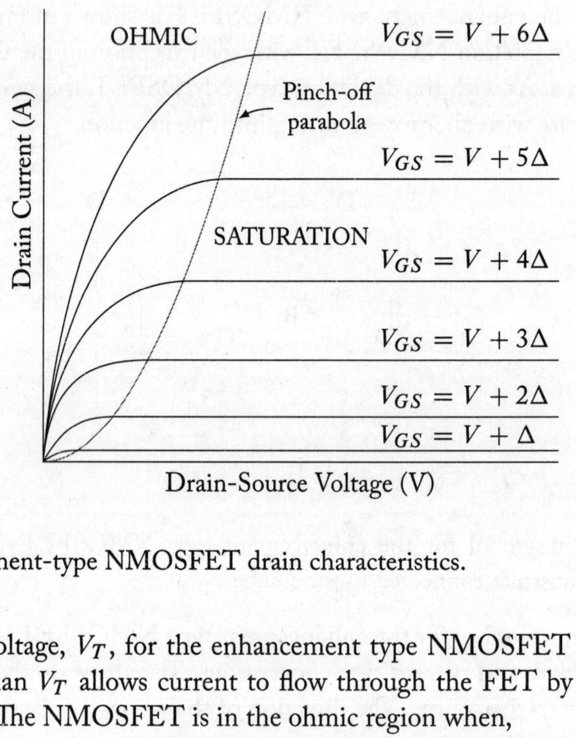

Figure 4.17: Enhancement-type NMOSFET drain characteristics.

The threshold voltage, V_T, for the enhancement type NMOSFET is a positive value. A positive V_{GS} greater than V_T allows current to flow through the FET by the formation of the drain-source channel. The NMOSFET is in the ohmic region when,

$$0 < V_{DS} \leq V_{GS} - V_T. \tag{4.27}$$

The pinch-off parabola, which delineated the boundary between the ohmic and saturation regions, is defined by:

$$V_{DS} = V_{GS} - V_T. \tag{4.28}$$

The saturation region lies beyond the pinch-off parabola:

$$V_{DS} \geq V_{GS} - V_T. \tag{4.29}$$

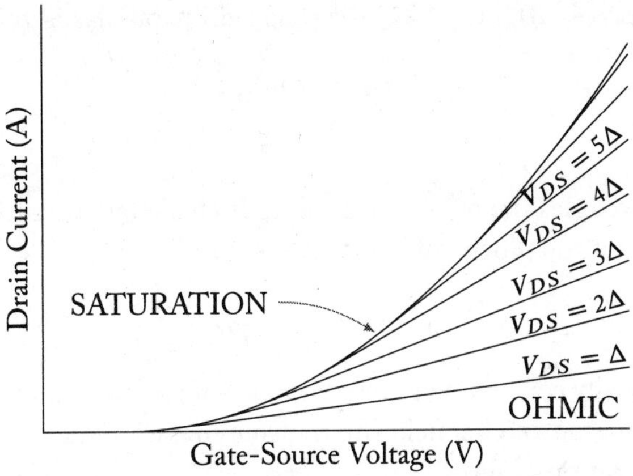

Figure 4.18: Enhancement-type NMOSFET transfer characteristics.

For the enhancement type NMOSFET operating in the ohmic or saturation regions,

$$I_D > 0 \qquad V_{GS} > V_T \qquad V_{DS} > 0.$$

Enhancement type NMOSFETs have a positive valued threshold voltage,

$$V_T > 0.$$

Ohmic Region

The ohmic region is that portion of the output characteristic curve between $V_{DS} = 0$ and the pinch-off parabola. The mathematical expression defining this region is,

$$0 < V_{DS} \leq V_{GS} - V_T. \tag{4.30}$$

The V-I relationship in this region is,

$$I_D = K\left[2\left(V_{GS} - V_T\right)V_{DS} - V_{DS}^2\right]. \tag{4.31}$$

The constant K is the trans conductance factor in units of amperes/volt^2. This transconductance factor is determined by the geometry of the FET, gate capacitance per unit area, and the surface mobility of the electrons in the n-channel.

For small values of V_{DS}, the drain current in Equation (4.31) is approximately,

$$I_D = K\left[2\left(V_{GS} - V_T\right)V_{DS}\right]. \tag{4.32}$$

Equation (4.32) is a linearly varying function of I_D and V_{DS} for constant V_{GS}. The output resistance in this region is the derivative of Equation (4.32) with respect to V_{DS} for constant V_{GS},

$$R_{DS}^{-1} = \frac{\partial I_D}{\partial V_{DS}} = 2K\left(V_{GS} - V_T\right). \tag{4.33}$$

By substituting Equation (4.28) into (4.31) the pinch-off parabolas is given by,

$$I_D = K V_{DS}^2. \tag{4.34}$$

Saturation Region

The saturation region occupies the region of the output characteristic curve of Figure 4.17 where $I_D > 0$ and to the right of the pinch-off parabola. That is,

$$V_{DS} \geq V_{GS} - V_T. \tag{4.35}$$

The drain current is virtually constant with respect to V_{DS} for a given V_{GS} in this region due to high conductivity in the drain-source channel. The transfer characteristic is obtained by substituting $V_{GS} - V_T$ for V_{DS} in Equation (4.34),

$$I_D = K (V_{GS} - V_T)^2. \tag{4.36}$$

The transfer characteristic Equation (4.35) is shown in Figure 4.18.

Cut-Off Region

The cut-off region is defined as,

$$V_{GS} < V_T. \tag{4.37}$$

In cut-off, the drain current is zero,

$$I_D = 0. \tag{4.38}$$

The FET is OFF in this region and does not conduct current. The region is used to implement the OFF state of a switch as described in Section 4.5.

The substrate potential affects the threshold current. In particular, for increasing negative substrate potential with respect to the source, the threshold voltage of the enhancement NMOSFET increases.

Example 4.4

An enhancement type NMOSFET with $V_T = 2\,\text{V}$ that conducts a current $I_D = 5\,\text{mA}$ for $V_{GS} = 4\,\text{V}$ and $V_{DS} = 5\,\text{V}$. What is the value of I_D for $V_{GS} = 3\,\text{V}$ and $V_{DS} = 6\,\text{V}$?

Solution:

First determine the region of operation at $V_{GS} = 4\,\text{V}$ and $V_{DS} = 5\,\text{V}$.

$$V_{GS} - V_T = 4\,\text{V} - 2\,\text{V} = 2\,\text{V}.$$

But $V_{DS} = 5\,\text{V}$ so,

$$V_{DS} > V_{GS} - V_T,$$

implying that the FET is in the saturation region.

The unknown quantity is the transconductance factor, K, of Equation (4.36),

$$I_D = K(V_{GS} - V_T)^2.$$

Solving for K,

$$\begin{aligned} K &= \frac{I_D}{(V_{GS} - V_T)^2} = \frac{5 \times 10^{-3}}{(4-2)^2} \\ &= 1.25 \times 10^{-3}\,\text{A/V}^2. \end{aligned}$$

With this information, the value of I_D for $V_{GS} = 3\,\text{V}$ and $V_{DS} = 6\,\text{V}$ can be determined. Again, the region of operation must be determined for the new operating parameters,

$$V_{GS} - V_T = 3\,\text{V} - 2\,\text{V} = 1\,\text{V}.$$

But $V_{DS} = 6\,\text{V}$ so,

$$V_{DS} > V_{GS} - V_T,$$

implying that the FET is again in the saturation region.

The current at I_D for $V_{GS} = 3\,\text{V}$ and $V_{DS} = 6\,\text{V}$ is,

$$\begin{aligned} I_D &= K\left(V_{GS} - V_T\right)^2, \\ &= 1.25 \times 10^{-3}\,(3-2)^2 \\ I_D &= 1.25\,\text{mA}. \end{aligned}$$

4.2.5 ENHANCEMENT TYPE PMOSFET

The circuit symbol for the enhancement type PMOSFET is shown in Figure 4.19. The symbol is similar to that of the enhancement NMOSFET with the exception of the direction of the arrow on the body (substrate) terminal. As with the depletion type PMOSFET, the symbol depicts a terminal representing the substrate with an arrow pointing into the junction.

A simplified circuit symbol for the enhancement type PMOSFET is shown in Figure 4.20 with the voltage and current sign conventions. The arrow on the circuit symbol points toward the junction to the source. The direction of the arrow corresponds to the direction of the current flow relative to the source terminal.

All terminal voltages are of opposite polarity from that of the enhancement NMOSFET. The polarity of the threshold voltage is opposite that of the enhancement NMOSFET. The enhancement type p-channel MOSFET voltages and currents have the following characteristics:

$$\begin{array}{lll} V_T < 0 & K > 0 & V_{SD} > 0 \\ V_{SG} > 0 & I_D < 0. & \end{array}$$

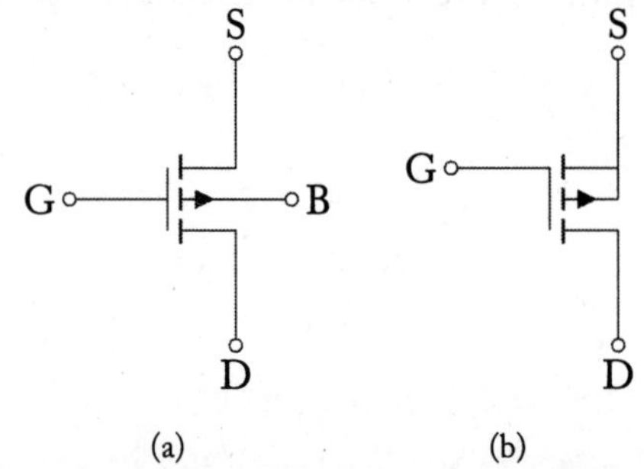

Figure 4.19: (a) Circuit symbol for the enhancement type PMOSFET. (b) Enhancement type PMOSFET with the substrate connected to the source.

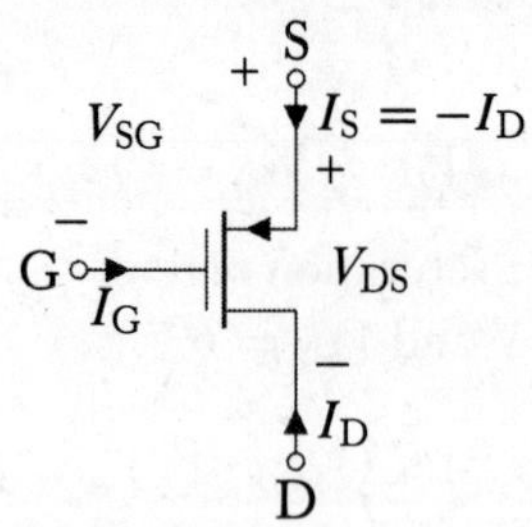

Figure 4.20: Simplified circuit symbol for the enhancement PMOSFET.

The conditions that identifies the regions of operation are,

$$\begin{aligned}&\text{Ohmic Region:}\quad 0 < V_{SD} < V_{SG} + V_T\\&\text{Saturation Region: } V_{SD} \geq V_{SG} + V_T\\&\text{Pinch-Off Parabola: } V_{SD} = V_{SG} + V_T.\end{aligned} \tag{4.39}$$

The enhancement PMOSFET characteristics are shown in Figure 4.21. The transfer characteristic of the enhancement PMOSFET is shown in Figure 4.22.

The threshold voltage, V_T, for the enhancement type PMOSFET is a negative value. A positive V_{SG} greater than $|V_T|$ allows current to flow through the FET by the formation of the drain-source channel.

4.3 THE FET AS A CIRCUIT ELEMENT

Using the terminal behavior of JFETs and MOSFETs, simple circuits using FETs can be developed. FETs are used in circuits as constant current devices, active loads, and voltage variable resistors, to name a few. MOSFETs are used extensively in digital circuits and have certain ad-

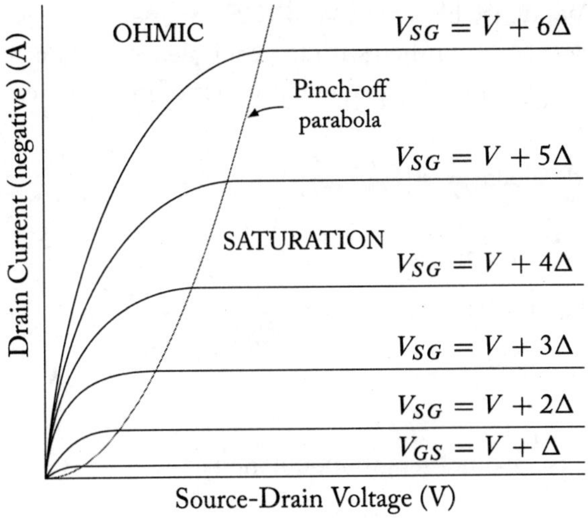

Figure 4.21: Enhancement-type PMOSFET drain characteristics.

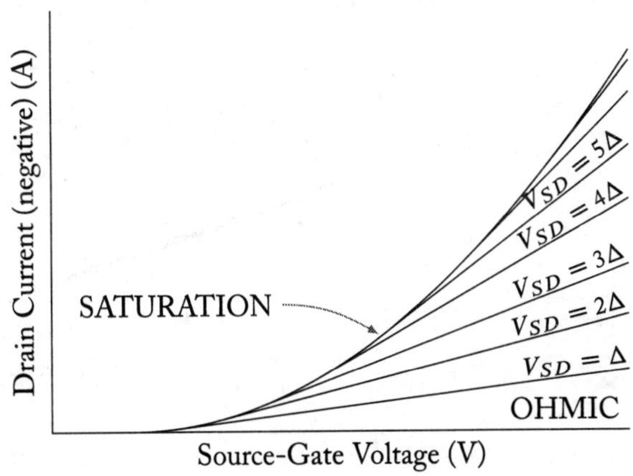

Figure 4.22: Enhancement-type PMOSFET transfer characteristics.

vantages over BJT designs. The SPICE model for the JFET and MOSFET are also developed in this section.

4.3.1 FET SPICE MODELS

When trying to convert the p-FET parameters discussed in this chapter to those used in SPICE models, it rapidly becomes apparent that the correspondence between the two sets of parameters is not clear. The correspondence between the SPICE parameter names and the FET parameter names used in this book is accomplished with the equivalences presented in this section.

In modeling JFETs, the primary parameter values of interest are:

$$\text{VTO} = \text{threshold voltage} = \pm|V_{PO}|$$

$$\text{BETA} = \text{transconductance coefficient} = \frac{I_{DSS}}{V_{PO}^2}$$

$$\text{LAMBDA} = \frac{1}{V_A} = \text{channel-length modulation}$$

V_A is the "early" voltag of the FET. The voltage V_A is that voltage (V_{DS} in the case of n-JFETs) that is the point of intersection of the $I_D = 0$ line and the extended line from the characteristic curves in saturation. Figure 4.23 provides a pictorial definition of the Early voltage.

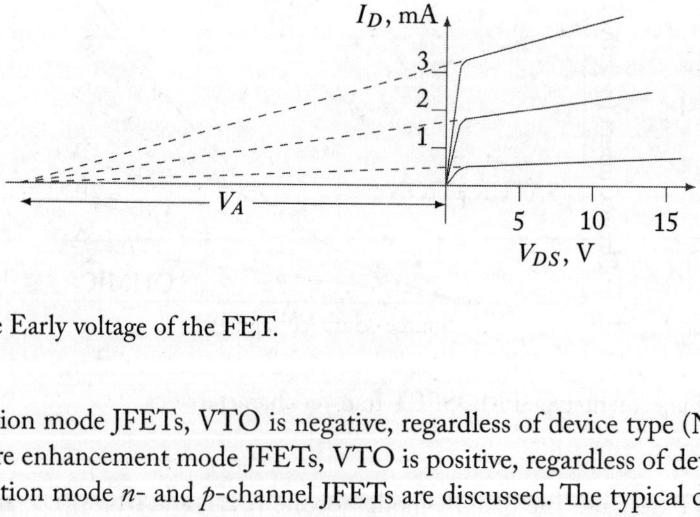

Figure 4.23: The Early voltage of the FET.

For depletion mode JFETs, VTO is negative, regardless of device type (NJF or PJF). For the extremely rare enhancement mode JFETs, VTO is positive, regardless of device type. In this book, only depletion mode n- and p-channel JFETs are discussed. The typical default values for the parameters are:

$$\text{VTO} = -2\,\text{V} \qquad \text{BETA} = 1\,\text{E} - 4\,\text{A/V}^2 \qquad \text{LAMBDA} = 0\text{V}^{-1}.$$

In modeling MOSFETs, the parameter values of interest are:

$$\begin{aligned}
\text{VTO} &= \text{zero-bias threshold voltage} = \pm|V_{PO}| \text{ for depletion MOSFETs} \\
&= \pm|V_T| \text{ for enhancement MOSFETs} \\
\text{KP} &= \text{transconductance coefficient} = \frac{2I_{DSS}}{V_{PO}^2} \text{ for depletion MOSFETs} \\
&= 2K \text{ for enhancement MOSFETs}^3 \\
\text{LAMBDA} &= \frac{1}{V_A} = \text{channel-length modulation}
\end{aligned}$$

An example of the .MODEL statement for an depletion NMOSFET is:

.MODEL MOS_TEST NMOS(VTO = −4 KP= 1.25E−3LAMBDA = 2E−6).

The default values for the parameters are:

$$\text{VTO} = 0\,\text{V} \qquad \text{BETA} = 2\text{E} - 5\,\text{A/V}^2 \qquad \text{LAMBDA} = 0\,\text{V}^{-1}.$$

VTO is positive (negative) for enhancement mode and negative (positive) for depletion mode n-channel (p-channel) devices:

VTO > 0 for enhancement NMOSFETs and depletion PMOSFETs
VTO < 0 for depletion NMSOFETs and enhancement PMOSFETs.

A summary of the FET parameter conversions to SPICE parameters is shown in Table 4.1.

4.3.2 FET AS A VOLTAGE VARIABLE RESISTOR

A voltage variable resistor (VVR) is a three terminal device where the resistance between two of the terminals is controlled by a voltage on the third. In the ohmic region, FETs demonstrate a variation in resistance, R_{DS}, described by Equations (4.9) and (4.33),

$$R_{DS}^{-1} = \frac{\partial I_D}{\partial V_{DS}} = \frac{2I_{DSS}}{V_{PO}}\left(\frac{V_{GS}}{V_{PO}} - 1\right)$$

$$R_{DS}^{-1} = \frac{\partial I_D}{\partial V_{DS}} = 2K\left(V_{GS} - V_T\right).$$

Written more conveniently as resistances, the drain-source resistances in the ohmic region of the n-JFET and of the depletion mode NMOSFET are:

$$R_{DS} = \frac{V_{PO}^2}{2I_{DSS}\left(V_{GS} - V_{PO}\right)}. \tag{4.40}$$

For the enhancement mode NMOSFET, the drain-source resistance is,

$$R_{DS} = \frac{1}{2K\left(V_{GS} - V_T\right)}. \tag{4.41}$$

[3]The channel width, W, and channel length, L, are here assumed to be equal (the default for SPICE). When converting manufacturer-specified MOSFET parameters to the parameters used in this text, the transconductance coefficient also depends on the ratio of those values: $K = 1/2KP(W/L)$.

Table 4.1: Summary of FET SPICE parameter conversions

FET type	SPICE parameter	Value
JFET	VTO	$-\lvert V_{PO}\rvert$
	BETA	I_{DSS}/V_{PO}^2
	LAMBDA	$1/V_A$
Depletion MOSFET	VTO	$\mp\lvert V_{PO}\rvert$
	KP	$2I_{DSS}/V_{PO}^2$
	LAMBDA	$1/V_A$
Enhancement MOSFET	VTO	$\pm\lvert V_T\rvert$
	KP	$2K$ (see footnote 3)
	LAMBDA	$1/V_A$

The n-channel JFET is used as an example in this section. Operation of n-JFETs in the ohmic region implies that the drain-source voltage is held small. Figure 4.24 illustrates typical n-JFET and resistor characteristics. The slope of V_{DS}/I_D is a function of V_{GS}, and the drain-source resistance R_{DS} is controlled by V_{GS}.

Figure 4.25 is the circuit diagram for a simple voltage controlled voltage divider using an n-JFET VVR to provide a means of voltage-controlling the divider ratio. As V_{GS} is changed from zero to V_{PO}, R_{DS} is changed.

The output voltage is,

$$V_{OUT} = V_{IN}\frac{R_L\left(1+R_{DS}^{-1}R_L\right)^{-1}}{R+R_L\left(1+R_{DS}^{-1}R_L\right)^{-1}}. \tag{4.42}$$

When V_{GS} approaches V_{PO}, R_{DS} approaches zero and the FET causes no signal attenuation. If $R_L \gg R_{DS}$, Equation (4.40) can be simplified to,

$$V_{OUT} = V_{IN}\frac{1}{1+RR_{DS}}. \tag{4.43}$$

Using Equation (4.9), The output voltage is expressed as a function of V_{GS},

$$V_{OUT} = \frac{1}{1+RR_{DSO}^{-1}\left[\left(\frac{V_{PO}-V_{GS}}{V_{PO}}\right)\right]}, \tag{4.44}$$

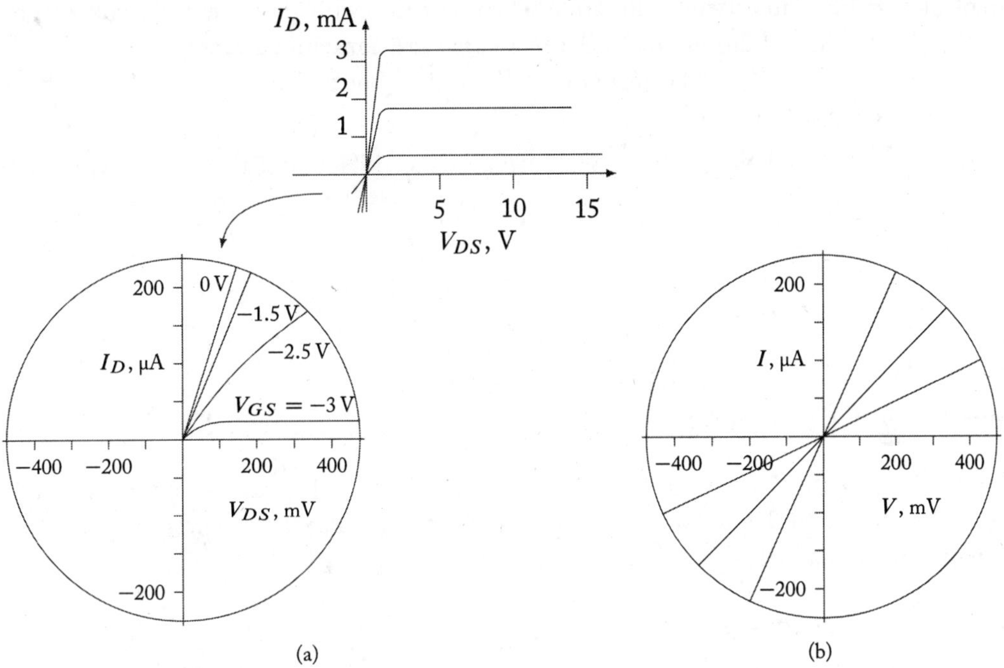

Figure 4.24: Comparison of n-JFET and resistor characteristics.

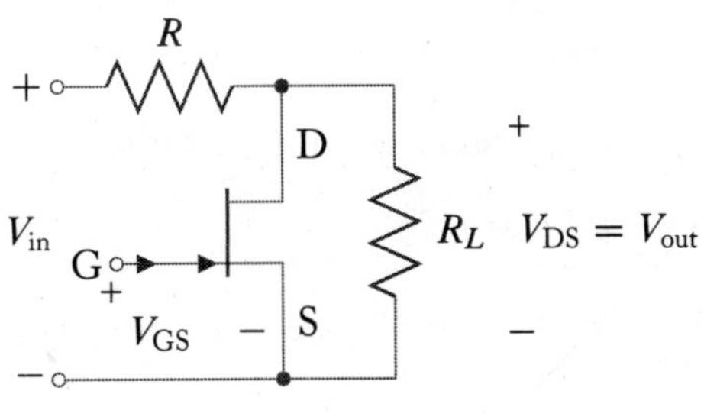

Figure 4.25: Simple voltage-controlled voltage divider.

where

$$R_{DSO} = \frac{\partial V_{GS}}{\partial I_D} \quad \text{at } V_{GS} = 0 \text{ and } V_{DS} = 0.$$

4.3.3 n-JFET AS A CONSTANT-CURRENT SOURCE

The ideal constant-current source would supply a given current to a load independent of the voltage across the load. In such a case, the output resistance of the source is infinite. The drain current of the JFET approaches saturation when operated with the gate-drain voltage greater than V_{PO}. Under this condition, the FET can be used as a constant current source.

One way to force V_{GD} to be greater than V_{PO} is to tie the gate of the n-JFET directly to the source as shown in Figure 4.26.

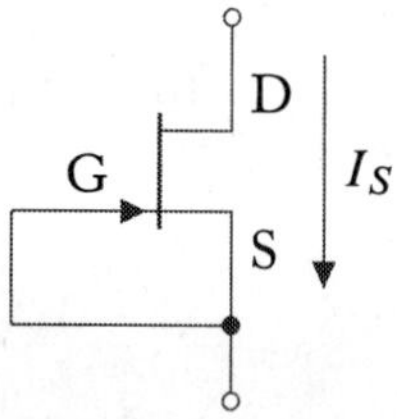

Figure 4.26: n-JFET constant current source.

In Figure 4.26, $V_{GS} = 0$, therefore in saturation, $V_{DS} = -V_{PO}$ and the drain current $I_D = I_{DSS}$. As long as the FET remains in saturation, it will provide a constant current flowing through the drain-source channel.

4.3.4 FET INVERTER

The methods used to analyze FET circuits are similar to that used to solve BJT circuits in Chapter 3. Three choices are available to analyze FET circuits. They are:

- Use the set of FET equations in Sections 4.1 and 4.2 along with additional circuit-dependent equations to obtain a numerical solution
- Use empirical V-I curves and obtain a graphical solution
- Use SPICE to obtain the solution[4]

As in BJT circuit analysis, the choice of technique used to analyze the circuit is strongly dependent on the complexity of the FET circuit.

Example 4.5
Consider an n-JFET inverter shown in Figure 4.27. The JFET characteristics are:

$$I_{DSS} = 8\,\text{mA}, \quad \text{and} \quad V_{PO} = -4\,\text{V}.$$

Find the drain current I_D and the drain-source voltage V_{DS} when $V_{GS} = -2\,\text{V}$.

[4]SPICE is the computer analysis tool of preference to the authors. Other computer simulation packages will produce similar results and often use the same computational engine as SPICE.

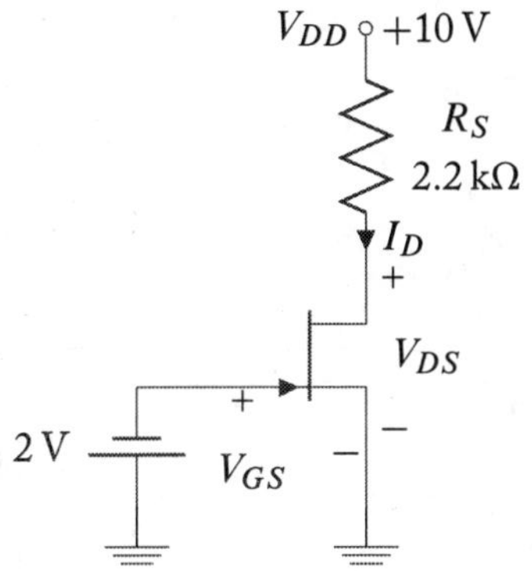

Figure 4.27: n-JFET inverter.

Solution #1 (Using characteristic equations for the n-JFET):

In order to find V_{DS}, an additional equation dependent on circuit topology is required. The loop equation needed is,

$$V_{DD} = I_D R_D + V_D.$$

Assume that the JFET is operating in saturation. The drain current can then be calculated from Equation (4.11):

$$I_D = I_{DSS}\left(1 - \frac{V_{GS}}{V_{PO}}\right)^2 = 8 \times 10^{-3}\left(1 - \frac{-2}{-4}\right)^2 = 2\,\text{mA}.$$

The drain-source voltage can then be obtained:

$$V_{DS} = V_{DD} - I_D R_D = 10 - (2 \times 10^{-3})(2200) = 5.6\ \text{V}.$$

To confirm that the n-JFET is operating in saturation, the condition for saturation is checked:

$$\begin{aligned} &V_{DS} \geq V_{GS} - V_{PO} \\ &5.6 \geq -2 - (-4) \\ &5.6 \geq 2. \end{aligned}$$

The saturation region assumption is verified and the calculations are valid.

Solution #2 (Graphical method):

The n-JFET output curves to perform a load-line analysis of the circuit must first be plotted. In the load-line method of analysis, the solution (operating point) of the n-JFET circuit is intersection of the straight line representing the constant load (R_D) and the curve for the desired V_{GS}. The values of I_D and V_{DS} corresponding to the intersection of the load line and the V_{GS} curve is the solution. The graphical solution of this Example is shown in Figure 4.5. The intersection

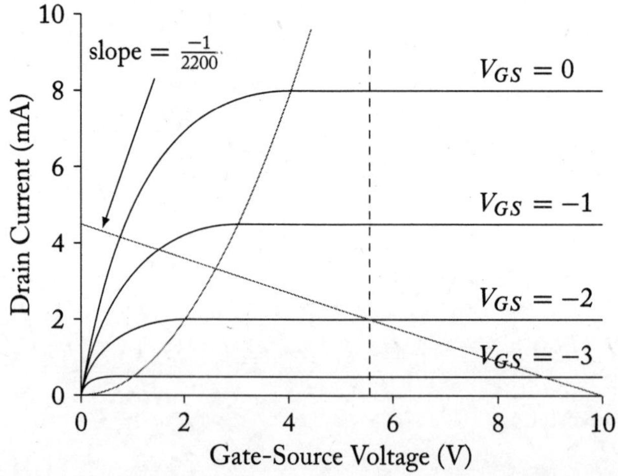

of the load-line and the $V_{GS} = -2$ V line occurs at $I_D = 2$ mA, and $V_{DS} = 5.6$ V, confirming solution #1. Also note that the solution lies in the saturation region.

The graphical technique also provides solutions for this circuit with other values of the source connected to the gate of the *n*-JFET. The intersection of the load-line and the plotted curves for these values of V_{GS} are:

V_{GS} (V)	I_D (mA)	V_{DS} (V)
−4	0.0	10
−3	0.5	8.9
−2	2.0	5.6
−1	3.7	1.8
0	4.0	1.2

Notice that the intersections of the load-line and the JFET curves for $V_{GS} = 0$ V and $V_{GS} = -1$ V occur in the ohmic region.

If other gate-source voltage values are chosen, the analysis must be expanded in one of two ways:

- additional JFET curves must be plotted, or
- the solutions must be obtained by interpolation between the curves.

Solution #3 (SPICE Solution):

The SPICE solution is shown in Figure 4.28. This solution is in the form of a Multisim DC operating point solution using a probe. The JFET SPICE parameters of interest are VTO $= -4$ and BETA $= 500 \times 10^{-6}$.

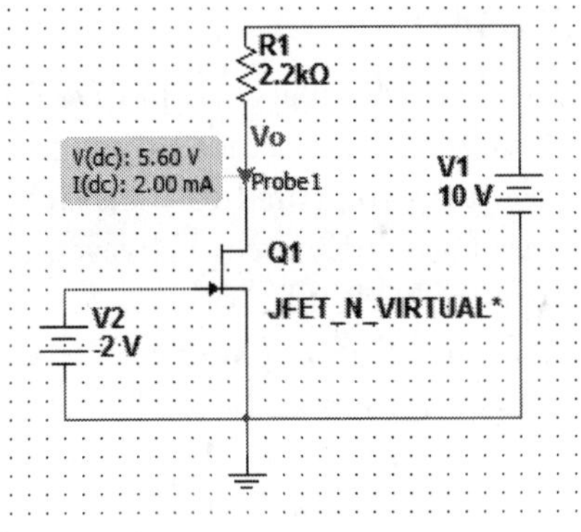

Figure 4.28: SPICE analysis of n-JFET inverting amplifier.

The SPICE results match those of the analytical and graphical approaches shown in Solution #1 and #2. For the SPICE simulation, the output current and voltage have also been calculated for additional values of the voltage at the gate of the JFET using DC sweep analysis. A plot of the resultant output voltage as a function of the input gate-source voltage is shown in Figure 4.29. Notice that for low values of input ($V_{GS} \approx -4$ V) the output is high ($V_{DS} \approx 10$ V): for high value of the input ($V_{GS} \approx 0$ V) the output is low ($V_{DS} \approx 0$ V). The output forms a digital inverse of the input. In the region between these two extremes the output is roughly a negative multiple of the input: a circuit of this topology forms a small-signal inverting amplifier. FET amplifiers are explored in Chapter 5 (Book 2).

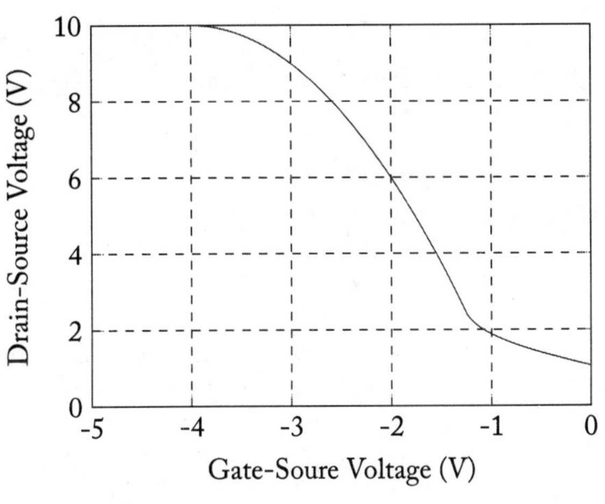

Figure 4.29: SPICE analysis: output voltage vs. input voltage.

4.3.5 FET AS AN ACTIVE LOAD

In some applications, a FET active load to replace the passive resistor used as a load in an inverting FET amplifier may be desirable. For instance in integrated circuit fabrication, passive resistors consume large areas of the chip compared to transistors. Both enhancement and depletion NMOSFET active loads are commonly used. However, the load-line characteristics are quite different between the two types of NMOSFETs.

An enhancement NMOSFET inverting amplifier with an enhancement NMOSFET load is shown in Figure 4.30. In this circuit, Q1 acts as the load and Q2 is the driver.

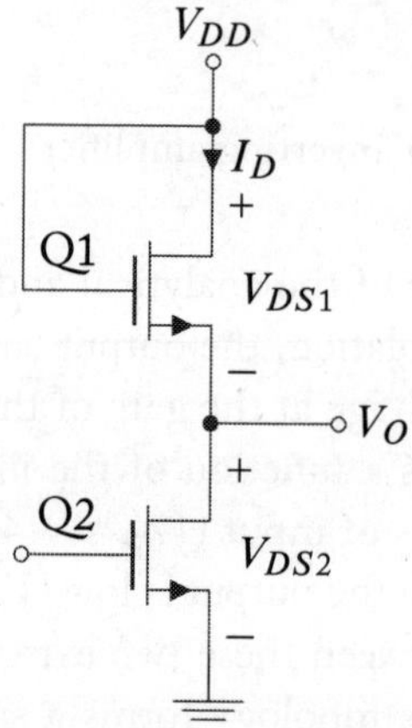

Figure 4.30: Enhancement NMOSFET inverting amplifier. with enhancement NMOSFET active load.

By studying the circuit, several interesting relationships can be found. The output voltage V_O is,

$$V_O = V_{DS2}.$$

Since the gate of Q1 is connected to its own drain,

$$V_{GS1} = V_{DS1}.$$

Because the Q2 gate-source and drain-source voltages are equal, Q2 is in saturation since THE difference between the gate-source and threshold voltages will always be greater than the drain-source voltage,

$$V_{DS} \geq V_{GS} - V_T.$$

The relationship between the two NMOSFETs is readily apparent using elementary circuit analysis. The sum of the voltages V_{DS1} and V_{DS2} must always equal the source voltage V_{DD},

$$V_{DS1} + V_{DS2} = V_{DD}.$$

The load-line created by Q1 can be found by replacing V_{DS1} with $V_{DD} - V_{DS2}$,

$$\begin{aligned} I_{D1} &= K\,(V_{GS1} - V_T)^2 \\ &= K\,[(V_{DD} - V_{DS2}) - 2]^2 . \end{aligned} \tag{4.45}$$

The Q1 load-line superimposed on the Q2 characteristic equation is shown in Figure 4.31. The curves were created using

$$K = 1/2KP = 625 \times 10^{-6} \qquad V_T = 2\,\text{V} \qquad V_{DD} = 10\,\text{V}.$$

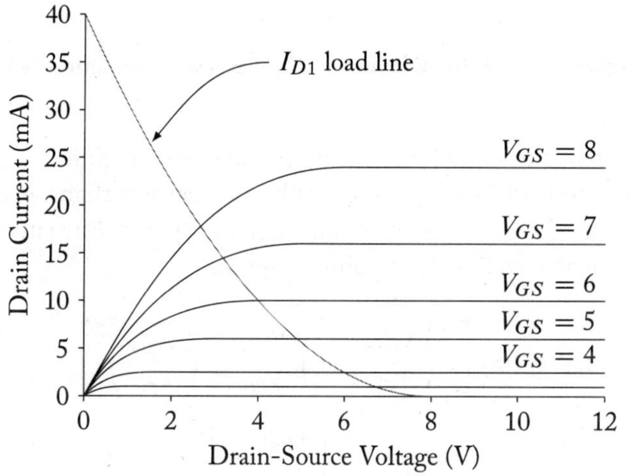

Figure 4.31: Enhancement NMOSFET inverting amplifier with enhancement NMOSFET load.

The load-line ends at

$$V_{DD} - V_{T1} = 8\,\text{V} \qquad I_D = K(V_{DD} - V_T)^2 = 40\,\text{mA}.$$

If the load line is plotted using Equation (4.45), the load-lines curves back up at $V_{DS} > V_{DD} - V_T$. In real circuits, the line ends at $V_{DS} = V_{DD} - V_T$.

Another common configuration is the enhancement NMOSFET driver with a depletion NMOSFET active load shown in Figure 4.32.

Several circuit relationships are evident from studying the circuit in Figure 4.32,

$$V_O = V_{DS2} \qquad V_{GS1} = 0 \qquad V_{DS1} + V_{DS2} = V_{DD}.$$

The depletion NMOSFET Q1 is in saturation only when

$$V_{DS1} = V_{DD} - V_{DS2} \geq V_{GS1} - V_{PO} \geq -V_{PO}.$$

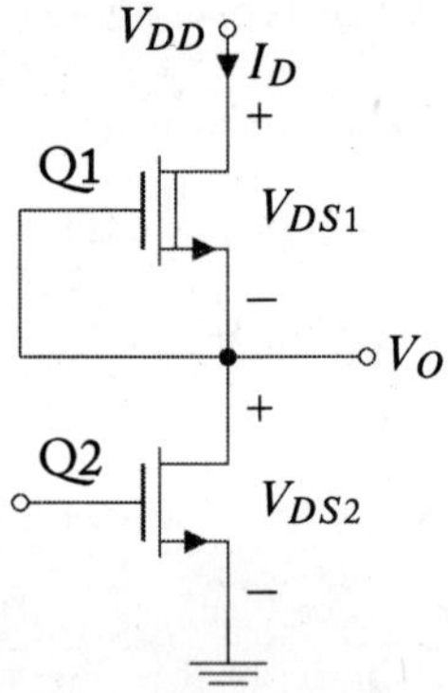

Figure 4.32: Enhancement NMOSFET inverting amplifier with depletion NMOSFET active load.

Since $V_{GS} = 0$, the current through Q1 is I_{DSS} in the saturation region.

As V_{DS2} increases toward V_{DD}, V_{DS1} eventually becomes less than $-V_{PO}$ causing Q1 to operate in the ohmic region. As Q1 enters its ohmic region, the load-line curves down toward V_{DD} following the characteristic equation of the ohmic region,

$$\begin{aligned} I_D &= I_{DSS}\left[2\left(\frac{V_{GS1}}{V_{PO}} - 1\right)\frac{V_{DS1}}{V_{PO}} - \left(\frac{V_{DS1}}{V_{PO}}\right)^2\right] \\ &= I_{DSS}\left[-2\frac{V_{DS1}}{V_{PO}} - \left(\frac{V_{DS1}}{V_{PO}}\right)^2\right] \\ &= I_{DSS}\left[-2\frac{V_{DD} - V_{DS2}}{V_{PO}} - \left(\frac{V_{DD} - V_{DS2}}{V_{PO}}\right)^2\right]. \end{aligned} \tag{4.46}$$

The resulting graphical solution is shown in Figure 4.33.

4.3.6 CMOS INVERTER

Complementary symmetry MOS (CMOS) circuits are fabricated with both enhancement NMOSFETs and enhancement PMOSFETs on the same chip. The main advantage of CMOS technology is the ability to design circuits with essentially zero DC power dissipation. Power is dissipated only during switching transitions. The CMOS inverter is shown in Figure 4.34. In Figure 4.34 Q1, the enhancement PMOSFET, is the load and Q2, the enhancement NMOSFET is the driver.

Qualitatively, the operation of the CMOS inverter is simple. When $V_I = 0$, $V_{SG1} = V_{DD}$, and $V_{GS2} = 0$. When the gate-source voltage of the enhancement NMOSFET is zero, Q2 is cut-off. Therefore, $V_O = V_{DD}$.

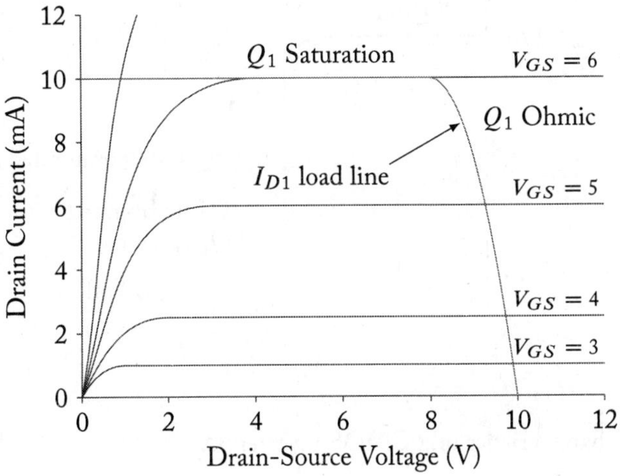

Figure 4.33: Enhancement NMOSFET inverting amplifier.

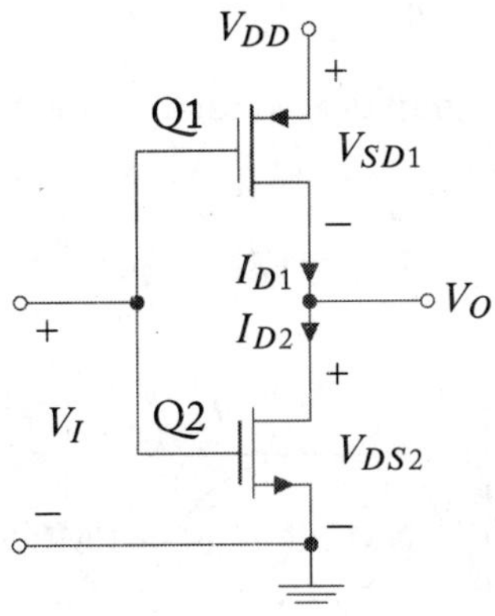

Figure 4.34: CMOS inverter.

When $V_I = V_{DD}$, $V_{GS2} = V_{DD}$ and $V_{SG1} = 0$. Since the source-gate voltage of the Q1 the enhancement PMOSFET is zero, Q1 is cut-off. Therefore, $V_O = 0$.

Note that at logic zero and logic one outputs ($V_O = 0$ and V_{DD}), there is no current flowing though the circuit. Therefore, the CMOS inverter only dissipates power when transitioning between the two logic levels.

A SPICE simulation clearly shows the transfer function of the CMOS inverter. The transfer function is shown in Figure 4.35.

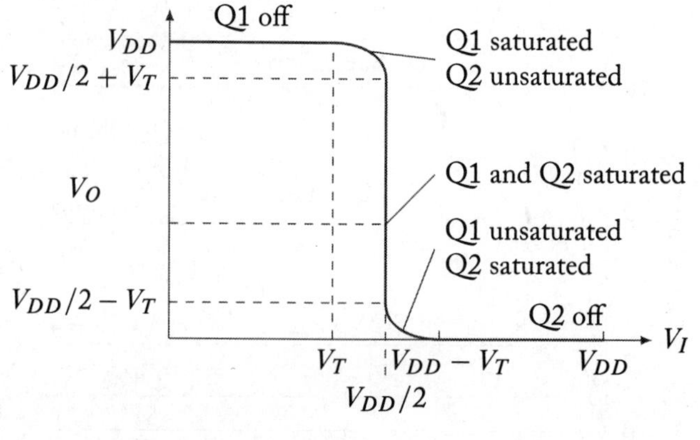

Figure 4.35: Transfer characteristics of a CMOS inverter.

Load-line analysis of the CMOS circuit is similar to other active load analysis. The circuit yields the following relationships,

$$V_O = V_{DS2} \qquad V_1 = V_{GS1} = V_{GS2}.$$

Using the condition for saturation, a graphical solution of the CMOS inverter can be found. The load-lines are shown in Figure 4.36.

$$V_{DS} \geq V_{GS} - V_T.$$

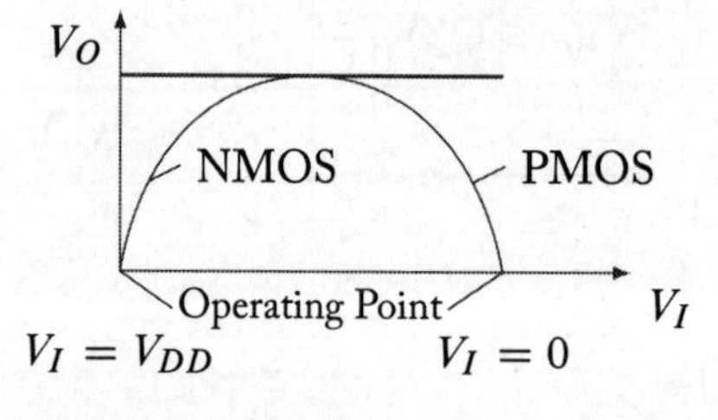

Figure 4.36: Load-line analysis of a CMOS inverter.

4.4 REGIONS OF OPERATIONS IN FETS

FET operation has been seen to fall into three regions of useful operation. The regions are described by the state of the drain-source channel controlled by the gate voltage. The three regions of interest are shown in Figure 4.37. Briefly, the three regions of operation are:

1. The cut-off region is defined as that region where the gate voltage disallows charge flow in the drain-source channel. The FET essentially looks like an open circuit. Applications for this region are primarily in switching and digital logic circuits.

2. The ohmic region is defined by a gradual increase in charge flow in the drain-source channel, the rate of which is controlled by the gate voltage. Applications for this region include the use of FETs as voltage variable resistors.

3. The saturation region. This region is defined by constant charge flow in the drain-source channel without regard to the drain-source voltage. The amount of constant current flow is regulated by the gate voltage. The region is commonly used for amplification with the modulation of the gate voltage and for constant current source applications.

In addition to these three regions, there is an additional region of severe consequences: the breakdown region. FETs are not manufactured to withstand extended use in the breakdown region and will exhibit catastrophic thermal runaway and destruction. To insure proper operation of the devices, manufacturers' specifications on maximum voltages must be heeded.

Due to the JFET fabrication process, the drain and source can, in most cases, be interchanged when the JFET is used as a circuit element without affecting the desired operating characteristics. This property (drain-source reciprocity) is not common to all types of FETs.

The characteristics of the FETs in the different regions are shown in Table 4.2.

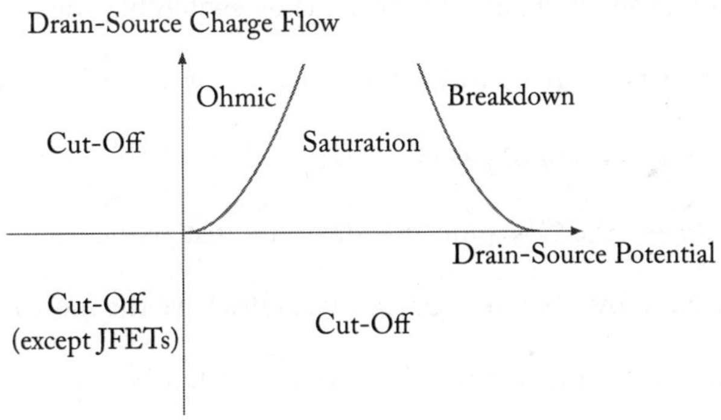

Figure 4.37: FET regions of operation.

4.5 THE FET AS AN ANALOG SWITCH

Both Bipolar Junction and Field Effect Transistors can be used as switches in a wide variety of analog electronic applications. Each type provides a great advantage over mechanical switches in both speed, reliability, and resistance to deterioration. FETs are the more common choice due to the inherent symmetry of FETs and the undesirable offset voltage (V_{CE} at $I_C = 0$) that is present in BJTs. The offset voltage of BJTs, typically on the order of a few millivolts, can produce significant errors in the transmission of low-level analog signals. Another advantage of FET switches is the high gate input impedance and the consequent low load that the voltage control port presents to control electronics.

Among the many possible electronic applications of solid-state analog switches are the following:

- Sample-and-hold circuits
- switchable gain amplifiers
- switched-capacitor filters
- digital-to-analog conversion
- signal gating and squelch control
- chopper stabilization of amplifiers

In addition, multiple switches connected to share a common output form a multiplexer: a common building block for analog and digital systems.

The basic function of a switch is to electrically isolate or connect two sections of a circuit. In order to accomplish that functionality, an ideal analog switch has the following design goals:

- in the ON state, it passes signals without attenuation or distortion
- in the OFF state, signals are not passed.
- the transitions between the ON and OFF states are instantaneous.

In order to accomplish the above design goals, real switches have the more realistic specifications:

- very low ON resistance compared to other circuit resistances
- very high OFF resistance compared to other circuit resistances
- low leakage currents in the OFF state
- low capacitive and/or inductive effects

FETs satisfy these specifications satisfactorily for many applications. Control signals applied to the gate-source port of a FET will vary the drain-source port between ON and OFF switch positions. Reasonably low ON resistance combined with extremely high OFF resistance make the FET ideal as a voltage controlled analog switch element. In Section 4.3.2 it was seen

Table 4.2: FET characteristics

FET Type	Region: Ohmic	Region: Saturation	Region: Cut-Off
n-JFET	$0 < V_{DS} \le V_{GS} - V_{PO}$ $I_D = I_{DSS}\left[2\left(\frac{V_{GS}}{V_{PO}} - 1\right)\frac{V_{DS}}{V_{PO}} - \left(\frac{V_{DS}}{V_{PO}}\right)^2\right]$	$V_{DS} \ge V_{GS} - V_{PO}$ $I_D = I_{DSS}\left(1 - \frac{V_{GS}}{V_{PO}}\right)^2$	$V_{GS} < V_{PO}$ $I_D = 0$
p-JFET	$0 < V_{SD} \le V_{SG} + V_{PO}$ $I_D = I_{DSS}\left[2\left(\frac{V_{GS}}{V_{PO}} - 1\right)\frac{V_{DS}}{V_{PO}} - \left(\frac{V_{DS}}{V_{PO}}\right)^2\right]$	$V_{SD} \ge V_{SG} + V_{PO}$ $I_D = I_{DSS}\left(1 - \frac{V_{GS}}{V_{PO}}\right)^2$	$V_{SG} < -V_{PO}$ $I_D = 0$
Enhancement NMOSFET	$0 < V_{DS} \le V_{GS} - V_T$ $I_D = K[2(V_{GS} - V_T)V_{DS} - {V_{DS}}^2]$	$V_{DS} \ge V_{GS} - V_T$ $I_D = K(V_{GS} - V_T)^2$	$V_{GS} < V_T$ $I_D = 0$
Depletion NMOSFET In depletion mode: $I_D > 0;\ V_{PO} < V_{GS} \le 0;\ V_{DS} > 0$ In enhancement mode: $I_D > 0;\ V_{GS} > 0;\ V_{DS} > 0$	$0 < V_{DS} \le V_{GS} - V_{PO}$ $I_D = I_{DSS}\left[2\left(\frac{V_{GS}}{V_{PO}} - 1\right)\frac{V_{DS}}{V_{PO}} - \left(\frac{V_{DS}}{V_{PO}}\right)^2\right]$	$V_{DS} \ge V_{GS} - V_{PO}$ $I_D = I_{DSS}\left(1 - \frac{V_{GS}}{V_{PO}}\right)^2$	$V_{GS} < V_{PO}$ $I_D = 0$
Enhancement PMOSFET	$0 < V_{SD} \le V_{SG} + V_T$ $I_D = K[2(V_{GS} - V_T)V_{DS} - {V_{DS}}^2]$	$V_{SD} \ge V_{SG} + V_T$ $I_D = K(V_{GS} - V_T)^2$	$V_{SG} < -V_T$ $I_D = 0$
Depletion PMOSFET In depletion mode: $I_D < 0;\ -V_{PO} < V_{SG} \le 0;\ V_{SD} > 0$ In enhancement mode: $I_D < 0;\ V_{SG} > 0;\ V_{SD} > 0$	$0 < V_{SD} \le V_{SG} + V_{PO}$ $I_D = I_{DSS}\left[2\left(\frac{V_{GS}}{V_{PO}} - 1\right)\frac{V_{DS}}{V_{PO}} - \left(\frac{V_{DS}}{V_{PO}}\right)^2\right]$	$V_{SD} \ge V_{SG} + V_{PO}$ $I_D = I_{DSS}\left(1 - \frac{V_{GS}}{V_{PO}}\right)^2$	$V_{SG} < -V_{PO}$ $I_D = 0$

that the drain-source resistance, R_{DS}, for a FET is highly dependent on the gate-source voltage.[5] The drain-source resistance expression for depletion type FETs is

$$R_{DS} = \frac{V_{PO}^2}{2I_{DSS}\left(V_{GS} - V_{PO}\right)}, \qquad V_{GS} \geq V_{PO}. \tag{4.47}$$

For enhancement type FETs,

$$R_{DS} = \frac{1}{2K\left(V_{GS} - V_T\right)}, \qquad V_{GS} \geq V_T. \tag{4.48}$$

Variation in the control signal, V_{GS}, can change this resistance from a the range of a few ohms to many Megohms. A simple application of an analog switch using a single FET is shown in Figure 4.38. Such simple analog switches typically use enhancement mode FETs although it is possible to form a switch with depletion mode FETs.[6] In order to keep the switch in the OFF state for all values of the input voltage, v_s, the control voltage, v_c, must be less than the minimum input signal value plus the threshold voltage of the FET:

$$v_{c(OFF)} < v_{s(\min)} + V_T. \tag{4.49}$$

Similarly, to keep the switch in the ON state for all values of the input voltage, the control voltage must be greater than the maximum input signal value plus the threshold voltage of the FET:[7]

$$v_{c(ON)} > v_{s(\max)} + V_T. \tag{4.50}$$

Equations (4.49) and (4.50) provide expressions for that absolute limits of ON and OFF control voltages. In order to ensure good switch performance, it is necessary to provide a design margin beyond these absolute limits.

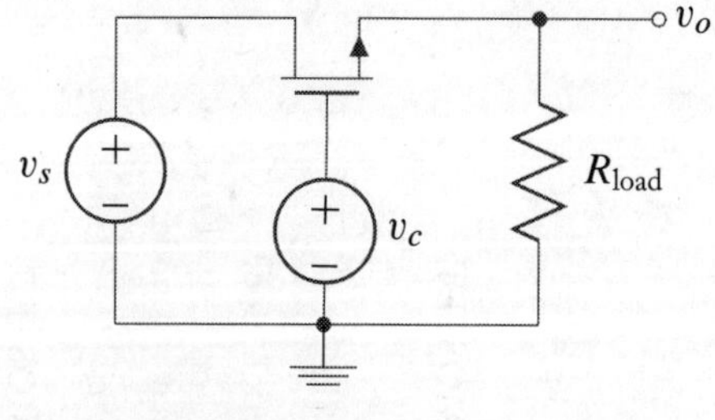

Figure 4.38: A Simple FET analog switch application.

[5]For simplicity of presentation, only the n-channel FET equations are given. Shown are the equations for drain-source resistance for depletion mode and enhancement mode FETs respectively.
[6]Enhancement mode FETs will be used in the discussions of this section.
[7]Care must be takes so that the FET gate breakdown voltage is not exceeded when choosing control voltages. Typical gate breakdown voltages are in excess of 25 V.

Example 4.6
The FET Analog Switch of Figure 4.38 is used on the output of an OpAmp whose power supplies are set at ± 15 VDC. These voltages are chosen as the control voltages for the switch. If the threshold voltage for the FET is 2 V, what range of output voltages will be properly controlled by the switch?

Solution:

The minimum signal voltage is given by Equation (4.49):

$$v_{s(\text{min})} > v_{c(OFF)} - T,$$

or

$$v_{s(\text{min})} > -15\,\text{V} - 2\,\text{V} = -17\,\text{V}.$$

But, for this case the signal voltage is limited by the output of the OpAmp to ≈ -15 V.

The maximum signal voltage is given by Equation (4.50):

$$v_{s(\text{max})} < v_{c(ON)} - V_T,$$

or

$$v_{s(\text{max})} < 15\,\text{V} - 2\,\text{V} = +13\,\text{V}.$$

Thus, the absolute maximum signal range is limited range to:

$$-15\,\text{V} < v_s < +13\,\text{V}.$$

Good design practice would place tighter limits the on the signal. These limits are functions of whatever additional design specifications may apply to the particular application of this switch.

One of the problems associated with the simple FET analog switch of Figure 4.38 is that the ON resistance of the switch is not constant as the input signal varies. Variation in ON resistance can be a serious limitation in some circuit applications as it can cause distortion of the analog signal. Variation in the input signal will cause the voltage at the source of the FET to vary. If the gate terminal of the FET is set at a specific control voltage level, the input signal variation therefore produces a variation in V_{GS} for the FET and consequently a variation in the drain-source resistance of the switch. Determination of the switch resistance as a function of input signal level requires the solution of two non-linear equations. For this particular circuit the equations reduce to:

$$R_{DS} = \frac{V_s}{i} - R_{\text{load}}, \tag{4.51}$$

and

$$R_{DS} = \frac{1}{V_{c(ON)} - iR_{\text{load}} - V_T}. \tag{4.52}$$

Computer simulation or the application of load line techniques are perhaps the two best methods to calculate the variation of switch resistance with input. A typical graph[8] of the switch resistance as a function of input signal is given in Figure 4.39. Notice that while the switch resistance is fairly small throughout most of the input signal range ($24\,\Omega < R_{DS} < 100\,\Omega$ as $-15\,\text{V} < v_s < 6.6\,\text{V}$), it increases dramatically as the level of the input signal approaches the positive control voltage. This drastic increase in switch resistance can greatly diminish the utility of such a simple switch. The OFF resistance of this switch is typically very high ($\geq 20\,\text{M}\Omega$) and the OFF performance of such a switch is limited only by a very small leakage current in the FET.

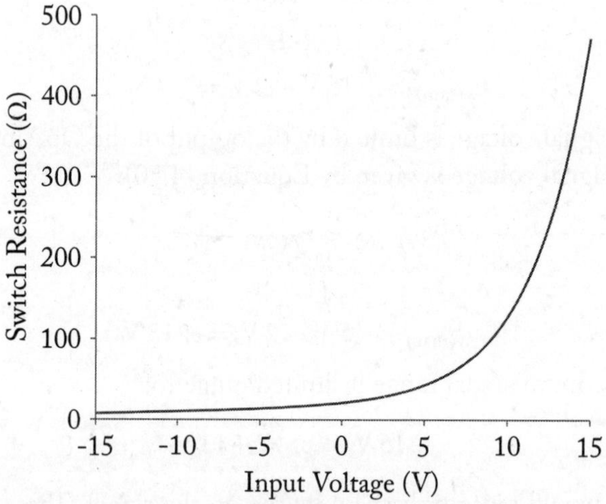

Figure 4.39: Switch resistance as a function of input signal.

A common approach to designing a better analog switch using FETs is shown in Figure 4.40. This switch is a parallel combination of a two switches: one constructed with an *n*-channel FET, the other with a *p*-channel FET. The triangular symbol is an inverter: it reverses the controlsignal levels so that the *p*-channel FET will operate with the *n*-channel FET.

It has been shown that the *n*-channel FET in this configuration will have low ON resistance for inputs signals near the negative limits. The *p*-channel FET will have low ON resistance for inputs signals near the positive limits. Since the total ON resistance of this parallel switch is the parallel combination of the two individual switch resistances, the switch has nearly constant ON resistance throughout the range of possible input signals. This near-constant resistance property is demonstrated in Figure 4.41.[9] In this example, the switch has an ON resistance of approximately $22.2\,\Omega$ throughout most of the possible range of input voltage:

$$v_s \leq v_{C(ON)} - V_T. \tag{4.53}$$

[8]Figure 4.39 was generated using the circuit parameters of Example 4.6, $R_{\text{load}} = 1\,\text{k}\Omega$, and $K = 1.5\,\text{mA/V}^2$.
[9]Figure 4.41 was generated using the same circuit and FET parameters as in Figure 4.39.

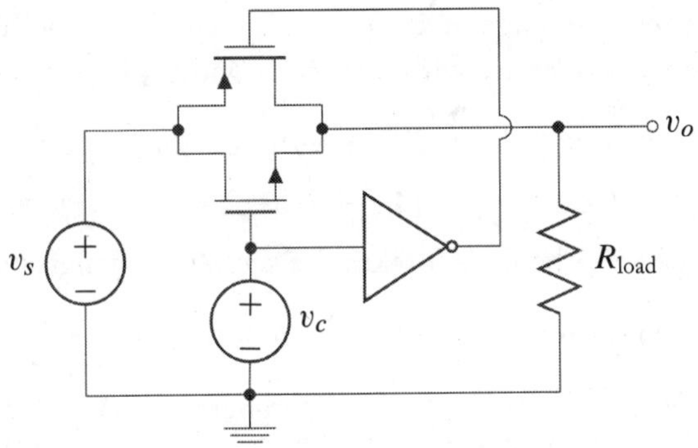

Figure 4.40: Parallel CMOS analog switch.

When the input voltage level exceeds the constraint of Equation (4.53), the input resistance drops slightly. In addition, the OFF resistance of this switch is extremely high: the OFF performance of this switch is limited by a very small leakage current through the FETs. The OFF resistance *suffers a significant degradation* when the limit of Equation (4.53) is exceeded: Input signals levels must be limited to these constraints.

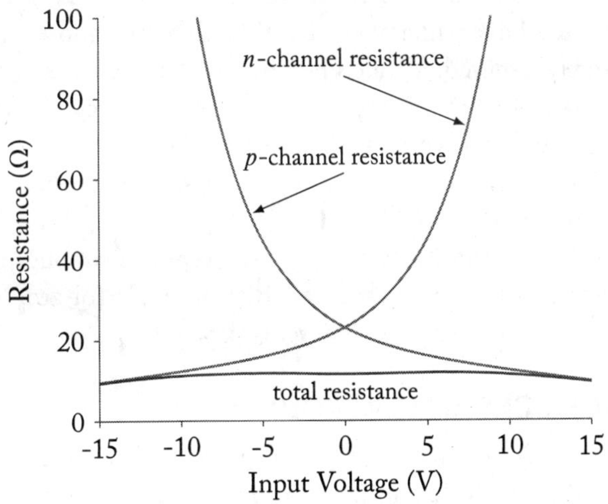

Figure 4.41: Complementary switch resistance as a function of input signal.

FET analog switches have specific limitations that are usually described in the specifications provided by the manufacturer. Among the most common of the limitations are:

- *Analog Output Leakage Current*—the algebraic sum of currents from the power supplies, ground, input signal, and control signal through a OFF switch.

- *Analog Voltage Range*—the range of analog voltage amplitudes with respect to ground over which the analog switch operates within the ON and OFF specifications.
- ON *Resistance* & ON *Resistance Variation*—the resistance of the switch over the analog voltage range.
- *Output Switching Times*—the time it takes the switch to change states.
- *Switch Current*—the maximum amount of current that can be fed through the switch.

In addition to the MOSFET switches described here, *n*-channel JFETs are used for analog switches. In order to maintain a depletion mode JFET switch in the ON state a rather complicated electronic control circuit is necessary. This control circuitry is usually fabricated on the same semiconductor chip as the switch and consists of both bipolar and JFET devices. These switches have a very constant ON resistance over the entire analog signal range. The disadvantage of these switches is their relatively high cost.

4.6 BIASING THE FET

The selection of an appropriate quiescent operating point for a FET is determined by conditions similar to those for a BJT. Here, the quiescent conditions are the zero-input DC values of the FET drain current, I_D, and the terminal voltage differences, V_{GS}, and V_{DS}. In this section, several bias circuits for FETs are examined. While bias circuitry can be used to put the FET in any of its regions of operation, the focus of this section is the saturation region. The saturation region is, of course, significant as the region where FET amplification occurs. While the examples in this section use *n*-channel FETs, *p*-channel FETs are biased in a similar manner except for a change in polarity of the voltage supplies.

The emphasis in this section in on biasing using voltage supplies and resistors. It is also common to bias transistors using current sources. Biasing FETs for amplifier applications using current sources is discussed extensively in Chapter 6 (Book 2).

4.6.1 THE SOURCE SELF-BIAS CIRCUIT

The Source self-bias circuit shown in Figure 4.42 is particularly useful in biasing JFETs and depletion mode FETs of other types. In this application, the external gate resistor, R_g serves to tie the voltage at the gate of the FET to ground (the gate current is essentially zero). This resistor is typically chosen to be a very large value (often on the order of a few Megohms). For a specified drain current, I_D, the gate-source voltage, V_{GS}, can be determined from the FET transfer characteristic in either analytic or graphical form. R_s is then determined from V_{GS} and I_D:

$$R_s = -\frac{V_{GS}}{I_D}. \tag{4.54}$$

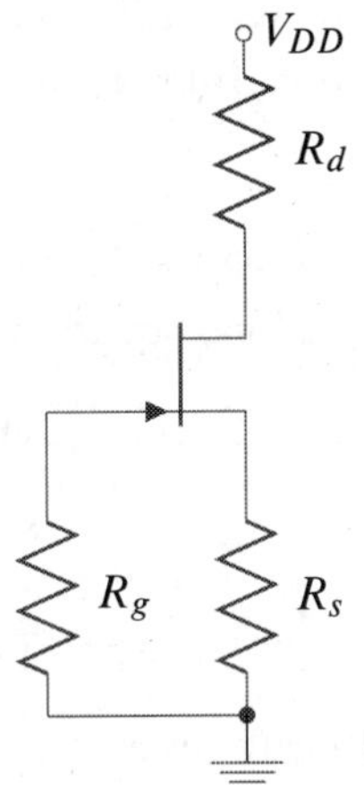

Figure 4.42: A source self-bias circuit.

Example 4.7
An n-channel JFET with the following characteristics:

$$V_{PO} = -3.5\,\text{V} \qquad \text{and} \qquad I_{DSS} = 10\,\text{mA}$$

is placed in the source self-bias circuit of Figure 4.42. Quiescent conditions of

$$I_D = 5\,\text{mA} \qquad \text{and} \qquad V_{DS} = 5\,\text{V},$$

are desired. Determine the bias resistors necessary to establish the quiescent conditions if the power supply voltage is: $V_{DD} = 15\,\text{V}$.

Solution:

The quiescent conditions for the JFET distinctly imply that it is in saturation:

$$V_{DS} \geq V_{GS} - V_{PO}.$$

Therefore the Equation (4.11) is the determining factor in finding V_{GS}:

$$I_D = I_{DSS}\left[1 - \frac{V_{GS}}{V_{PO}}\right]^2.$$

Substitution of values yields:

$$\frac{5}{10} = \left[1 - \frac{V_{GS}}{-3.5}\right]^2 \quad \text{or } V_{GS} = 3.5\left(\pm\frac{1}{\sqrt{2}} - 1\right) = -1.025\,\text{V},\ -5.975\,\text{V}.$$

The value between 0 V and V_{PO} is the only valid solution to Equation (4.11): the other is a spurious result caused by taking the square root. Thus, $V_{GS} = -1.025$ V. The resistor R_S can now be determined:

$$R_S = -\frac{-1.025\,\text{V}}{5\,\text{mA}} = 205\ \Omega\,.$$

The resistor R_D is determined by writing a loop equation around the drain-source loop:

$$I_D R_S + V_{DS} + I_D R_D - V_{DD} = 0,$$

or

$$R_d = \frac{V_{DD} - V_{DS}}{I_D} - R_s = \frac{15 - 5}{5\,\text{m}} - 205 = 1.795\ \text{k}\Omega\,.$$

The Resistor R_g is typically chosen arbitrarily large:

$$R_g = 1\ \text{M}\Omega\,.$$

One drawback of the source self-bias circuit is that the quiescent conditions are sensitive to variation in the FET parameters V_{PO} and I_{DSS}. The restriction of Equation (4.54),

$$V_{GS} = -I_D R_d\,,$$

coupled with the FET characteristic equations potentially leads to wide variation in quiescent condition. This variation is best described graphically as in Figure 4.43. In the circuit of Example 4.7, a FET with the following parameter variations is used.

$$8\,\text{mA} < I_{DSS} < 12\,\text{mA} \qquad\qquad -4\,\text{V} < V_{PO} < -3\,\text{V}.$$

If the design procedure results of Example 4.7 are used, the resultant ranges in quiescent conditions due to FET parameter variation can be read off Figure 4.43. They are:

$$-1.20\,\text{V} < V_{GS} < -0.85\,\text{V}\ \text{and}\ 4.13\,\text{mA} < I_D < 5.87\,\text{mA}.$$

The variation in FET parameters has caused a change in the two quiescent quantities of more than $\pm$ 17% from the nominal quiescent conditions calculated in Example 4.7. In many cases it is necessary to control the quiescent point to a greater degree than is possible with the source self-bias circuit.

4.6.2 THE FIXED-BIAS CIRCUIT

In addition to sensitivity to FET parameter variation, the source self-bias circuit mandates overly restrictive constraints on the external bias resistors in many application. As will be seen in Chapter 5 (Book 2), the resistors, R_d and R_s, play important roles in determining amplifier gain and

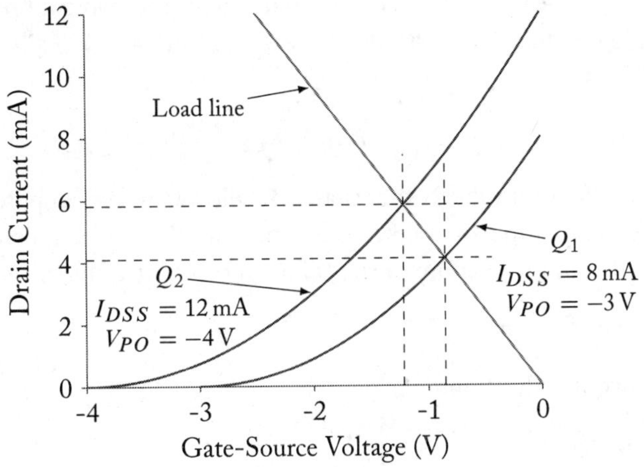

Figure 4.43: The change in quiescent point due to FET parameters variation.

output impedance. It is a rare occurrence when quiescent conditions are more significant than these two amplifier performance factors. The restrictive nature of the source self-bias circuit is centered at the holding of the FET gate terminal at ground potential. Removing that restriction greatly improves the versatility of a bias scheme. The Fixed-Bias circuit of Figure 4.44 is a simple depletion-mode FET bias circuit that allows the gate to be at voltages other than ground.

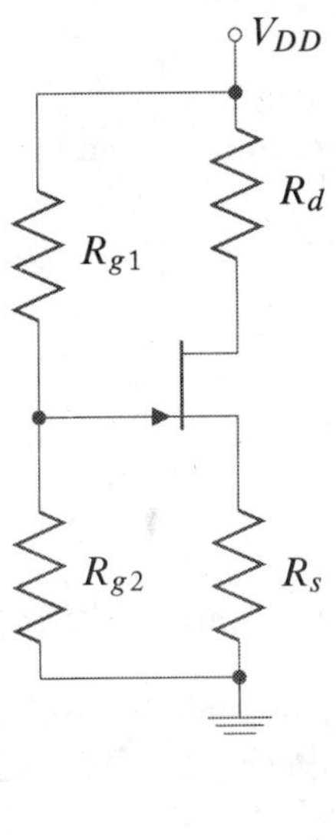

Figure 4.44: A fixed-bias circuit.

Since the gate current in the FET is essentially zero, the two gate resistors R_{g1} and R_{g2} set the voltage at the gate at any desired value between ground and V_{DD}:

$$V_G = \frac{R_{g2}}{R_{g1} + R_{g2}} V_{DD}. \tag{4.55}$$

Setting the FET gate voltage at values other than zero allows for a wider possible choice of source and drain resistances to accomplish specific quiescent conditions. Design choices for these resistors are based primarily on one or more of the following specific criteria:

- Amplification specifications
- Output resistance specifications
- Quiescent point stability

The first two of these design criteria will be discussed in Chapter 5 (Book 2): quiescent point stability involves reducing the variation in drain current due to FET parameter variation. The relationship of drain current to the external resistors in a fixed-bias circuit is determined by the voltage across the drain resistor:

$$I_D = \frac{V_S}{R_s} = \frac{V_G - (V_G - V_S)}{R_s} = \frac{V_G - V_{GS}}{R_s}, \tag{4.56}$$

where V_G is given by Equation (4.55). The variation of I_D due to variation of the FET parameter, V_{GS}, is inversely proportional to the value of the source resistor, R_s. While other design criteria may put upper limits on R_s, quiescent point stability indicates that a large value for R_s is desirable. The basic load line interpretation of this stability principle is described in Figure 4.45.

Example 4.8

An n-channel JFET with the following nominal characteristic parameters:

$$V_{PO} = -3.5\,\text{V} \qquad \text{and} \qquad I_{DSS} = 10\,\text{mA},$$

is to be biased with quiescent conditions of

$$I_D = 5\,\text{mA} \qquad \text{and} \qquad V_{DS} = 5\,\text{V}.$$

The FET is subject to parameter variation:

$$-4\,\text{V} < V_{PO} < -3\,\text{V} \qquad \text{and} \qquad 8\,\text{mA} < I_{DSS} < 12\,\text{mA}.$$

Determine the bias resistors necessary to establish the quiescent conditions so that the drain current will not vary more than $\pm$ 4% due to the FET parameter variation. The power supply is given to be: $V_{DD} = 20\,\text{V}$.

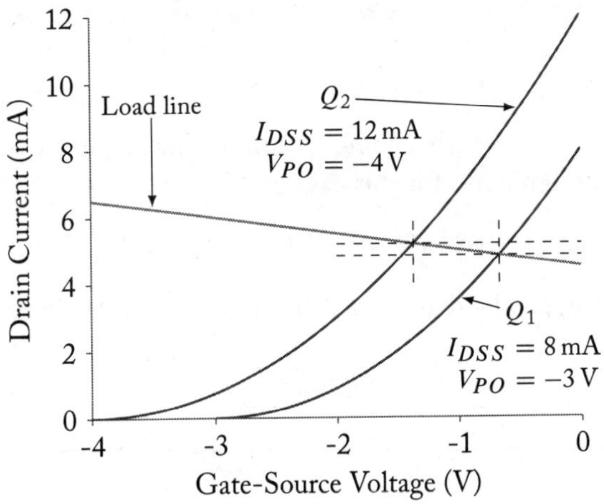

Figure 4.45: Change in quiescent point due to FET paramater variation, fixed bias circuit.

Solution:

The nominal quiescent conditions for the FET are identical to those of Example 4.7. Thus, the value of V_{GS} is obtained in the same manner and is:

$$V_{GS(\text{nominal})} = -1.025\,\text{V}.$$

The extreme values of V_{GS} can be obtained either graphically, as in Figure 4.45, or analytically using Equation (4.11):

$$I_D = I_{DSS}\left[1 - \frac{V_{GS}}{V_{PO}}\right]^2.$$

These extreme values are found to be:

$$V_{GS(\text{min})} = -1.37\,\text{V}\{I_{D(\text{max})} = 5.2\,\text{mA}\}$$
$$V_{GS(\text{max})} = -0.676\,\text{V}\{I_{D(\text{min})} = 4.8\,\text{mA}\}.$$

The minimum value of R_s that will satisfy the stability requirements can be calculated as the *maximum* value of

$$R_{s(\text{min})} = \max\left\{\frac{\Delta V_{GS}}{\Delta I_D}\right\}.$$

For this particular case,

$$R_{s(\text{min})} = \max\left\{\frac{-1.025 + 1.367}{0.0002} \text{ or } \frac{-1.025 + 0.676}{-0.0002}\right\} = 1.75\,\text{k}\Omega.$$

If R_s is chosen to be this minimum value, the nominal voltage at the source and gate of the FET are:

$$V_s = I_{D(\text{nominal})} R_s = 8.75\,\text{V}\; V_G = V_S + V_{GS(\text{nominal})} = 7.725\,\text{V}.$$

The gate voltage can be obtained with a combination of a large variety of resistors. One pair that will satisfy the minimal design goals for this case is:

$$R_{g1} = 1.62\,\text{M}\Omega \qquad \text{and} \qquad R_{g2} = 1.02\,\text{M}\Omega.$$

The remaining resistor, R_d, can be determined from the drain-source voltage requirement as:

$$R_d = \frac{V_{DD} - V_{DS}}{I_D} - R_s = \frac{20 - 5}{5\,\text{m}} - 1750 = 1.25\,\text{k}\Omega.$$

4.6.3 BIASING ENHANCEMENT MODE FETS

The fixed-bias circuit presented in the preceding section is also useful in biasing enhancement mode FETs. The only significant difference in the procedure necessary to determining resistance values to properly bias the FET is in the initial determination of the quiescent conditions. V_{GS} in enhancement FETs carries the opposite sign as for depletion FETs and a different relationship between V_{GS} and I_D exists. The proper expression for the FET gate-source voltage in the saturation region is given by Equation (4.36):

$$I_D = K\left(V_{GS} - V_T\right)^2.$$

Bias stability can be achieved by following the procedure outlined in Example 4.8.

The source self-bias circuit of Figure 4.42 is not useful for enhancement FETs. However, several other possibilities for biasing of this type of FET exist. In addition to the fixed-bias circuit of Figure 4.44, two other common bias circuits for enhancement-mode FETs are given in Figure 4.46. Both circuits offer some bias stabilization with the circuit of Figure 4.46b offering superior stability and flexibility in resistor value choice. The connection in these two circuits between the gate and drain through resistor R_f provides a signal path between input and output when the FET is used in amplifier applications. This path provides advantages and disadvantages that will be discussed in Chapter 8 (Book 2).

4.7 CONCLUDING REMARKS

The Field Effect Transistor has been described in this chapter as a highly useful device with three basic regions of operation: the saturation, ohmic, and cut-off regions. Entrance into each of these regions is controlled by two voltages: the gate-source voltage and the drain-source voltage.

The large-signal characterization of a FET is non-linear. While it can be modeled through local linearization, these linear models are highly dependent on the point of linearization. Therefore, unlike the BJT, the FET cannot easily be described with a unique, simple linear model in

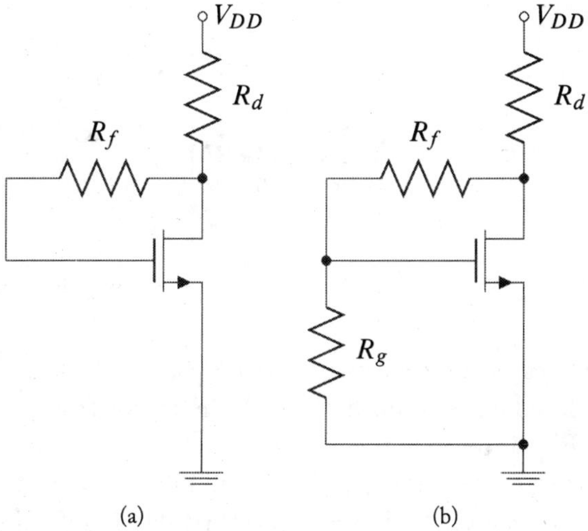

Figure 4.46: Additional bias circuits for enhancement-mode FETs.

each of its regions of operation. The design engineer must rely on a set of non-linear analytic expressions for FET description.

Switching applications, including logic gate applications, are achieved with the FET in transitioning between the cut-off and ohmic region. Linear applications take place in the saturation region. While a variety of applications have been examined in this chapter, the possibilities for circuitry using FETs extend far beyond what has been shown here. Many additional FET linear applications and several non-linear applications will be examined in later chapters. Additional restrictions placed on FET circuit design by frequency response limitations will be discussed in Chapter 9 (Book 3).

Summary Design Example

In order to provide bi-directional current to a DC motor, the motor is connected to an "H-driver" circuit. The basic topology of such a circuit contains four switches as shown below.

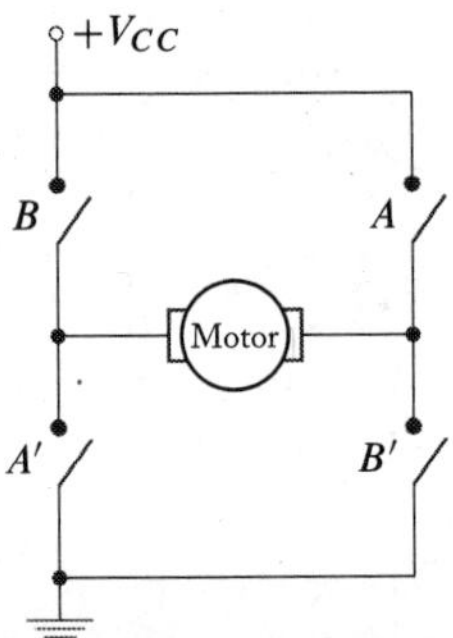

The H-driver switches operate in counteracting pairs: only one pair is closed at a time. When the A-A' pair is closed, current flows through the motor from right to left (in the above diagram): closing the B-B' pair reverses the current flow.

It is necessary to design a circuit to control a small DC motor bi-directionally. Control signals will be provided by standard CMOS logic levels (HIGH $\approx$ 5 V, LOW $\approx$ 0 V). The voltage and current ratings for the motor are:

voltage—3 V to 15 V.

maximum current—300 mA.

Design an H-driver to provide proper power to the motor when control signals are applied.

Solution:

The switches in an H-driver can be either mechanical or solid state. Because of the small currents being switched and the complexity necessary for control of mechanical switches, solid state switching is a good choice for this design. FET switches are the preferred choice in most solid state switching applications.

The best FET H-driver is essentially formed from two counteracting CMOS gates (with appropriate, high-current FETs) and a buffer FET. The basic topology is shown below.

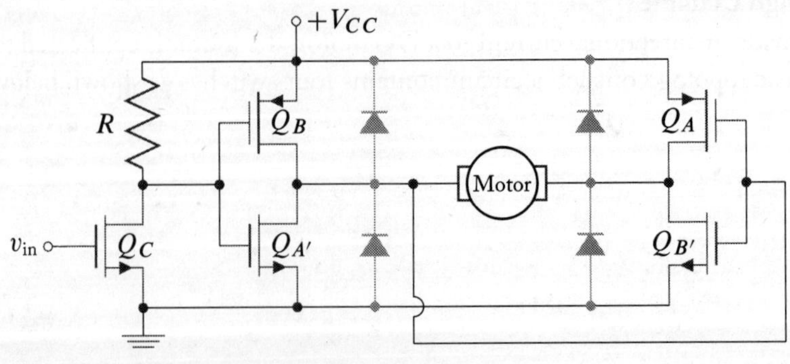

Each complementary pair of FETs (Q_A–$Q_{B'}$ & Q_B–$Q_{A'}$) acts as a counteracting switch and the interconnection ensures that Q_A & $Q_{A'}$ and Q_B & $Q_{B'}$ act simultaneously. The buffer

FET, Q_C, provides 0 V and $+V_{CC}$ to ensure accurate switching. The switching FETs must be enhancement-mode FETs with a threshold voltage significantly less than the minimum applied voltage and with maximum current capability in excess of 300 mA. Q_C must also be a similar enhancement-mode FET, but with smaller current capability. R should be small compared to the input resistance of $Q_{A'}$ and Q_B, but large to avoid wasting power.

A summary of appropriate design choices is:

Q_A & Q_B— p-channel, enhancement MOSFETs, $V_T \approx -1\,\text{V}, I_{D(\text{max})} \approx -0.5\,\text{A}$

$Q_{A'}$ & $Q_{B'}$— n-channel, enhancement MOSFETs, $V_T \approx 1\,\text{V}, I_{D(\text{max})} \approx 0.5\,\text{A}$

Q_C— n-channel, enhancement signal MOSFET, $V_T \approx 1\,\text{V}$,

R— 100 kΩ.

It is common when switching loads that have a significant inductive component (i.e., motors) to shunt the output of a switch with a reverse biased diode (shaded in the diagram). This diode provides a current path during the switching transition. For such a small motor, these diodes are probably not necessary. Many switching FETs have diodes incorporated into their design.

4.8 PROBLEMS

4.1. An n-channel JFET is described by the following parameters:

$$I_{DSS} = 4.5\,\text{mA}$$
$$V_{PO} = -3.6\,\text{V}$$

(a) If the JFET is in saturation, what gate-to-source voltage, V_{GS}, is necessary to achieve a drain current of 2.6 mA?

(b) What is the minimum V_{DS} that will satisfy the conditions stated in part a)?

(c) If $V_{DS} = 2\,\text{V}$, what gate-to-source voltage is necessary to achieve the same drain current?

4.2. A p-channel JFET is described by the following parameters:

$$I_{DSS} = 4.0\,\text{mA}$$
$$V_{PO} = -2.8\,\text{V}$$

(a) If the JFET is in saturation, what source-to-gate voltage, V_{SG}, is necessary to achieve a drain current of 1.8 mA?

(b) What is the minimum V_{SD} that will satisfy the conditions stated in part a)?

(c) If $V_{SD} = 1.5\,\text{V}$, what gate-to-source voltage is necessary to achieve the same drain current?

4.3. The parameters for a given JFET are:

$$I_{DSS} = 7.5\,\text{mA}$$
$$V_{PO} = -4\,\text{V}$$

The JFET is to be biased at

$$I_D = 2\,\text{mA}$$
$$V_{DS} = 6\,\text{V}$$

with the circuit topology as shown.

Determine the Values of R_D and R_S to complete this design if $V_{DD} = 20\,\text{V}$.

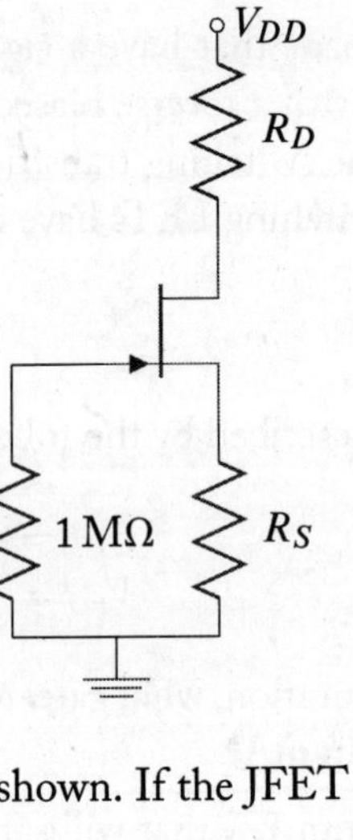

4.4. Given the n-channel JFET circuit shown. If the JFET is described by

$$V_{PO} = 2.5\,\text{V} \quad \text{and} \quad I_{DSS} = 4\,\text{mA},$$

find I_D and V_{DS}.

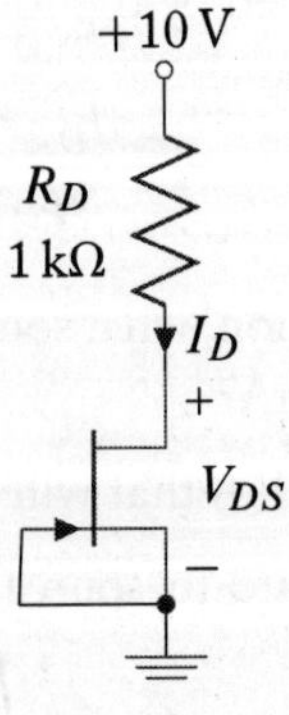

4.5. Complete the design of the n-channel JFET circuit shown for $I_D = 3\,\text{mA}$ and $V_{DS} = 5\,\text{V}$. The JFET parameters are:

$$I_{DSS} = 7\,\text{mA} \quad \text{and} \quad V_{PO} = 2.5\,\text{V}.$$

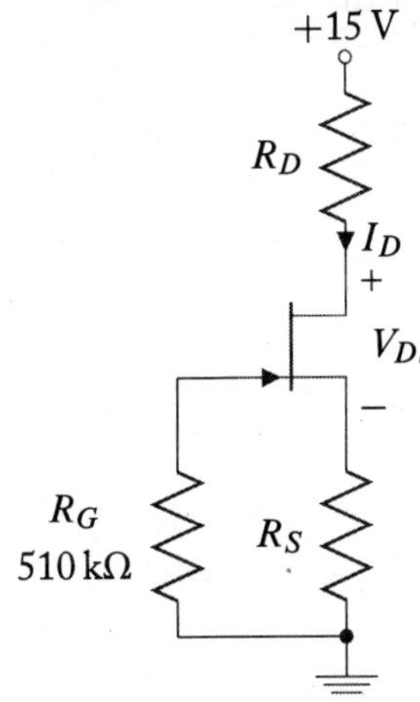

4.6. An n-channel, depletion type MOSFET is described by the following parameters:

$$I_{DSS} = 8.2\,\text{mA}$$
$$V_{PO} = -3.1\,\text{V}$$

(a) If the NMOSFET is in saturation, what gate-to-source voltage, V_{GS}, is necessary to achieve a drain current of 4.0 mA?

(b) What is the minimum V_{DS} that will satisfy the conditions stated in part a)?

(c) If $V_{DS} = 2\,\text{V}$, what gate-to-source voltage is necessary to achieve the same drain current?

(d) What is the output resistance of the NMOSFET at the conditions of part c)?

4.7. An n-channel MOSFET has the following characteristics:

$$V_{PO} = 3\,\text{V},\ V_A = 170\,\text{V},\ \text{and}\ I_{DSS} = 8\,\text{mA}.$$

(a) Determine the minimum drain-source voltage, V_{DS}, for the MOSFET to be in saturation.

(b) Determine the output resistance of the MOSFET in the ohmic region when $V_{GS} = 1.5\,\text{V}$.

(c) Determine the output resistance of the MOSFET in the saturation region when $I_D = 2\,\text{mA}$.

4.8. Assume an n-channel depletion type MOSFET with parameters:

$$I_{DSS} = 10\,\text{mA}$$
$$V_{PO} = -5\,\text{V}$$

Determine V_o and I_D for the circuit shown.

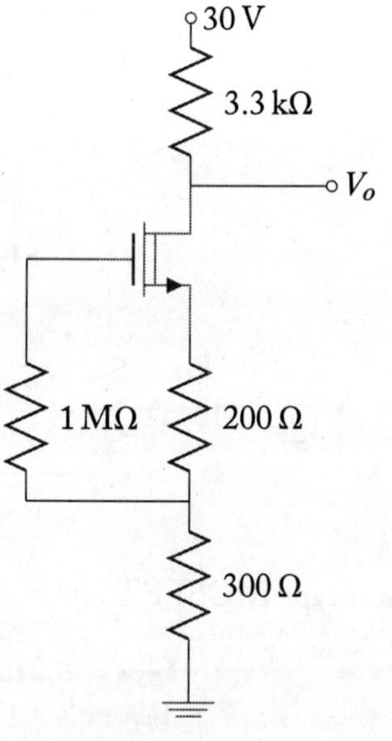

4.9. Find the drain current, drain-source voltage, and gate-drain voltage for the given circuit. Assume $K = 0.25\,\text{mA/V}^2$ and $V_T = 2.5\,\text{V}$.

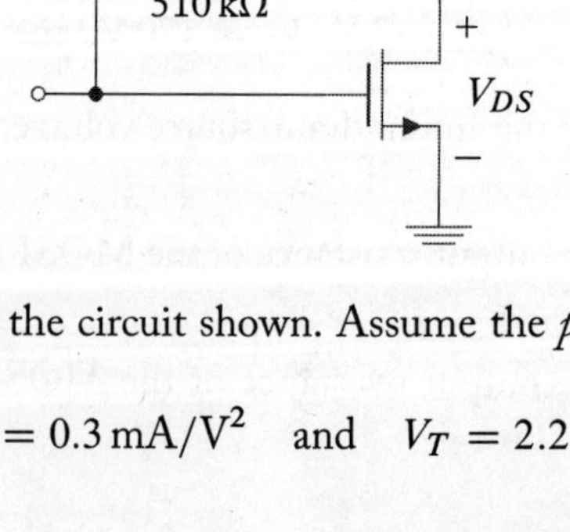

4.10. Determine the Q-point for the circuit shown. Assume the p-channel MOSFET is described by:

$$K = 0.3\,\text{mA/V}^2 \quad \text{and} \quad V_T = 2.2\,\text{V}$$

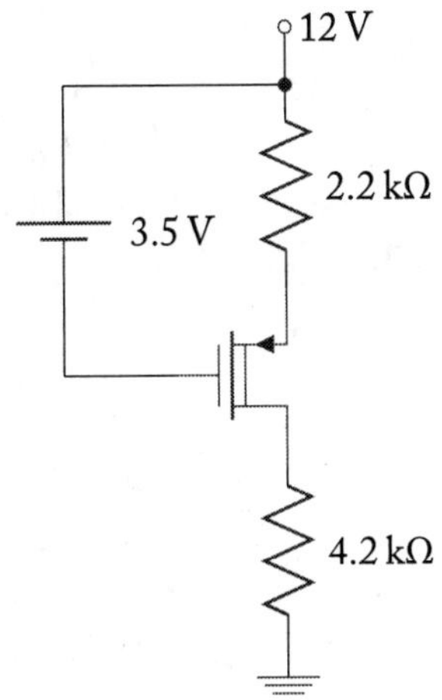

4.11. An n-channel, enhancement type MOSFET is described by the following parameters:

$$I_{DSS} = 2.4\,\text{mA/V}^2$$
$$V_T = 1.2\,\text{V}$$

(a) If the NMOSFET is in saturation, what gate-to-source voltage, V_{GS}, is necessary to achieve a drain current of 4.0 mA?

(b) What is the minimum V_{DS} that will satisfy the conditions stated in part a)?

(c) If $V_{DS} = 1.2\,\text{V}$, what gate-to-source voltage is necessary to achieve the same drain current?

(d) What is the output resistance of the NMOSFET at the conditions of part c)?

4.12. For the n-channel MOSFET circuit shown, determine the drain current, I_D, and drain-source voltage, V_{DS}, using

(a) The analytical method (using equations)

(b) Load-line analysis. Use SPICE to arrive at the transistor characteristic curves.

The MOSFET parameters are:

$$V_T = 1.5\,\text{V},\ V_A = 170\,\text{V},\ \text{and}\ K = 1.2\,\text{mA/V}^2.$$

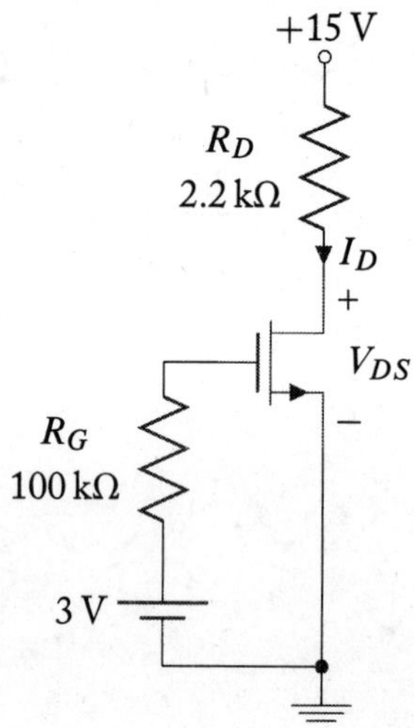

4.13. Design a circuit to bias an n-channel depletion MOSFET with $I_{DSS} = 8$ mA and $V_{PO} = -4$ V using the "bootstrapping" configuration shown. The design specifications require that $I_D = 4$ mA and $V_{DS} = 3$ V. The "bootstrapping" bias technique is used to preserve the high input resistance of the circuit.

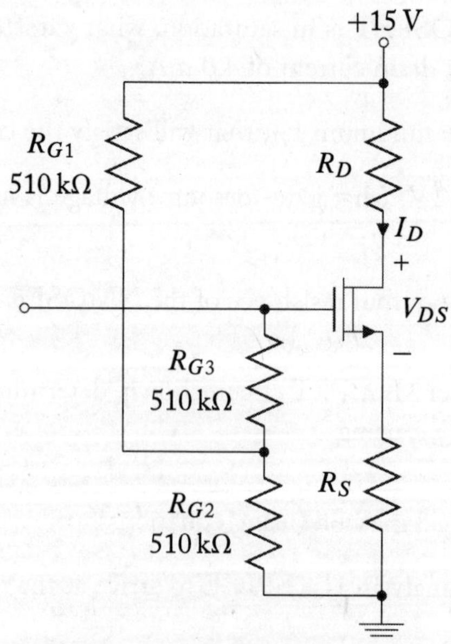

4.14. Complete the design of the n-channel depletion MOSFET circuit shown so that

$$I_D = 2\,\text{mA} \quad \text{and} \quad V_{DS} = 4\,\text{V}.$$

The MOSFET parameters are:

$$I_{DSS} = 5\,\text{mA} \quad \text{and} \quad V_{PO} = 3\,\text{V}.$$

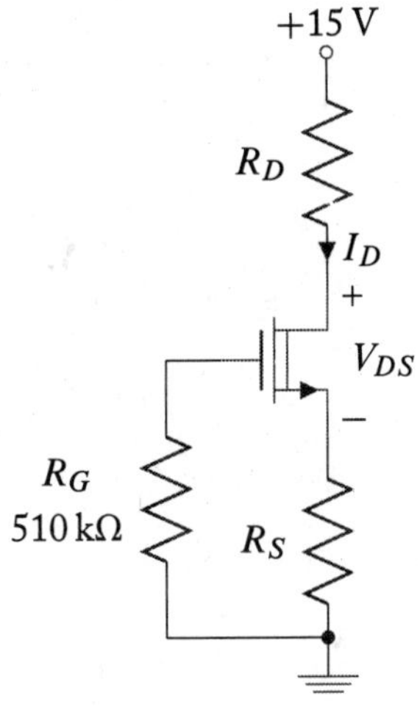

4.15. For the circuit shown: the n-channel JFET is described by:

$$I_{DSS} = 8\,\text{mA}, \quad \text{and} \quad V_{PO} = -6.5\,\text{V}.$$

(a) Complete the design of the circuit to achieve a Q-point of:

$$I_D = 6\,\text{mA and } V_{DS} = 3\,\text{V}$$

by determining R_{G1} and R_{G2} for $R_{G1}//R_{G2} = 54\,\text{k}\Omega \pm 5\,\text{k}\Omega$. Calculate V_{GS} and V_{DS}.

(b) The JFET is replaced by a p-channel JFET with $I_{DSS} = 8\,\text{mA}$ and $V_{PO} = 4\,\text{V}$. Draw the p-channel JFET circuit diagram so that the FET is biased in the saturation region. Find the Q-point for the JFET in saturation (V_{DS}, I_D and V_{GS}) using the resistor values found in part a).

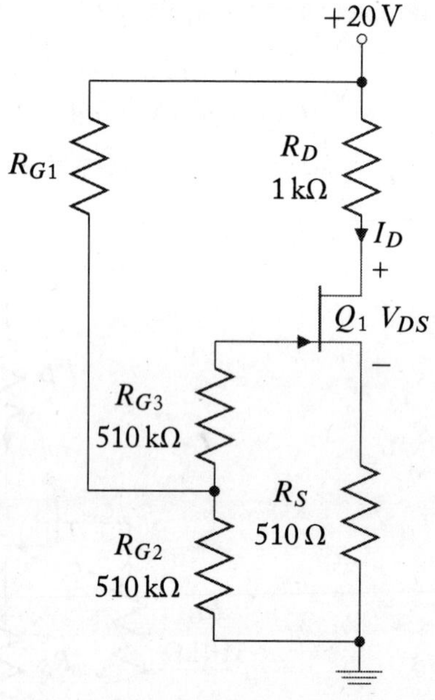

4.16. Complete the design of the n-channel enhancement MOSFET circuit shown so that

$$I_D = 2\,\text{mA}.$$

The MOSFET parameters are:

$$K = 1.3\,\text{mA/V}^2 \quad \text{and} \quad V_T = 2\,\text{V}.$$

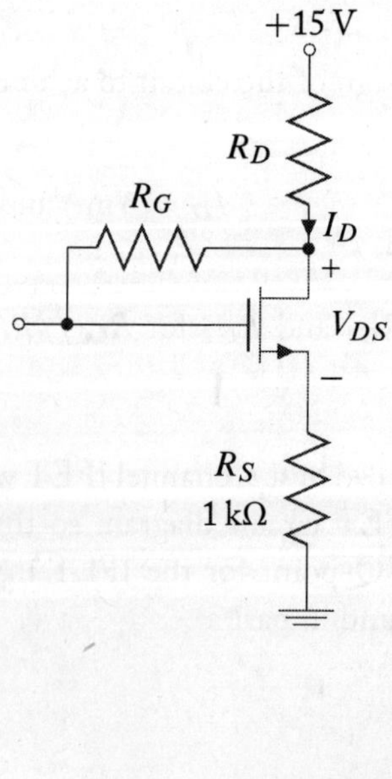

4.17. Use SPICE to generate the transistor characteristic curves for a p-channel depletion MOSFET with parameters:

$$V_{PO} = 4\,\text{V}$$
$$I_{DSS} = -7\,\text{mA}$$

over the range $0 \leq V_{SD} \leq 15\,\text{V}$.

Using the curves generated, determine the FET drain current and V_{DS} in the circuit shown.

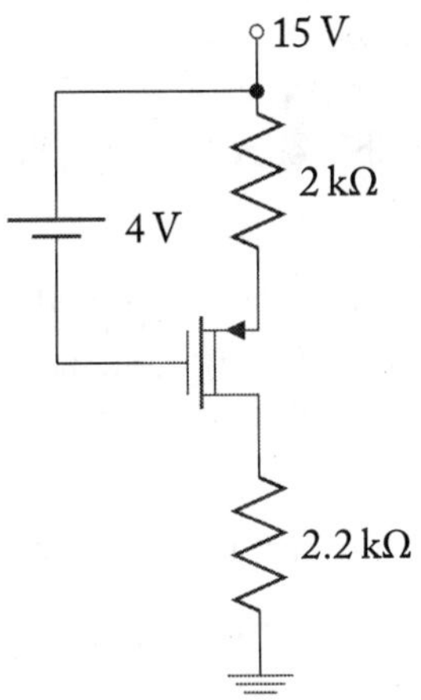

4.18. For the p-channel MOSFET circuit shown, determine the drain current, I_D, and drain-source voltage, V_{DS}, using

(a) The analytical method (using equations)

(b) Load-line analysis. Use SPICE to arrive at the transistor characteristic curves.

The MOSFET parameters are:

$$V_{PO} = 2.5\,\text{V},\ V_A = 150\,\text{V},\ \text{and}\ I_{DSS} = -10\,\text{mA}.$$

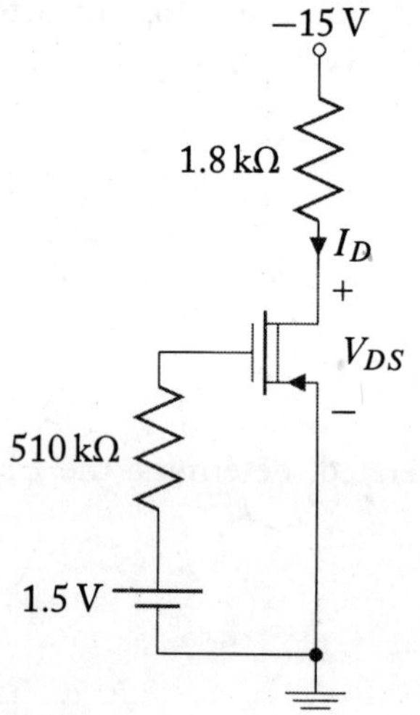

4.19. For the p-channel MOSFET circuit shown, determine the drain current, I_D, and drain-source voltage, V_{DS}, using

(a) The analytical method (using equations)

(b) Load-line analysis. Use SPICE to arrive at the transistor characteristic curves.

The MOSFET specifications are:

$$V_{PO} = 3.5\,\text{V},\ V_A = 150\,\text{V},\ \text{and}\ I_{DSS} = -10\,\text{mA}.$$

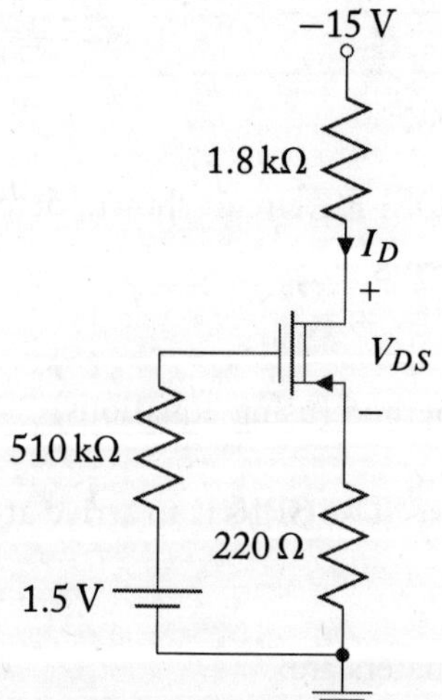

4.20. Plot the input and output characteristics of an n-channel enhancement MOSFET with

$K = 1.2 \times 10^{-3}\,\text{A/V}^2$ and $V_T = 3.5\,\text{V}$. Design a circuit so that the MOSFET is biased at $V_{DS} = 5\,\text{V}$ and $I_D = 1\,\text{mA}$ with $V_{DD} = 10\,\text{V}$.

4.21. Plot the input and output characteristics of an *n*-channel enhancement MOSFET with $K = 1.2 \times 10^{-3}\,\text{A/V}^2$ and $V_T = 3.5\,\text{V}$. Design a circuit so that the MOSFET is biased at $V_{DS} = 3.5\,\text{V}$ and $I_D = 3\,\text{mA}$ with $V_{DD} = 10\,\text{V}$. Compare the region of FET operation with that of the previous problem.

4.22. Plot the input and output characteristics of a *p*-channel enhancement MOSFET with $K = 1.2 \times 10^{-3}\,\text{A/V}^2$ and $V_T = -2\,\text{V}$. Design a circuit so that the MOSFET is biased at $V_{SD} = 6\,\text{V}$ and $I_D = -1\,\text{mA}$ with $V_{DD} = -12\,\text{V}$.

4.23. Plot the input and output characteristics of a *p*-channel enhancement MOSFET with $K = 1.2 \times 10^{-3}\,\text{A/V}^2$ and $V_T = -2\,\text{V}$. Design a circuit so that the MOSFET is biased at $V_{SD} = 3.5\,\text{V}$ and $I_D = -3\,\text{mA}$ with $V_{DD} = -12\,\text{V}$. Compare the region of FET operation with that of the previous problem.

4.24. For the circuit shown, the MOSFET is described by:

$$I_{DSS} = 8\,\text{mA},\ I_D = 4\,\text{mA},\ V_{DS} = 8\,\text{V},\ \text{and}\ V_{PO} = -5\,\text{V}.$$

(a) Find R_D, and R_{G1} and R_{G2} for $R_G = R_{G1}//R_2 = 1\,\text{M}\Omega$.

(b) The MOSFET is replaced by one with parameters:

$$I_{DSS} = 10\,\text{mA} \quad \text{and} \quad V_{PO} = -6\,\text{V}.$$

Find the new Q-point.

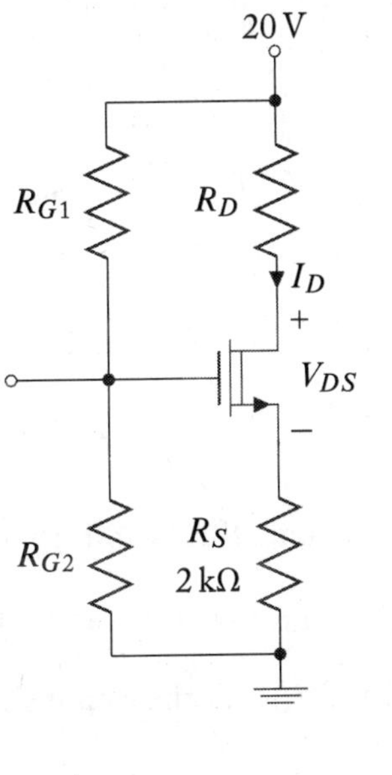

4.25. The output current of a FET current sources can be adjusted through the use of a resistor at the source of the FET as shown. If the JFET parameters are given as:

$$I_{DSS} = 5\,\text{mA} \quad V_{PO} = -2.5\,\text{V}$$

(a) Determine the resistor value to obtain an output current, I_D, of 2 mA.

(b) Determine the resistor value to obtain an output current, I_D, of 3 mA.

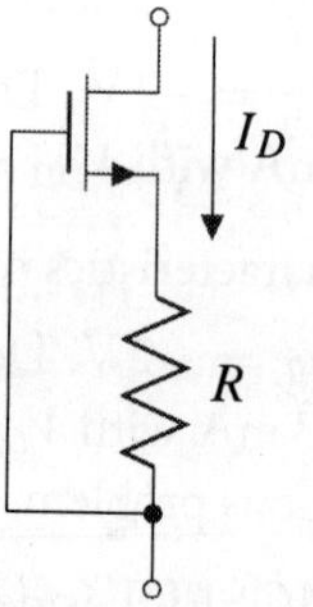

4.26. Current sources can be realized with p-channel as well as n-channel devices. If the p-channel MOSFET parameters are given as:

$$I_{DSS} = -5\,\text{mA}$$
$$V_{PO} = 2.0\,\text{V}$$

(a) Determine the resistor value to obtain an output current, I_O, of 2 mA.

(b) Determine the resistor value to obtain an output current, I_O, of 3 mA.

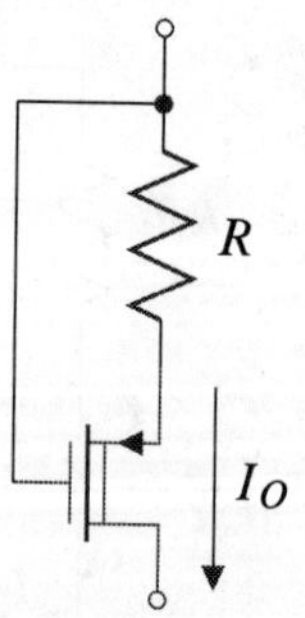

4.27. Given two identical n-channel JFETs with parameters,

$$I_{DSS} = 6\,\text{mA} \quad \text{and} \quad V_{PO} = -2.5\,\text{V},$$

determine V_{DS2}, V_{DS2}, and I_{D2} for the circuit shown.

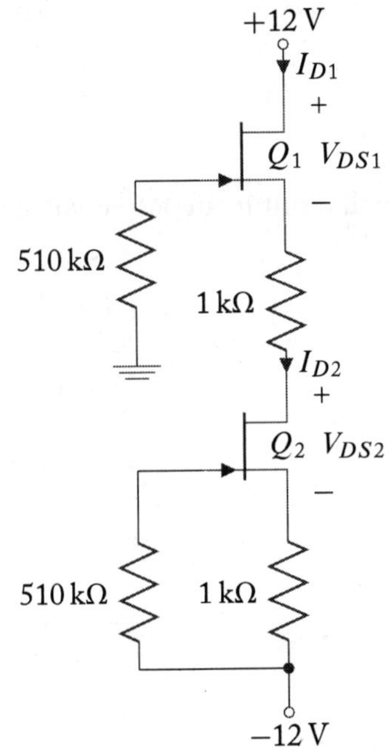

4.28. Find the output voltage V_O, of the circuit shown, for the following FET choices:

(a) Given two identical n-channel MOSFETs with parameters,

$$V_T = 2.5\,\text{V} \quad \text{and} \quad K = 0.15\,\text{mA/V}^2.$$

(b) Given two different n-channel MOSFET with parameters:

$$V_{T1} = 1.5\,\text{V},\, V_{T1} = 3\,\text{V},\, \text{and}\, K = 0.15\,\text{mA/V}^2.$$

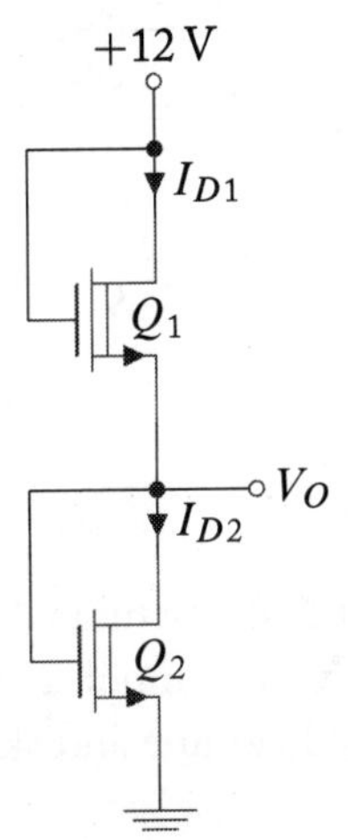

4.29. Given the voltage division circuit shown. The MOSFET is described by:

$$K = 1\,\text{mA/V}^2,\ V_T = 2\,\text{V and } V_A = 100\,\text{V}.$$

What input voltage will result in an output voltage of 1.0 V?

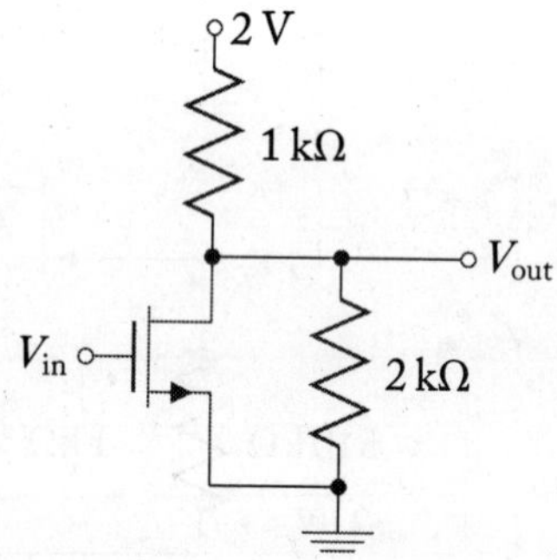

4.30. Given the voltage division circuit shown. The MOSFET is described by:

$$I_{DSS} = 4\,\text{mA},\ V_{PO} = -2\,\text{V and } V_A = 100\,\text{V}.$$

What input voltage will result in an output voltage of 1.0 V?

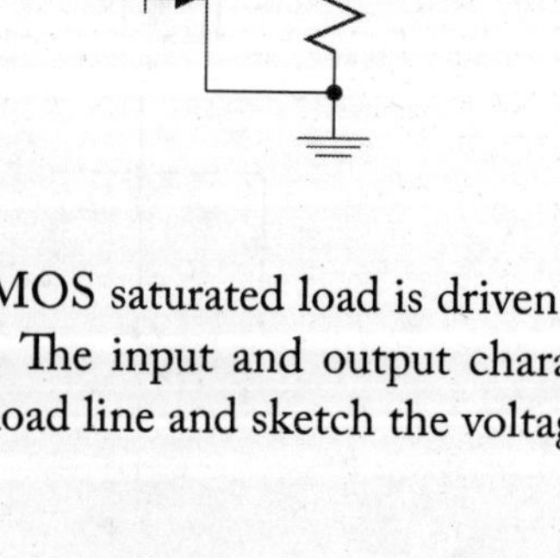

4.31. An n-channel enhancement MOS saturated load is driven with an n-channel enhancement MOS driver as shown. The input and output characteristics are shown in Figure 4.17 and 4.18. Show the load line and sketch the voltage transfer characteristic (V_O vs. V_I).

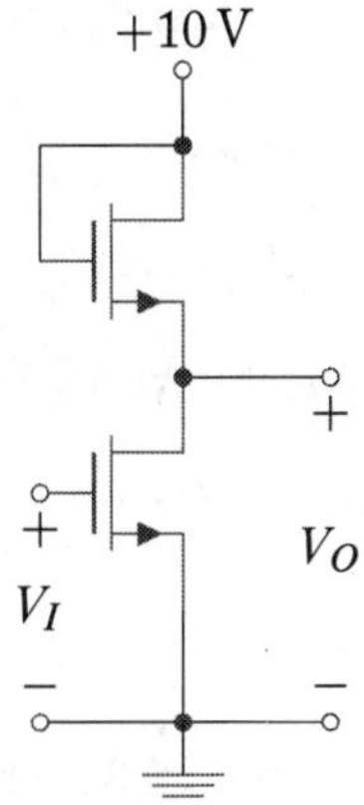

4.32. Use SPICE to determine the ON and OFF resistance the analog FET switch shown in Figure 4.38 as a function of the input voltage, v_s. The FET is described by the parameters:

$$K = 3.0\,\text{mA/V}^2,\ V_T = 2.5\,\text{V and } V_A = 100\,\text{V}.$$

Assume the load resistance, $R = 100\,\Omega$, the input voltage exists over the range, $0\,\text{V} < v_s < +10\,\text{V}$, and the control voltage, v_c, switches between 0 V and +10 V.

4.33. Use SPICE to determine the ON and OFF resistance the analog FET switch shown in Figure 4.40 as a function of the input voltage, v_s. The FETs are described by the parameters:

$$K = 3.0\,\text{mA/V}^2,\ |V_T| = 2.5\,\text{V and } |V_A| = 100\,\text{V}$$

Assume the load resistance, $R = 200\,\Omega$, the input voltage exists over the range, $0\,\text{V} < v_s < +10\,\text{V}$, and the control voltage, v_c, switches between 0 V and +10 V.

4.34. An n-channel MOSFET with the following nominal characteristic parameters:

$$V_{PO} = -2\,\text{V},\ I_{DSS} = 8\,\text{mA}$$

is to be biased with quiescent conditions of:

$$I_D = 3\,\text{mA} \quad \text{and} \quad V_{DS} = 4\,\text{V},$$

using resistors and a single 20 V power supply. The FET is subject to parameter variation:

$$-2.5 < V_{PO} < -1.5 \quad \text{and} \quad 7\,\text{mA} < I_{DSS} < 10\,\text{mA}.$$

Determine the bias resistors necessary to establish quiescent conditions so that the drain current will not vary more than ±2.5% due to the FET parameter variation. Verify that your design meets specifications using SPICE.

4.35. The n-channel MOSFET in the circuit shown is described by the parameters:

$$V_T = 1.5\,\text{V} \quad \text{and} \quad K = 0.5\,\text{mA/V}^2.$$

Determine I_D and V_{DS} analytically. Verify your results using SPICE.

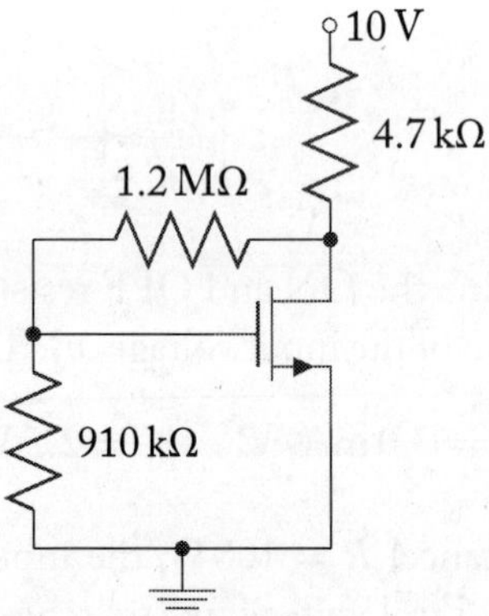

4.36. Design a bias circuit to achieve a quiescent condition of $I_D = 6\,\text{mA}$ and $V_{DS} = 6\,\text{V}$ using a single 20 V power supply and a FET with the following parameters:

$$V_{PO} = -2\,\text{V} \quad \text{and} \quad I_{DSS} = 12\,\text{mA}.$$

It is required that the drain of the FET be connected to the 20 V supply through a 2 kΩ resistor ($R_D = 2\,\text{k}\Omega$).

4.37. Design an FET constant current source for the circuit configuration shown. Determine all resistance values so that

$$V_{CE} = 3\,\text{V} \quad \text{and} \quad I_C = 5\,\text{mA}.$$

The following components are available:

npn BJT: 2N2222
Depletion NMOSFET: $I_{DSS} = 5\,\text{mA}$, $V_{PO} = -3.5\,\text{V}$.

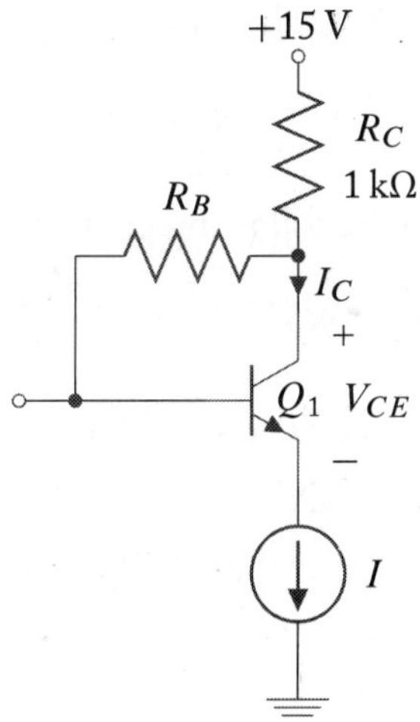

4.38. Design an inverting DC OpAmp amplifier with a voltage variable gain. The maximum required gain magnitude is 10. The following components are available:

n-JFET:	$I_{DSS} = 8\,\text{mA},\ V_{PO} = -4\,\text{V}$
Near-ideal OpAmp:	μA741CN
Power Supply:	0 to ±15 V
Standard Value Resistors	

Show complete analysis of the design.

4.39. An n-channel MOSFET with the following nominal characteristic parameters,

$$V_T = 2\,\text{V} \quad \text{and} \quad K = 1.25\,\text{mA/V}^2$$

is required to be biased at the following quiescent conditions:

$$I_D = 2\,\text{mA} \quad \text{and} \quad V_{DS} = 3\,\text{V}.$$

The FET is subjected to the following parameter variations:

$$1 < V_T < 3\,\text{V} \quad \text{and} \quad 1.00\,\text{mA/V}^2 < K < 1.5\,\text{mA/V}^2.$$

The power supply voltage provided is +12 V.

Design a circuit to bias the FET so that the drain current does not vary more than ±3% due to the FET parameter variations. Verify that the design meets the stability specification using SPICE.

4.40. A p-channel JFET with the following nominal characteristic parameters,

$$V_{PO} = 3\,\text{V} \quad \text{and} \quad I_{DSS} = -12\,\text{mA}$$

is required to be biased at the following quiescent conditions:

$$I_D = -2\,\text{mA} \quad \text{and} \quad V_{DS} = 5\,\text{V}.$$

The JFET is subjected to the following parameter variations:

$$1.5 < V_{PO} < 4.5\,\text{V} \quad \text{and} \quad -10\,\text{mA} < I_{DSS} < -14\,\text{mA}.$$

The power supply voltage provided is −24 V.

Design a circuit to bias the FET so that the drain current does not vary more than ±5% due to the FET parameter variations. Verify that the design meets the stability specification using SPICE.

4.9 REFERENCES

[1] ——, *Analog IC Data Book,* Precision Monolithics, Inc., Santa Clara, 1990

[2] ——, *Integrated Circuits Data Book*, Silconix Inc., Santa Clara, 1988

[3] ——, *PSpice Reference Manual*, MicroSim Corp., Irvine, 1989

[4] Baliga, B. J. and Chen, D. Y., editors, *Power Transistors: Device Design and Applications*, IEEE Press, New York, 1984.

[5] Colclaser, R. A. and Diehl-Nagle, S., *Materials and Devices for Electrical Engineers and Physicists*, McGraw-Hill Book Company, New York, 1985.

[6] Ghausi, M. S., *Electronic Devices and Circuits: Discrete and Integrated,* Holt, Rinehart and Winston, New York, 1985.

[7] Gray, P. R., and Meyer, R. G., *Analysis and Design of Analog Integrated Circuits*, 3rd. Ed., John Wiley & Sons, Inc., New York, 1993.

[8] Horowitz, P., and Hill, W., *The Art of Electronics,* 2nd. Ed. Cambridge University Press, Cambridge, 1992.

[9] Millman, J., *Microelectronics, Digital and Analog Circuits and Systems*, McGraw-Hill Book Company, New York, 1979.

[10] Sedra, A. S. and Smith, K. C., *Microelectronic Circuits*, 3rd. Ed., Holt, Rinehart, and Winston. Philadelphia, 1991.

[11] Tuinenga, P., *SPICE: A Guide to Circuit Simulation and Analysis Using PSpice*, 2nd. Ed., Prentice Hall, Englewood Cliffs, 1992.

Authors' Biographies

Thomas F. Schubert, Jr., and Ernest M. Kim are colleagues in the Electrical Engineering Department of the Shiley-Marcos School of Engineering at the University of San Diego.

THOMAS F. SCHUBERT, JR.

Thomas Schubert received BS, MS, and PhD degrees in Electrical Engineering from the University of California at Irvine (UCI). He was a member of the first engineering graduating class and the first triple-degree recipient in engineering at UCI. His doctoral work discussed the propagation of polarized light in anisotropic media.

Dr. Schubert arrived at the University of San Diego in August, 1987 as one of the two founding faculty of its new Engineering Program. From 1997–2003, he led the Department as Chairman, a position that became Director of Engineering Programs during his leadership tenure. Prior to coming to USD, he was at the Space and Communications Division of Hughes Aircraft Company, the University of Portland, and Portland State University. He is a Registered Professional Engineer in the State of Oregon.

In 2012, Dr. Schubert was awarded the Robert G. Quinn Award by the American Society of Engineering Education "in recognition of outstanding contributions in providing and promoting excellence in engineering experimentation and laboratory instruction."

ERNEST M. KIM

Ernest Kim received his B.S.E.E. from the University of Hawaii at Manoa in Honolulu, Hawaii in 1977, an M.S.E.E. in 1980 and Ph.D. in Electrical Engineering in 1987 from New Mexico State University in Las Cruces, New Mexico. His dissertation was on precision near-field exit radiation measurements from optical fibers.

Dr. Kim worked as an Electrical Engineer for the University of Hawaii at the Naval Ocean Systems Center, Hawaii Labs at Kaneohe Marine Corps Air Station after graduating with his B.S.E.E. Upon completing his M.S.E.E., he was an electrical engineer with the National Bureau of Standards in Boulder, Colorado designing hardware for precision fiber optic measurements. He then entered the commercial sector as a staff engineer with Burroughs Corporation in San Diego, California developing fiber optic LAN systems. He left Burroughs for Tacan/IPITEK Corporation as Manager of Electro-Optic Systems developing fiber optic CATV hardware and systems. In 1990 he joined the faculty of the University of San Diego. He remains an active consultant in radio frequency and analog circuit design, and teaches review courses for the engineering Fundamentals Examination.

Dr. Kim is a member of the IEEE, ASEE, and CSPE. He is a Licensed Professional Electrical Engineer in California.

Index